Contributions in Mathematical Physics

A Tribute to Gerard G. Emch

Volumes containing conference proceedings, workshop lectures, and collection of invited articles in area of mathematics are published in this series.

The aim is to provide an overview of the development through surveys, recent trends, problems and their current status as well as historical background.

All contributions for publication should be thoroughly reviewed before submitting to us.

We need manuscripts ready for direct reproduction. All articles must be prepared in an uniform format. Necessary instructions for format etc., are available from the publisher.

Already Published Volumes

Algebra-Some Recent Advances: Edited by I. B. S. Passi

Number Theory: Edited by R. P. Bambah, V. C. Dumir, R. J. Hans-Gill

Number Theory and Discrete Mathematics. Edited by A. K. Agarwal, Bruce C. Berndt, Christian F. Krattenthaler, Gray L. Mullen, K. Ramachandra and Michel Waldschmidt

Current Trends in Number Theory: Edited by S.D. Adhikari, S.A. Katre, B. Ramakrishnan

Advances in Algebra and Geometry: Edited by C. Musili

A Tribute to C. S. Seshadri: Perspectives in Geometry and Representation Theory: Edited by V. Lakshmibai, V. Balaji, V. B. Mehta, K. R. Nagaraj, K. Paranjape. P. Sankaran and R. Sridharan

Elliptic Curves, Modular Forms and Cryptography: Edited by A. K. Bhandari, D. S. Nagaraj, B. Ramakrishnan, T. N. Venktaraman

History of the Mathematical Sciences: Edited by Ivor Gratan-Guinness and B. S. Yadav

Transformation Groups: Symplectic Torus Actions and Toric Manifolds: Edited by Goutam Mukherjee

Contributions in Mathematical Physics

A Tribute to Gerard G. Emch

Edited by
S. Twareque Ali
Kalyan B. Sinha

HINDUSTAN
BOOK AGENCY

Editors:

S. Twareque Ali
Department of Mathematics and Statistics
Concordia University
1455, de Maisonneuve Blvd. West
Montreal, Quebec
Canada H3G1 M8

stali@mathstat.concordia.ca

Kalyan B. Sinha
Jawaharlal Nehru Centre for Advanced Scientific Research
P.O. Jakkur
Bangalore 560064
India

kbs_jaya@yahoo.co.in

Published By

Hindustan Book Agency (India)
P 19 Green Park Extension
New Delhi 110 016
India

email: hba@vsnl.com
http://www.hindbook.com

ISBN-10: 81-85931-79-8
ISBN-13: 978-81-85931-79-1

Contents

The Publications of Gérard G. Emch

Books (Author)

(with C. Liu) The Logics of Thermostatistical Physics, (xiv + 703 pp.), Springer-Verlag, Berlin-Heidelberg, 2002.

Mathematical and Conceptual Foundations of 20th-Century Physics, (x + 550 pp.), Elsevier North-Holland Mathematical Studies, No. **100**, Amsterdam, 1984.

Algebraic Methods in Statistical Mechanics and Quantum Field Theory, (xiv + 333 pp.), J. Wiley-Interscience, New York, 1972. **Russian** transl. G. A. Zaitsev, 1975.

Books (Editor)

(with R. Sridharan and M.D. Srinivas) Contributions to the History of Indian Mathematics, Hindustan Book Agency (India), New Delhi, 2005.

(with S.T. Ali, A. Odzijewicz, M. Schlichenmaier and S.L. Woronowicz) Twenty Years of Bialowieza: A Mathematical Anthology, Aspects of Differential Geometric Methods in Physics, World Scientific Monogrtaph series in Mathematics, Vol. 8, World Scientic Publishers, Singapore, 2005.

(with J. Mehra and A.S. Wightman) Philosophical Reflections and Syntheses, (vii + 632 pp.), Vol. VI of The Collected Works of Eugene Paul Wigner, Springer, Heidelberg, 1995; 2nd printing, 1997.

(with G. Hegerfeldt and L. Streit, eds.) On Klauder's Path: a Field Trip, (xxi + 263 pp.), World Scientific Publishers, Singapore, 1994.

Journal Articles and/or Book Chapters

1. Le caratère de Saint–Exupéry, *Ann. Collège de Gen'eve* **14** (1955) 54–58.
2. Contribution à l'étude de $Cr(H_20)_6^{+++}$, Travail de diplôme, Université de Genève, 1959.
3. (with R. Lacroix) Résonance paramagnétique du complexe $Cr(H_2O)_6^{+++}$, *Arch. des Sciences* **13** (1960) 157–158.
4. (with R. Lacroix) Résonance paramagnétique de l'ion Cr^{++-}, *Helv. Phys. Acta* **33** (1960) 1021–1031.
5. Représentations du groupe de Lorentz, *Publ. Sem. Phys. Math., Geneva* (1961) 1–41.

6. (with R. Lacroix) Covalence et résonance paramagnétique, *Helv. Phys. Acta.* **35** (1962) 592–616.
7. (with C. Piron) Note sur les symétries en théorie quantique, *Helv. Phys. Acta.* **35** (1962) 542–543.
8. (with C. Piron) On the Symmetries in Quantum Theory, *J. Math. Phys.* **4** (1963) 469–473.
9. A propos d'une mécanique quantique quaternionienne, *Helv. Phys. Acta.* **36** (1963) 499.
10. Mécanique quantique quaternionienne et Relativité restreinte I, *Helv. Phys. Acta* **36** (1963) 739–769.
11. Mécanique quantique quaternionienne et Relativité restreinte II, *Helv. Phys. Acta* **36** (1963) 770–788.
12. Sur la généralité des équations maitresses quantiques, *Helv. Phys. Acta* **37** (1964) 270–283.
13. Coarse-graining in Liouville Space and Master Equations, *Helv. Phys. Acta.* **37** (1964) 532–544.
14. On the Markov Character of the Master Equations, *Helv. Phys. Acta* **38** (1965) 164–171.
15. (with C. Favre) Coarse-graining in Liouville Space and Ergodicity, *Publ. Sem. Phys. Math., Geneva* (1965) 1–32.
16. (with J.M. Jauch) Structures logiques et mathématiques en physique quantique, *Dialectica* **19** (1965) 259–279.
17. Representations of the Lorentz Group in Quaternionic Quantum Mechanics, *in Lectures in Theoretical Physics* **VIIa** *(= Boulder Summer School 1964),* W.E. Brittin, ed., University of Colorado Press (1965) 1–36.
18. (with M. Guenin) A Gauge Invariant Formulation of the BCS Model, *J. Math. Phys.* **7** (1966) 915–921.
19. Rigorous Results in Non-Equilibrium Statistical Mechanics, *in Lectures in Theoretical Physics* **VIIIa** *(= Boulder Summer School 1965),* W.E. Brittin, ed., University of Colorado Press (1966) 65–99.
20. A Non-Markovian Model for the Approach to Equilibrium, *J. Math. Phys.* **7** (1966) 1198–1206.
21. On the Definition of States in Quantum Statistical Mechanics, *J. Math. Phys.* **7** (1966) 1413–1420.

22. A Common Interpretation of Phase Transitions in Various Models, Journ Math. Phys. **8** (1967) 13–18.

23. On the Origin of Phase Transitions in Various Models, Techn. Notes in Appl. Math., University of Maryland, **BN–433** (1955)

24. van der Waals Wiggles, Maxwell Rule and Temperature Dependent Excitations, *J. Math. Phys.* **8** (1967) 19–25.

25. (with H.J.F. Knops and E.J. Verboven) On the Extension of Invariant Partial States in Statistical Mechanics, *Commun. Math. Phys.* **7** (1968) 164–172.

26. Projective representations of Groups in Quaternionic Hilbert Spaces, NSF Conference on Orthomodular Lattices and Empirical Logic, University of Massachusetts, Amherst (1968)

27. (with R. Herman) A Generalization of Hugenholtz's Theorem on Types, Notices Amer. Math. Soc. **15** (1968) 219.

28. (with G.L. Sewell) Non-Equilibrium Statistical Mechanics of Open Systems, *J. Math. Phys.* **9** (1968) 946–958.

29. (with K.B. Sinha) An Adaptation of Powers' Non-Interaction Theorem to the case of Boson Fields, Bull. Amer. Phys. Soc. **14** (1968) 86.

30. (with H.J.K. Knops and E.J. Verboven) On Partial Weakly Clustering States with an Application to the Ising Model, *Commun. Math. Phys.* **8** (1968) 300–314.

31. (with H.J.K. Knops and E.J. Verboven) Ergodic States and Phase Transitions, *Journ. Phys. Soc. Japan (Suppl.)* **26** (1969) 301–303.

32. (with H.J.K. Knops and E.J. Verboven) The Breaking of Euclidean Symmetry with an Application to the Theory of Crystallization, *J. Math. Phys.* **11** (1970) 1655–1668.

33. (with H.J.K. Knops) Pure Thermodynamical Phases as KMS States. *J. Math. Phys.* **11** (1970) 3008–3018.

34. (with C. Radin) Time evolution in Generalized Ising Models, Bull. Amer. Phys. Soc. **15** (1970) 33.

35. (with C. Radin) Relaxation of Local Deviations from Equilibrium. *J. Math. Phys.* **12** (1971) 2013–2046.

36. (Notes by J. Dufty) An Introduction to C^*-algebraic Methods in Many-Body Problems, NSF Summer School, Bozeman, Montana, 1971.

37. An Introduction to C^*-algebraic Methods in Physics, *Adv. Chem. Phys.* **22** (1972) 315–364.

38. On Quantum Measurement Processes, *Helv. Phys. Acta* **45** (1972) 1049–1056.

39. (with J.C. Wolfe) A Model for Dissipative Behavior in Nonlinear Quantum Optics, *J. Math. Phys.* **13** (1972) 1236–1243.

40. The C^*-algebraic Approach to Phase Transitions, Chapter 4 in Volume 1 of *Phase Transitions and Critical Phenomena,* C. Domb and M.S. Green, eds., Academic Press, London (1972) 137–175.

41. Diffusion, Einstein Formula, and Mechanics, *J. Math. Phys.* **14** (1973) 1775–1783.

42. (with S.T. Ali) Fuzzy Observables in Quantum Mechanics, *J. Math. Phys.* **15** (1974) 176–182.

43. (with J.C. Wolfe) C^*-algebraic Formalism for Coarse-graining I, General Theory, *J. Math. Phys.* **15** (1974) 1343–1347.

44. (with J.C. Wolfe) C^*-algebraic Formalism for Coarse-graining II, Momentum Coarse-graining of Fermi Systems in Finite Volume, *J. Math. Phys.* **15** (1974) 1348–1350.

45. (with J.C. Wolfe) C^*-algebraic Formalism for Coarse-graining III, Momentum Coarse -graining of Fermi Systems in Infinite Free Space, *J. Math. Phys.* **15** (1974) 1351–1365.

46. Positivity of the K-entropy on Non-Abelian K-Flows, *Z. Wahrscheinlichkeitstheorie u. verw. Gebiete* **29** (1974) 241–252.

47. Quantum Generalizations of the Kolmogorov Entropy, Workshop on Dynamical Systems, Repino, 1974 79–131.

48. Minimal K-Flow Associated to a Quantum Diffusion Process, *in Physical Reality and Mathematical Description, a volume in honor of Josef M. Jauch;* C.P. Enz and J. Mehra, eds., Reidel, Dordrecht (1974) 477–493.

49. (with C.W. Burckardt) Discipline and Interdisplinarity, Polyrama, EPF-Lausanne, **21** (1974) 2.

50. (with S.T. Ali) Covariant Conditional Expectations in Quantum Probability Theory, *Rep. ICTP, Trieste* (1974) 1–17.

51. Non-Abelian Special K-Flows, *J. Funct. Analysis* **19** (1975) 1–12.

52. An Algebraic Approach to the Theory of K-Flows and K-Entropy, *in Mathematical Problems in Theoretical Physics, H. Araki, ed., Springer LNP* **39** *Heidelberg* (1975) 315–318.

53. (with Ph. Martin) A Rigorous Model Sustaining van Hove's Phenomenon, *Helv. Phys. Acta.* **48** (1975) 59–78.

54. (with B. Whitten-Wolfe) A Mechanical Quantum Measuring Process, *Helv. Phys. Acta.* **49** (1976) 45–55.

55. Non-Equilibrium Quantum Statistical Mechanics, *in Current Problems in Elementary Particle and Mathematical Physics, Acta. Phys. Austriaca Suppl.* **XV** *(= Schladmig Winter School 1975),* P. Urban, ed., Springer, Wien (1976)

56. Algebraic K-Flows, *in Dynamical Systems in Mathematical Physics, M. Keane, ed., Astérisque (Soc. Math. de France)* **40** (1976) 63–65.

57. Generalized K-Flows, *Commun. Math. Phys.* **49** (1976) 191–215.

58. A Dilation Problem in Non-equilibrium Statistical Mechanics, *in Statistical Mechanics and Statistical Methods, a volume in honor of Elliott W. Montroll,* U. Landman, ed., Plenum Press, New York (1977) 313–329.

59. Minimal Dilations of CP-Flows, *in C^*-Algebras and Applications to Physics, H. Araki and R.V. Kadison, eds., Springer LNM* **650**, *Heidelberg* (1978) 156–161.

60. Phase Transitions, Approach to Equilibrium and Structural Stability, *in Group Theoretical Methods in Physics, P. Kramer and A. Rieckers, eds., Springer LNP* **79**, *Heidelberg* (1978) 223–246.

61. Stochasticity and Irreversibility in Infinite Mechanical Systems, *in Mathematical Problems in Theoretical Physics, G. dell'Antonio, S. Doplicher, and G. Jona-Lasino, eds., Springer LNP* **80**, *Heidelberg* (1978) 426–432.

62. (with S. Albeverio and J.-P. Eckmann) Quasi–free Generalized K-Flows, *Reports on Mathematical Physics* **13** (1978) 73–85.

63. (with K.B. Sinha) Weak Quantization in a Non-perturbative Model, *J. Math. Phys.* **20** (1979) 1336–1340.

64. (with J.C. Varilly) On the Standard Form of the Bloch Equation, *Lett. Math. Phys.* **3** (1979) 113–116.

65. (with J.C. Varilly) Some Remarks on Dilating Semigroups of Completely Positive Mappings, *Reports on Math. Phys.* **18** (1980) 97–102.

66. An Algebraic Approach for Spontaneous Symmetry Breaking in Quantum Statistical Mechanics, Chapter 5 *in Groups, Systems and Many-body Physics,* P. Kramer and M. Dal Cin, eds., Vieweg, Brauschweig/Wiesbaden (1980) 246–284.

67. Irreversibility, Decay and Asymptotic Dynamics, *J. Quant. Chem.* **XVII** (1980) 107–120.

68. Some Mathematical Problems in Non-equilibrium Statistical Mechanics, *in Advances in Functional Analysis, Holomorphy and Approximation Theory, S. Machado, ed., Springer LNM* **843**, *Heidelberg* (1981) 264–295.

69. Von Neumann's Uniqueness Theorem Revisited, *in Mathematical Analysis and Applications, 2 volumes in honour of Laurent Schwartz*; L. Nachbin, ed., Academic Press, New York (1981) *Adv. Math. Suppl. Studies* **7A** 361–368.

70. Quantum Measurement Processes: A Gedanken Experiment, *in Quantum Theory and the Structures of Time and Space, 4, L. Castell, M. Drieschner, and C.F. von Weizsaecker, eds., Carl Hanser Verlag, Munchen* (1981) 137–151.

71. Prequantization and KMS Structures, *Intern'l J. Theor. Phys.* **20** (1981) 891–903.

72. (with W.P.C. King) Faithful Normal States on JBW-algebras, *in Proceedings of the 1980 Summer Institute on Operator Algebras and Applications, AMS Proc. of Symposia in Pure Mathematics, R.V. Kadison, ed.,* **38** *Part 2* (1982) 305–307.

73. Quantum and Classical Mechanics on Homogeneous Riemannian Manifolds, *J. Math. Phys.* **23** (1982) 1785–1791.

74. Stochasticity in Non-equilibrium Statistical Mechanics, *in Stochastic Processes in Quantum Theory and Statistical Physics, S. Albeverio, ed., Springer LNP* **173**, *Heidelberg* (1982) 147–153.

75. A Geometric Dequantization Programme for the Solution of the Dirac Problem, *International J. Theor. Phys.* **22** (1983) 397–420.

76. Classical Implications of the Quantum Statistical Interpretation, *in New Horizons of Quantum Chemistry, 5, P.O. Lowdin & B. Pullman, eds., Reidel Publ., Dordrecht* (1983) 5.

77. A Derivation of the Wigner-Moyal Correspondence and its Extension beyond Flat Spaces, *in Quantum Theory and the Structures of Space and Time, 5, L. Castell & C.F. von Weiszaecker, eds., Carl Hanser Verlag, Munchen* (1983) 19–33.

78. Review of "Angular Momentum in Quantum Physics: Theory and Application, The Racah-Wigner Algebra in Quantum Theory, by L.C. Biedenharn and J.D. Louck", Bull. Amer. Math. Soc. **10** (1984) 150–152.

79. (with S. DeBièvre) Relativistic Particle Scattering, *Commun. Math. Phys.* **101** (1985) 539–557.

80. (with G.C. Hegerfeldt) New Classical Properties of Quantum Coherent States, *J. Math. Phys.* **27** (1986) 2731–2737.

81. (with S.T. Ali) Geometric Quantization: Modular Reduction Theory and Coherent States, *J. Math. Phys.* **27** (1986) 2936–2943.

82. (with Sungpyo Hong) Time–like Geodesic Flows on Lorentz Manifolds, *Ergodic Th. & Dynam. Sys.* **7** (1987) 175–192.

83. KMS Structures in Geometric Quantization, *Contemporary Mathematics (AMS)* **62** (1987) 175–186.

84. (with A.J. Rica da Silva) Algebraic aspects of geometric quantization, Abstract AMS **9** (1988) 375, # 845-53-167; see also (with A.J. Rica da Silva) The Weyl Group for Curved Manifolds in Classial and Quantum Theories; *and* On an Algebraic Property of Geometric Quantization for the Weyl Group, Physdics preprints, University of Florida, 1989.

85. Geometric Quantization: Regular Representations and Modular Algebras, *in Group Theoretical Methods in Physics, Moscow, V.V. Dodonov and V.I. Man'ko, eds., Springer LNP 382, Heidelberg* (1991) 356–364.

86. Kolmogorov Flows, Dynamical Entropies and Mechanics, *in Quantum Probability and Related Topics, VII, L. Accardi, K.R. Parthasarathy and K.B. Sinha, eds., World Scientific, Singapore* (1992) 125–137.

87. (with P.E. Ehrlich) The Conjugacy Index and Simple Astigmatic Focusing, *in Contemporary Mathematics* **127***: Geometry, Physics and Nonlinear PDE's, V. Oliker and A. Treibergs, eds., AMS, Providence, RI* (1992) 27–39.

88. (with P.E. Ehrlich) Quasi-Time Functions in Lorentzian Geometry, *in Geometric Analysis and Nonlinear Partial Differential Equations, I.J. Bakelman, ed., Marcel Dekker, New York* (1992) 203–212.

89. (with P.E. Ehrlich) Gravitational Waves and Causality, *Rev. Math. Phys.* **4** (1992) 163–221.

90. (with P.E. Ehrlich) Geodesic and Causal Behavior of Gravitational Plane Waves: Astigmatic Conjugacy, *in Proc. of Symposia in Pure Mathematics* **54**, *Part 2, R. Greene and S.T. Yau, eds., AMS, Providence RI* (1993) 203–209.

91. Chaos in Classical and Quantum Theories, *in Symposium on the Foundations of Modern Physics, Cologne, P. Busch, P. Lahti and P. Mittelstaedt, eds., World Scientific Publ. Singapore* (1993) 176–177.

92. Models vs. Models, *Proc. Oxford Wigner Symposium 1993; Intern'l Journ. Modern Phys. B* (to appear); reproduced as Appendix E in Emch and Liu (see below).

93. The Philosophy of E.P. Wigner, *in Classical and Quantum Systems, Foundations and Symmetries, Wigner Symposium, H.-D. Doebner, W. Scherer and F. Schroeck, eds. World Scientific, Singapore* (1993) 2–8.

94. Central Extensions for the Weyl CCR in curved Space, *Proc. 19th International Colloquium on Group Theoretical Methods in Physics, Salamanca,* M.A. del Olmo, M. Santander and J. Mateos Guilarte, eds., Anales de Fisica, Monografias, Ciemat/RSEF, Madrid (1993) 426–429.

95. (with H. Narnhofer, W. Thirring and G.L. Sewell) Anosov Actions on Non-Commutative Algebras, *J. Math. Phys.* **35** (1994) 5582–5599.

96. (with G.C. Hegerfeldt and L. Sreit) Preface (p.ix) in *On Klauder's Path: a Field Trip,* G.G. Emch, G.C. Hegerfeldt and L. Streit, eds., World Scientific Publishers, Singapore, 1994. 149–158.

97. (with A. El Gradechi and S.T. Ali) Modular Algebras in Geometric Quantization, *J. Math. Phys.* **35** (1994) 6237–6243.

98. Annotation (pp. 1–28), in *Philosophical Reflections and Syntheses, Vol. VI of The Collected Works of Eugene Paul Wigner,* J. Mehra and A.S. Wightman, eds., Springer, Heidelberg, 1995; 2nd printing, 1997.

99. (with J. Bertrand and G. Rideau) The Cohomology of the Classical and Quantum Weyl CCR on Curved spaces, *Lett. Math. Phys.* **34** (1995) 149–158.

100. Modular Structures in Geometric Quantization, *in Quantization, Coherent States and Complex Structures, J.P. Antoine et al., eds., Plenum Press, New York* (1995) 33–42.

101. De l'ubiquité de la condition KMS, *in Modern Group Theoretical Methods in Physics,* J. Bertrand et al., eds., Kluwer Academic Publ., Dordrecht, The Netherlands (1995) 115–128.

102. Comments on a Recent Paper by S. Adler on Projective Group Representations in Quaternionic Hilbert Spaces, *J. Math. Phys.* **37** (1996) 6582–6585.

103. (with S.L. Adler) A Rejoinder on Quaternionic Projective Representations, *J. Math. Phys.* **38** (1997) 4758–4762.

104. Chaotic Dynamics in Non-Commutative Geometry, *in Quantizations, Deformations and Coherent States,* J.P. Antoine et al., eds., Polish Scientific Publ., Warsaw (1997) 33–42.

105. Beyond Irreducibility and Back, *Rep. Math. Phys.* **40** (1997) 187–193.

106. Foundations of Quantum Mechanics: Building on Von Neumann's Heritage, *Intern. J. Quantum Chemistry* **65** (1997) 379–387.

107. On the Need to Adapt de Finetti's Probability Interpretation to QM, *in Quantum Probability, Banach Center Publications* **43** (1998) 157–166.

108. (with I. Peter) Quantum Anosov Flows: A new Family of Examples, *J. Math. Phys.* **39** (1998) 4513–4539.

109. Wavelets from Coherent states, Part I *Proc. Intern'l Soc. Opt. Eng. Conference: Wavelets Applications VI* H. Szu, ed. **3733** (1999) 34–43 [for Part II, see J. Klauder. *ibid.*]

110. (with A. Emch–Dériaz) Madame du Châtelet, a Fair presentation of Newton's *Principia. Proc. Xe Congrès International des Lumières, Dublin, 1999.*

111. (with A.Z. Jadcyk) Weak Projective Representations, Quaternions and Monopoles, *Stochastic Processes, Physics and Geometry: New Interplays. A CMS/AMS Volume in Honor of Sergio Albeverio Conf. Proc. Canad. Math. Soc.* **29** (2000) 157–164.

112. Review of "John von Neumann and the Foundations of Quantum Physics, M. Rédei and M. Stöltzner, eds., Kluwer Academic Publisher, Dordrecht," *Foundations of Physics* **32** (2002) 981–985.

113. Review of "Mathematical Topics between Classical and Quantum mechanics, by N. P. Landsman, Springer Monographs in Mathematics, New York, Springer-Verlag, 1998" *Studies in History and Philosophy of Modern Physics* **33** (2002) 148–150.

114. On Wigner's different usages of models, *Electronic Proceedings of the Wigner Centenial Conference, held in Pécs 2002,* Paper No. 59, 1–4 (2003); to appear also in *Acta Physica Hungarica: Heavy Ion Physics.*

115. Is there a Quantum de Finetti Programme? in *Foundations of Probability and Physics-2, Ser. Math. Modelling in Phys., Eng., and Cogn. Sci.*, **5**, Växjö Univ. Press (2003) 159–178.

116. The Dynamics of Modelling, in *Time, Quantum & Information, Festschrift zum 90. Geburtstag von Prof. C.F. von Weizsäcker,* L. Castell and O. Ischebeck, eds., Springer-Verlag, Heidelberg, 2003, pp. 243–259.

117. Quantum Measurement: a bridge to Classical Physics, in *Quantum Information and Computation,* E. Donkor, A.R. Pirich and H.E. Brandt, eds., Intern. Soc. Opt. Eng. (SPIE) Proceedings 5105 (2003), pp. 255–264.

118. Interactive modelling, *J. Stat. Phys.* **116** (2004) 17–28.

119. Review of "Quantum Mechanics and its Emergent Macrophysics, by G.L. Sewell, Princeton University Press, Princeton and Oxford, 2002", *Studies in History and Philosophy of Modern Physics* **35**(2004) 129–133.

120. So ... what was the question? *J. of Nonlinear Math. Phys.* **11** (2004) *Supplement: Geometrical Methods in Physics,* S.T. Ali et al. eds.; pp. 1–8.

121. (with S.T. Ali, A. Odzijewicz, M. Schlichenmaier and S.L. Woronowicz) Preface (p. vi–x) in *Twenty Years of Bialowieza: A Mathematical Anthology, Aspects of Differential Geometric Methods in Physics,* S.T. Ali, G. G. Emch, A. Odzijewicz, M. Schlichenmaier and S.L. Woronowicz, eds., World Scientific Monogrtaph series in Mathematics, Vol. 8, World Scientic Publishers, Singapore, 2005.

122. Probabilistic issues in statistical mechanics, *Stud. Hist. Phil. Modern Phys.* **36** (2005) 303–322.

123. (with C. Liu) Explaining quantum aneous symmetry braking, *Stud. Hist. Phil. Modern Phys.* **36** (2005) 137–163.

124. Not what models are, but what models do, in *Foundations of Probability and physics 3,* A. Khrennikov, ed. *Amer. Inst. Phys. Proceedings* **750** (2005) 76–86

125. Contributions of Indian Mathematicians to Quantum Statistics, in *Contributions to the History of Indian Mathematics,* G.G. Emch, R. Sridharan aand M.D. Srinivas, eds., Hindustan Book Agency (India), New Delhi, 2005, pp. 261–288.

126. (with A. Emch–Dériaz) On Newton's French Translator: How faithful was Madame du Châtelet? in *Emilie du Châtelet, rewriting Enlightenment philosophy and science,* J. Zinnser and J.C. Hayes, ed.; *Studies in Voltaire and the Eighteenth Century,* Oxford (2006), pp. 226–251.

127. (with A. Emch–Dériaz) Mathématicienne, et comment, *in Madame Du Châtelet, La femme des Lumières,* E. Badinter & D. Muzerelle, eds. Bibliothèque Nationale de France, Paris (2006); pp. 90–94.

128. (with A. Emch–Dériaz) Emilie du Châtelet, Minerve des Lumières, *in Pour la science: Les génies de la science, Dossier N*^{o} 27, mai-juillet 2006; pp. 10–15.

129. Quantum Statistical Physics, Chapter 10 in *Handbook of the Philosophy of Science, Philosophy of Physics,* J. Butterfield, and J. Earman,eds., Elsevier, 2007, pp. 1075–1182.

Preface

This volume is a tribute to Gérard G. Emch, on the occasion of his seventieth birthday. To all of us who have put down their appreciation in writing here, Gérard Emch has been a teacher, mentor and friend. This collection of articles, some inspired by Gérard's own work, is a sequel to the special session, dedicated to him, at the XXVth Workshop on Geometrical Methods in Physics, held in Bialowieza, Poland, from July 2 - 8, 2006. A large number of Gérard's students, friends and collaborators were invited to that meeting, but of course, not everyone was able to attend. We did however, receive expressions of good wishes from them all. The call for papers also went out to a longer list of names than just the ones represented here. In presenting this volume to Gérard, we offer him our collective good wishes and our thankfulness for having shared with us all that he has - scientifically, professionally and in friendship.

Scientifically, Gérard's work spanned five decades, including in its gamut the foundations of quantum mechanics, the C* -algebraic approach to quantum mechanics and field theory, equilibrium and non-equilibrium statistical mechanics, quantum ergodic theory and in more recent years the history and philosophy of science. It is less well known that Gérard started his scientific career working on paramagnetic resonance in certain compounds. However, the scientific atmosphere of the Département de Physique Théorique in the University of Geneva of those days, under the influence of Joseph M. Jauch was probably overwhelming and Gérard gravitated to more theoretical and fundamental issues of quantum physics. Collaborating with Jauch and Piron, Gérard did some pioneering work on quantum logic (particularly on quaternionic quantum mechanics), symmetries and representations of the Lorentz group. Along with these, there were already signs of Gérard getting interested in quantum statistical mechanics - both equilibrium and non-equilibrium. In the interim period, the celebrated paper of Haag and Kastler on the C*-algebraic approach to quantum mechanics and quantum field theory made its expected impact on Gérard and his early batch of graduate students at the University of Rochester (including the editors here). This led to a series of publications on phase transition and partial states, on the one hand, and on coarse-graining and the quantum master equation on the other, by Gérard with his students and other collaborators. As a natural sequel followed a number of significant papers by Gérard on K-flows, as well as on the question of dilation of irreversible dissipative dynamics represented by completely positive semi-groups on *-algebras. Many of the research areas pioneered by Gérard are now under intense development. With one of the present editors and other students, Gérard also studied the relationship between geometric quantization and generalized coherent states before launching himself into an entirely new area of research, viz., geometry and theory of general relativity. The new millennium saw Gérard getting interested in history and philosophy of quantum mechanics and thermodynamics; there were several papers with Antoinette Emch-Dériaz (a professor of history at the University of Florida, Gainesville) amongst others and a book on the logic of thermodynamics that Gérard co-authored in this period.

This is a large canvas of scientific creativity, displaying considerable energy and originality on the part of Gérard in his distinguished scientific career.

Human relationships are multi-faceted and complex. It would be hard to describe our indebtedness to Gérard and all the ways in which we had come to appreciate him. For those among us, who had been his students, his influence has in many ways shaped the conduct of our professional and scientific lives. Looking back on those years, what stands out most in our thoughts is the very personal relationship that Gérard and Antoinette nurtured with his students. Some of his students and associates remember fondly the delightful hospitality of the Emch's, the initiation into the art of savouring good wine and of hand-grinding Turkish Coffee (ouch!). Unforgettable also is Gérard's sense of humour and ready wit, acerbic at times in the old days, however mellowing like good wine with age!

All of us who have collaborated with or just known Gérard over the last four decades and have cherished this association, wish him the very best in the years to come.

S. Twareque Ali
Montreal, Canada

Kalyan B. Sinha
Bangalore, India.

Berezin-Toeplitz Quantization over Matrix Domains

S. Twareque Ali and M. Engliš

We explore the possibility of extending the well-known Berezin-Toeplitz quantization to reproducing kernel spaces of vector-valued functions. In physical terms, this can be interpreted as accommodating the internal degrees of freedom of the quantized system. We analyze in particular the vector-valued analogues of the classical Segal-Bargmann space on the domain of all complex matrices and of all normal matrices, respectively, showing that for the former a semi-classical limit, in the traditional sense, does not exist, while for the latter only a certain subset of the quantized observables have a classical limit: in other words, in the semiclassical limit the internal degrees of freedom disappear, as they should. We expect that a similar situation prevails in much more general setups.

Dedicated to Gerard G. Emch on the occasion of his 70th birthday

1. Introduction

Let Ω be a symplectic manifold, with symplectic form ω, and $\mathcal{H}$ a subspace of $L^2(\Omega, d\mu)$, for some measure μ, admitting a reproducing kernel K. For $\phi \in C^\infty(\Omega)$, the *Toeplitz operator* T_ϕ with symbol ϕ is the operator on $\mathcal{H}$ defined by

$$T_\phi f = \mathbf{P}(\phi f), \qquad f \in \mathcal{H},$$

where $\mathbf{P} : L^2(\Omega, d\mu) \to \mathcal{H}$ is the orthogonal projection. Using the reproducing kernel K, this can also be written as

$$T_\phi f(x) = \int_\Omega f(y)\phi(y)K(x, y)\, d\mu(y).$$

It is easily seen that T_ϕ is a bounded operator whenever ϕ is a bounded function, and $\|T_\phi\|_{\mathcal{H}\to\mathcal{H}} \leq \|\phi\|_\infty$, the supremum norm of ϕ.

The first author would like to acknowledge support from the NSERC, Canada and FQRNT, Québec. Research of the second author was supported by GA ČR grant no. 201/03/0041 and AV ČR research plan no. AV0Z10190503.

Suppose now that both the measure μ and the reproducing kernel subspace $\mathcal{H}$ are made to depend on an additional parameter $h > 0$ (shortly to be interpreted as the Planck constant), in such a way that the associated Toeplitz operators $T_\phi^{(h)}$ on $\mathcal{H}_h$ satisfy, as $h \searrow 0$,

$$\|T_\phi^{(h)}\|_{\mathcal{H}_h \to \mathcal{H}_h} \to \|\phi\|_\infty, \tag{1.1}$$

and

$$\|T_\phi^{(h)} T_\psi^{(h)} - T_{\phi\psi}^{(h)}\|_{\mathcal{H}_h \to \mathcal{H}_h} \to 0, \tag{1.2}$$

$$\|\tfrac{2\pi}{ih}[T_\phi^{(h)}, T_\psi^{(h)}] - T_{\{\phi,\psi\}}^{(h)}\|_{\mathcal{H}_h \to \mathcal{H}_h} \to 0 \tag{1.3}$$

(where $\{\cdot, \cdot\}$ is the Poisson bracket with respect to ω), and, more generally,

$$T_\phi^{(h)} T_\psi^{(h)} \approx \sum_{j=0}^{\infty} h^j \, T_{C_j(\phi,\psi)}^{(h)} \qquad \text{as } h \to 0,$$

for some bilinear differential operators $C_j : C^\infty(\Omega) \times C^\infty(\Omega) \to C^\infty(\Omega)$, with $C_0(\phi, \psi) = \phi\psi$ and $C_1(\phi, \psi) - C_1(\psi, \phi) = \frac{i}{2\pi}\{\phi, \psi\}$. Here the last asymptotic expansion means, more precisely, that

$$\left\| T_\phi^{(h)} T_\psi^{(h)} - \sum_{j=0}^{N} h^j T_{C_j(\phi,\psi)}^{(h)} \right\| = O(h^{N+1}) \quad \text{as } h \searrow 0, \qquad \forall N = 0, 1, 2, \dots. \tag{1.4}$$

One then speaks of the *Berezin-Toeplitz quantization.* Indeed, it is well known that the recipe

$$\phi * \psi := \sum_{j=0}^{\infty} h^j \, C_j(\phi, \psi)$$

then gives a *star-product* on Ω, and (1.2), (1.3) just amount to its correct semiclassical limit.

Observe that if we introduce the normalized reproducing kernels — or *coherent states* — by

$$k_y := \frac{K(\cdot, y)}{\|K(\cdot, y)\|_{\mathcal{H}}}, \qquad \text{i.e.} \qquad k_y(x) = \frac{K(x, y)}{K(y, y)^{1/2}},$$

and define the *Berezin transform* of an operator T on $\mathcal{H}$ by

$$\widetilde{T}(y) := \langle T k_y, k_y \rangle_{\mathcal{H}},$$

then (1.4) implies that as $h \to 0$,

$$\widetilde{T_\phi^{(h)} T_\psi^{(h)}} \approx \sum_{j=0}^{\infty} h^j \, \widetilde{T_{C_j(\phi,\psi)}^{(h)}}, \tag{1.5}$$

pointwise and even uniformly on Ω. Often, one also has a stronger version of (1.1), namely

$$\widetilde{T_\phi^{(h)}}(x) \to \phi(x) \qquad \forall x \in \Omega. \tag{1.6}$$

The simplest instance of the above situation is $\Omega = \mathbb{R}^{2n} \simeq \mathbf{C}^n$, with the standard (Euclidean) symplectic structure, and

$$\mathcal{H}_h = L^2_{\text{hol}}(\Omega, d\mu_h) \tag{1.7}$$

the Segal-Bargmann space of all holomorphic functions square-integrable with respect to the Gaussian measure $d\mu_h(z) := e^{-\|z\|^2/h}(\pi h)^{-n}\, dz$ (dz being the Lebesgue measure on $\mathbf{C}^n$). The space $\mathcal{H}_h$ admits the reproducing kernel $K_h(x, y) = e^{\langle x,y\rangle/h}$. The Berezin transform turns out to be given just by the familiar heat operator semigroup,

$$\widetilde{T_\phi^{(h)}}(x) = (\pi h)^{-n} \int_{\mathbf{C}^n} e^{-\|x-y\|^2/h} \phi(y)\, dy,$$

so that, by the stationary phase (WJKB) expansion (see Section 6 below; for a discussion of the (WJKB) approximation, see for example [Me], Chapter 7) [1]

$$\widetilde{T_\phi^{(h)}}(x) = \sum_{j=0}^{\infty} \frac{h^j}{j!} \Delta^j \phi(x). \tag{1.8}$$

Similarly,

$$\widetilde{T_\phi^{(h)} T_\psi^{(h)}}(x) = (\pi h)^{-2n} \int_{\mathbf{C}^n} \int_{\mathbf{C}^n} \phi(y)\psi(z) e^{(\langle x,y\rangle + \langle y,z\rangle + \langle z,x\rangle - \|x\|^2 - \|y\|^2 - \|z\|^2)/h}\, dy\, dz,$$

so by stationary phase again,

$$\widetilde{T_\phi^{(h)} T_\psi^{(h)}}(x) = \sum_{\alpha,\beta,\gamma} \frac{\partial^\alpha \overline{\partial}^{\alpha+\gamma} \phi(x)}{\alpha!\gamma!} \, \frac{\partial^{\beta+\gamma} \overline{\partial}^\beta \psi(x)}{\beta!} \, h^{|\alpha|+|\beta|+|\gamma|}. \tag{1.9}$$

[1] Throughout this paper, we are using the slightly nonstandard Laplacian $\Delta = \sum_j \partial^2/\partial z_j \partial \overline{z}_j$, which differs from the usual one by a factor of 4.

(Here the summation extends over all multiindices α, β, γ, i.e. n-tuples of nonnegative integers $\alpha = (\alpha_1, \dots, \alpha_n)$, etc., and we are using the usual multiindex notations $|\alpha| = \alpha_1 + \dots + \alpha_n$, $\alpha! = \alpha_1! \dots \alpha_n!$, $\partial^\alpha = \partial^{|\alpha|}/\partial x_1^{\alpha_1} \dots \partial x_n^{\alpha_n}$.) Inserting (1.8) and (1.9) into (1.5), we get formulas for the cochains C_j:

$$C_j(\phi, \psi) = \sum_{|\alpha|=j} \frac{1}{\alpha!} \partial^\alpha \phi \cdot \overline{\partial}^\alpha \psi. \tag{1.10}$$

The resulting star-product coincides, essentially, with the familiar Moyal product.

Other examples of Berezin-Toeplitz quantization include the unit disc $\mathbf{D}$ with the Poincaré metric, bounded symmetric domains, strictly pseudoconvex domains with metrics having reasonable boundary behaviour, or, provided one allows not only holomorphic functions but also sections of line bundles as elements of $\mathcal{H}_h$, all compact Kähler manifolds whose Kähler form is integral. In all these cases, the choice of the spaces (1.7) which works are the weighted Bergman spaces $\mathcal{H}_h = L^2_{\text{hol}}(\Omega, e^{-\Phi/h}\omega^n)$, where n is the complex dimension of Ω and Φ is a Kähler potential for ω (so, for instance, for the unit disc $\mathcal{H}_h = L^2_{\text{hol}}(\mathbf{D}, (1-|z|^2)^{(1/h)-2})$). See [KS], [BMS] or [AE] for the details and further discussion.

In this note, we explore the possibility of extending the above formalism to reproducing kernel spaces of *vector-valued* functions. In physical terms, this can be interpreted as accommodating the *internal degrees of freedom* of the quantized system.

In more concrete terms, this means that we again consider, for a given $h > 0$, a suitable measure μ_h on Ω, and a reproducing kernel subspace $\mathcal{H}_h \subset L^2_{\mathbf{C}^N}(\Omega, d\mu_h)$, where the subscript $\mathbf{C}^N$ indicates that we are now dealing with vector-valued functions taking values in $\mathbf{C}^N$ for some $N \geq 1$. The reproducing kernel K_h of $\mathcal{H}_h$ will thus now be a matrix-valued object, $K_h : \Omega \times \Omega \to \mathbf{C}^{N\times N}$. We can again consider, for each $\phi \in C^\infty(\Omega)$, the associated Toeplitz operators, and investigate the existence of the asymptotic expansion (1.4). In fact, we can now even allow *matrix-valued* symbols $\phi \in C^\infty_{\mathbf{C}^{N\times N}}(\Omega)$. By analogy with the scalar-valued case, one may again expect appropriate asymptotic expansions

$$\widetilde{T^{(h)}_\phi}(x) = \sum_{j=0}^{\infty} L_j\phi(x)\, h^j,, \tag{1.11}$$

$$\widetilde{T^{(h)}_\phi T^{(h)}_\psi}(x) = \sum_{j=0}^{\infty} M_j(\phi, \psi)(x)\, h^j, \tag{1.12}$$

with some differential and bidifferential operators L_j and M_j, respectively, whose comparison would yield (1.5), thus suggesting that (1.4) is also likely to hold.

Unfortunately, at this level of generality the results are negative: we show that for certain spaces $\mathcal{H}_h$ as above, which are quite natural generalizations of the Euclidean situation (1.7) to vector-valued functions, the semiclassical expansions (1.11), (1.12) fail to hold. More specifically, it seems that there are asymptotic expansions for $\widetilde{T_\phi^{(h)}}(x)$ and $\widetilde{T_\phi^{(h)} T_\psi^{(h)}}(x)$ in powers of h, but their coefficients do not depend only on the jets of ϕ and ψ at x, but also at other points (i.e. are not local operators); besides, in addition to integer powers of h, half-integer powers seem to also enter the picture. Finally, (1.1) and (1.6) may also break down.

However, it turns out that upon restricting to appropriate domains Ω and appropriate classes of functions ϕ, ψ, the situation can be saved completely: namely, the following picture emerges. The admissible functions ϕ can be identified with functions $f(d_1; d_2, \ldots, d_N)$ on $\mathbf{C} \times \mathbf{C}^{N-1}$ that are symmetric in the $N-1$ variables $d_2, \ldots, d_N$. Clearly, one can associate a function $u^\#$ of this form to any function $u : \mathbf{C} \to \mathbf{C}$ by the recipe $u^\#(d_1; d_2, \ldots, d_N) := u(d_1)$. We show that for any functions f, g of the above form, there exist uniquely determined functions u_r on $\mathbf{C}$, $r = 0, 1, 2, \ldots$, such that

$$\widetilde{T_f^{(h)} T_g^{(h)}} \approx \sum_{r=0}^{\infty} h^r \, \widetilde{T_{u_r^\#}^{(h)}}. \tag{1.13}$$

Further, the u_r are given by differential expressions involving f and g; and, finally, if f and g themselves are of the form $v^\#$ and $w^\#$, respectively, for some functions v, w on $\mathbf{C}$, then in fact $u_r = C_r(v, w)$ with the bidifferential operators C_r given by (1.10) for $n = 1$, i.e. $\Omega = \mathbf{C}$. This suggests the following interpretation: our quantum system has N internal degrees of freedom, which have no classical couterparts, so that only a subset of the quantized observables have a classical limit. In the semiclassical limit the internal degrees of freedom disappear, as they should. We conjecture that similar quantizations can be carried out by our method in much more general setups.

The paper is organized as follows. In Section 2, we introduce the spaces $\mathcal{H}_h$ of vector-valued functions on $\mathbf{C}^{N \times N}$, as well as their analogues for the subset of all normal matrices in $\mathbf{C}^{N \times N}$. These spaces have previously appeared in [AEG]. In Section 3, we define a generalization of the Berezin transform — which will now also be a matrix-valued object — and establish its basic properties. Sections 4 and 5 discuss the semiclassical asymptotic expansions (1.11)–(1.12) in the above two settings of the domains of all matrices and all normal matrices, respectively. In Section 6 we present our first result in the positive direction, by establishing an asymptotic expansion — which is, however, of a highly non-local nature — for $\widetilde{T_\phi^{(h)}}$ and $\widetilde{T_\phi^{(h)} T_\psi^{(h)}}$ in the case of the normal matrices. Finally, in Sections 7 and 8 we introduce our restricted class of observables ϕ and establish the asymptotic expansion (1.13).

2. The Domains and The Spaces

Our first domain is $\Omega = \mathbf{C}^{N\times N}$, with the measures

$$d\mu_h(Z) = e^{-\mathrm{Tr}(Z^*Z)/h}\,(\pi h)^{-N^2}\,dZ,$$

where dZ denotes the Lebesgue measure on $\mathbf{C}^{N\times N}$. The measures μ_h are normalized to be of total mass one. The functions $\Psi_j(Z) := Z^j$ satisfy

$$\int_\Omega Z^{*j}Z^k\,d\mu_h(Z) = \delta_{jk}h^k c_k I \tag{2.1}$$

for some numbers $c_k > 0$; see [AEG]. Explicitly, c_k are given by ([Gin], formula (1.40), and [Kri]) [2]

$$c_k = \begin{cases} \dfrac{(k+N+1)!}{N!(k+1)(k+2)} & \text{for } k \geq N-1, \\ \dfrac{(k+N+1)!}{N!(k+1)(k+2)} - \dfrac{N!}{(k+1)(k+2)(N-k-2)!} & \text{for } k < N-1, \end{cases}$$

that is,

$$c_k = \frac{1}{(k+1)(k+2)}\left[\prod_{j=1}^{k+1}(N+j) - \prod_{j=1}^{k+1}(N-j)\right]. \tag{2.2}$$

In particular, $c_0 = 1, c_1 = N, c_2 = N^2+1$, etc.

It follows that if $\chi_1, \ldots, \chi_N$ is the standard basis of $\mathbf{C}^N$, then the functions

$$\frac{Z^k\chi_j}{\sqrt{c_k h^k}}, \qquad j = 1, \ldots, N,\ k = 0, 1, 2, \ldots, \tag{2.3}$$

are orthonormal in $L^2_{\mathbf{C}^N}(\Omega, d\mu_h)$. Let $\mathcal{H}_h$ be the subspace spanned by these functions. Then the function

$$K_h(X,Y) = \sum_{k=0}^{\infty}\frac{X^kY^{*k}}{c_k h^k} \tag{2.4}$$

converges for all $X, Y \in \Omega$ and is the reproducing kernel of $\mathcal{H}_h$, in the sense that

$$\int_\Omega K_h(X,Y)f(Y)\,d\mu_h(Y) = f(X), \qquad \forall f \in \mathcal{H}_h,\ \forall X \in \Omega.$$

[2] The authors are grateful to M. Bertola for this result.

Our second domain will be the subset $\Omega_{\text{norm}} = \{Z \in \mathbf{C}^{N\times N} : \ Z^*Z = ZZ^*\}$ of all normal matrices in $\mathbf{C}^{N\times N}$. By the Spectral Theorem, any $Z \in \Omega_{\text{norm}}$ can be written in the form

$$Z = U^* D U, \tag{2.5}$$

with $U \in U(N)$ unitary and D diagonal; D is determined by Z uniquely up to permutation of the diagonal elements, and if the latter are all distinct and their order has been fixed in some way, then U is unique up to left multiplication by a diagonal matrix with unimodular elements. Consequently, there exists a unique measure $d\mu_h(Z)$ on Ω_{norm} such that

$$\int_{\Omega_{\text{norm}}} f(Z)\, d\mu_h(Z) = (\pi h)^{-N} \int_{U(N)} \int_{\mathbf{C}^N} f(U^* D U)\, e^{-\|D\|^2/h}\, dU\, dD \qquad \forall f,$$

where dU is the normalized Haar measure on $U(N)$, and dD is the Lebesgue measure on $\mathbf{C}^N$, where we are identifying the diagonal matrix $D = \text{diag}(d_1, \ldots, d_N)$ with the vector $d = (d_1, \ldots, d_N) \in \mathbf{C}^N$, and $\|D\|^2 = \|d\|^2 := |d_1|^2 + \cdots + |d_N|^2$. Again, one easily checks [AEG] that

$$\int_{\Omega_{\text{norm}}} Z^{*j} Z^k \, d\mu_h(Z) = \delta_{jk} k! h^k I, \tag{2.6}$$

so that the elements (2.3) are orthogonal also in $L^2_{\mathbf{C}^N}(\Omega_{\text{norm}}, d\mu_h)$, and we let $\mathcal{H}_h$ be the subspace spanned by them. The reproducing kernel is then given by

$$K_h(X, Y) = \sum_{k=0}^{\infty} \frac{X^k Y^{*k}}{k! h^k} \tag{2.7}$$

(with the series converging for all $X, Y \in \mathbf{C}^{N\times N}$), in the sense that

$$\int_{\Omega_{\text{norm}}} K_h(X, Y) f(Y)\, d\mu_h(Y) = f(X), \qquad \forall f \in \mathcal{H}_h,\ \forall X \in \Omega_{\text{norm}}.$$

Remark. At first sight, the most natural candidate for the vector-valued space $\mathcal{H}$ would seem to be the subspace $L^2_{\text{hol},\mathbf{C}^N}(\Omega, d\mu)$ of *all* holomorphic functions in $L^2_{\mathbf{C}^N}(\Omega, d\mu)$ (i.e. of all square-integrable $\mathbf{C}^N$-valued functions which depend holomorphically on the coordinates $z_{11}, \ldots, z_{NN}$ of the point $z \in \Omega$). However, this choice turns out to be too simple-minded: the reproducing kernel is then just $k(x, y)I$, where $k(x, y)$ is the reproducing kernel of the ordinary (scalar-valued) space $L^2_{\text{hol}}(\Omega, d\mu)$; the Toeplitz operator T_ϕ (to be introduced in the next section) is just the $N \times N$ matrix $[T_{\phi_{jk}}]_{j,k=1}^N$ of Toeplitz operators on $L^2_{\text{hol}}(\Omega, d\mu)$; and the Berezin transforms (also to be introduced in the next section) are just $\widetilde{T_\phi} =$

$[\widetilde{T_{\phi_{jk}}}]_{j,k=1}^{N}$ and $\widetilde{T_\phi T_\psi} = [\sum_{l=1}^{N} \widetilde{T_{\phi_{jl}} T_{\psi_{lk}}}]_{j,k=1}^{N}$. Thus, for instance, for the spaces (1.7) with $n = 1$ (i.e. on $\Omega = \mathbf{C}$), $\|T_\phi^{(h)} T_\psi^{(h)} - T_{\phi\psi}^{(h)}\| \to 0$ as $h \to 0$, while

$$\|\tfrac{2\pi}{ih}[T_\phi^{(h)}, T_\psi^{(h)}] - T_{[\![\phi,\psi]\!]}^{(h)}\| \to 0,$$

where

$$[\![\phi, \psi]\!] := \partial\phi \cdot \overline{\partial}\psi - \partial\psi \cdot \overline{\partial}\phi$$

(here $\partial, \overline{\partial}$ are applied individually to each element of a matrix, and the dot stands for matrix multiplication). From the physical point of view, this "matrix-valued Poisson bracket" seems to be a rather doubtful object, indicating that the spaces $L^2_{\mathrm{hol},\mathbf{C}^N}$ are probably not the right route to take.

3. The Berezin Transform

Let, quite generally, $\mathcal{H}$ be a reproducing kernel subspace of $\mathbf{C}^N$-valued functions in $L^2_{\mathbf{C}^N}(\Omega, d\mu)$, for some domain Ω and measure μ on it, with reproducing kernel K; thus K is a $\mathbf{C}^{N\times N}$-valued function on Ω and

$$f(X) = \int_\Omega K(X, Y) f(Y)\, d\mu(Y) \qquad \forall X \in \Omega,\ f \in \mathcal{H}. \tag{3.1}$$

In particular, for any $\chi \in \mathbf{C}^N$, the functions

$$K_{Y,\chi}(X) := K(X, Y)\chi$$

belong to $\mathcal{H}$, and

$$\langle f, K_{Y,\chi}\rangle_{\mathcal{H}} \equiv \int_\Omega K_{Y,\chi}^* f\, d\mu = \chi^* f(Y). \tag{3.2}$$

See [AAG] for more information on such spaces and their reproducing kernels.

Let T be an arbitrary bounded linear operator on $\mathcal{H}$. For any fixed $X \in \Omega$, the expression $\langle T K_{X,\chi}, K_{X,\chi'}\rangle$ is evidently linear in χ and χ'^*; thus there exists a unique $N \times N$ matrix $\widetilde{\widetilde{T}}(X)$ such that

$$\chi'^* \widetilde{\widetilde{T}}(X)\chi = \langle T K_{X,\chi}, K_{X,\chi'}\rangle.$$

We define the *Berezin transform* $\widetilde{T}$ of T by

$$\widetilde{T}(X) := K(X, X)^{-1/2} \widetilde{\widetilde{T}}(X) K(X, X)^{-1/2}.$$

That is,

$$\chi'^{*}\widetilde{T}(X)\chi = \langle TK_{X,K(X,X)^{-1/2}\chi}, K_{X,K(X,X)^{-1/2}\chi'}\rangle_{\mathcal{H}}.$$

At first sight, this definition may seem a little *ad hoc;* the reason behind it is that this seems to be the only way to make the following statements true.

Proposition 3.1. *$\widetilde{T}$ is a $\mathbf{C}^{N\times N}$-valued function on Ω satisfying*

(i) $\widetilde{T^*} = (\widetilde{T})^*$;

(ii) *if ϕ is a matrix-valued function on Ω such that $\phi K_{X,\chi} \in \mathcal{H}$ for all $X \in \Omega$ and $\chi \in \mathbf{C}^N$, then*

$$\widetilde{M_\phi}(X) = K(X,X)^{-1/2}\phi(X)K(X,X)^{1/2}$$

where $M_\phi f := \phi f$;

(iii) *in particular,* $\widetilde{I}(X) = I\ \forall X \in \Omega$;

(iv) $\|\widetilde{T}(X)\|_{\mathbf{C}^N\to\mathbf{C}^N} \le \|T\|_{\mathcal{H}\to\mathcal{H}},\ \forall X \in \Omega.$

If the elements of $\mathcal{H}$ are holomorphic functions, then also

(v) $\widetilde{T}(X) = 0\ \forall X$ *only if* $T = 0$.

Proof. (i) is immediate from the definition, while (ii) follows from the reproducing property (3.1), and (iii) is a trivial special case of (ii). To prove (iv), observe that, by (3.2),

$$\|K_{X,\chi}\|^2_{\mathcal{H}} = \langle K_{X,\chi}, K_{X,\chi}\rangle_{\mathcal{H}} = \chi^* K_{X,\chi}(X) = \chi^* K(X,X)\chi = \|K(X,X)^{1/2}\chi\|^2_{\mathbf{C}^N}.$$

Thus, for any $\chi, \chi' \in \mathbf{C}^N$,

$$\begin{aligned} |\chi'^{*}\widetilde{T}(X)\chi| &= |\langle TK_{K(X,X)^{-1/2}\chi}, K_{K(X,X)^{-1/2}\chi'}\rangle| \\ &\le \|T\|\ \|K_{X,K(X,X)^{-1/2}\chi}\|\ \|K_{K(X,X)^{-1/2}\chi'}\| \\ &= \|T\|\ \|\chi\|_{\mathbf{C}^N}\ \|\chi'\|_{\mathbf{C}^N}, \end{aligned}$$

and the assertion follows.

Finally, for (v), note that $\widetilde{T} \equiv 0$ implies $\widetilde{\widetilde{T}} \equiv 0$, i.e.

$$\langle TK_{Y,\chi'}, K_{X,\chi}\rangle = 0 \qquad \forall \chi, \chi' \tag{3.3}$$

whenever $X = Y$. If $\mathcal{H}$ consists of holomorphic functions, then $K(X,Y)$ is holomorphic in X and conjugate-holomorphic in Y; thus the left-hand side of (3.3) is holomorphic in X and conjugate-holomorphic in Y. It is well known that such functions are uniquely determined by their restriction to the diagonal $X = Y$; consequently, (3.3) holds for all X, Y, i.e.

$$\chi^*(TK_{Y,\chi'})(X) = 0 \qquad \forall \chi, \chi' \, \forall X, Y.$$

Hence $TK_{Y,\chi'} = 0$ for all Y and χ', and thus for any $f \in \mathcal{H}$

$$\chi'^*(T^* f)(Y) = \langle T^* f, K_{Y,\chi'} \rangle = \langle f, TK_{Y,\chi'} \rangle = 0,$$

i.e. $T^* f = 0$. Thus $T^* = 0$ and $T = 0$. ■

In analogy with the scalar-valued situation, we next define for any $\phi \in C^\infty_{\mathbf{C}^{N\times N}}(\Omega)$ the Toeplitz operator T_ϕ on $\mathcal{H}$ by the recipe

$$T_\phi f(X) := \int_\Omega K(X, Y)\phi(Y) f(Y) \, d\mu(Y),$$

or, equivalently,

$$T_\phi f = \mathbf{P}(\phi f),$$

where $\mathbf{P} : L^2_{\mathbf{C}^N}(\Omega, d\mu) \to \mathcal{H}$ is the orthogonal projection. Note that the last formula implies that $\|T_\phi\|_{\mathcal{H}\to\mathcal{H}} \leq \|\phi\|_\infty := \sup_{X\in\Omega} \|\phi(X)\|_{\mathbf{C}^N\to\mathbf{C}^N}$; in particular, by Proposition 3.1, also $\|\widetilde{T}_\phi(X)\|_{\mathbf{C}^N\to\mathbf{C}^N} \leq \|\phi\|_\infty$.

Proposition 3.2. *The following formulae hold:*

$$\widetilde{T}_\phi(X) = K(X, X)^{-1/2} \cdot \int_\Omega K(X, Y)\phi(Y)K(Y, X) \, d\mu(Y) \cdot K(X, X)^{-1/2};$$

$$\begin{aligned}\widetilde{T_\phi T_\psi}(X) = K(X, X)^{-1/2} & \\ \cdot \int_\Omega \int_\Omega & K(X, Y)\phi(Y)K(Y, Z)\psi(Z)K(Z, X) \, d\mu(Y) \, d\mu(Z) \\ & \cdot K(X, X)^{-1/2}.\end{aligned}$$

In particular, if ϕ is a multiplier of $\mathcal{H}$ (i.e. $\phi f \in \mathcal{H}$ whenever $f \in \mathcal{H}$), then

$$\widetilde{T}_\phi(X) = K(X, X)^{-1/2}\phi(X)K(X, X)^{1/2};$$

and, similarly, when ϕ and ψ^* are multipliers of $\mathcal{H}$, then

$$\widetilde{T_\phi T_\psi}(X) = K(X, X)^{-1/2}\phi(X)K(X, X)\psi(X)K(X, X)^{-1/2}.$$

4. Bad Behaviour: All Matrices

We now exhibit an example of some pathological phenomena, showing in particular that the straightforward generalizations of the expansions (1.11) and (1.12) cannot hold: first, apart from the integer powers of h, we will see that also $\sqrt{h}$ enters the picture; and second, instead of evaluations at X we get also contributions from other points. This applies to the case of the full matrix domain $\Omega = \mathbf{C}^{N\times N}$; for the normal matrices Ω_{norm}, we will show in the next section that at least the second of these pathologies still remains.

Theorem 4.1. *Consider the full matrix domain $\Omega = \mathbf{C}^{N\times N}$ with $N = 2$. Let $\mathcal{H}_h$ be the spaces from Section 2, with reproducing kernels K_h given by (2.4). Let X be the matrix*

$$X = \begin{pmatrix} 0 & 1 \\ 0 & 0 \end{pmatrix}.$$

Then:

(i) $\lim_{h\to 0} \widetilde{T_\phi^{(h)}}(X) = \begin{pmatrix} \frac{\phi_{11}(0)+\phi_{22}(0)}{2} & 0 \\ 0 & \phi_{22}(0) \end{pmatrix}.$

Note that the matrix on the right-hand side does not depend on the value of ϕ at X, but rather on its value at 0.

(ii) *If $\phi(Y) = \sqrt{2}y_{22}I$, where y_{22} denotes the* (2, 2)*-entry of Y, then*

$$\widetilde{T_\phi^{(h)}}(X) = X\sqrt{h} + O(h) \quad \text{as } h \to 0.$$

Proof. Since $X^2 = 0$, the series (2.4) for $K_h(X, Y)$ becomes simply

$$K_h(X, Y) = I + \frac{XY^*}{c_1 h} = I + \frac{XY^*}{2h}. \tag{4.1}$$

Thus

$$\begin{aligned}\widetilde{\widetilde{T_\phi^{(h)}}}(X) &= \int_\Omega K_h(X, Y)\phi(Y)K_h(Y, X)\,d\mu_h(Y) \\ &= \int_\Omega \left[\phi(Y) + \frac{XY^*\phi(Y)}{2h} + \frac{\phi(Y)YX^*}{2h} + \frac{XY^*\phi(Y)YX^*}{4h^2}\right] d\mu_h(Y).\end{aligned} \tag{4.2}$$

On the other hand, the change of variable $Z = Y/\sqrt{h}$ and Taylor's expansion imply that for any C^∞ function f on Ω,

$$\begin{aligned}\int_\Omega f(Y)\,d\mu_h(Y) &= \int_\Omega f(\sqrt{h}Z)\,d\mu_1(Z) \\ &= \sum_{j=0}^\infty \frac{1}{j!}\Delta^j f(0)\,h^j,\end{aligned} \tag{4.3}$$

where $\Delta = \sum_{j,k=1}^2 \partial^2/\partial z_{jk}\partial\overline{z}_{jk}$ is the Laplacian on $\mathbf{C}^{2\times 2}$. Applying this to (4.2), we therefore get

$$\begin{aligned}\widetilde{\widetilde{T_\phi^{(h)}}}(X) = \phi(0) &+ O(h) \\ &+ \frac{X[\Delta(Y^*\phi(Y))(0) + O(h)]}{2} \\ &+ \frac{[\Delta(\phi(Y)Y)(0) + O(h)]X^*}{2} \\ &+ X\Big[\frac{\Delta(Y^*\phi(Y)Y)(0)}{4h} + O(1)\Big]X^*.\end{aligned}$$

Now $\widetilde{T_\phi^{(h)}}(X) = K(X,X)^{-1/2}\widetilde{\widetilde{T_\phi^{(h)}}}(X)K(X,X)^{-1/2}$; note that

$$K_h(X,X)^{-1/2} = \begin{pmatrix}\sqrt{\frac{2h}{2h+1}} & 0 \\ 0 & 1\end{pmatrix},$$

and thus, for any matrix $A = \begin{pmatrix} a_{11} & a_{12} \\ a_{21} & a_{22}\end{pmatrix}$,

$$\begin{aligned}K(X,X)^{-1/2}AK(X,X)^{-1/2} &= \begin{pmatrix}\frac{2h}{2h+1}a_{11} & \sqrt{\frac{2h}{2h+1}}a_{12} \\ \sqrt{\frac{2h}{2h+1}}a_{21} & a_{22}\end{pmatrix}, \\ K(X,X)^{-1/2}XAK(X,X)^{-1/2} &= \begin{pmatrix}\frac{2h}{2h+1}a_{21} & \sqrt{\frac{2h}{2h+1}}a_{22} \\ 0 & 0\end{pmatrix}, \\ K(X,X)^{-1/2}AX^*K(X,X)^{-1/2} &= \begin{pmatrix}\frac{2h}{2h+1}a_{12} & 0 \\ \sqrt{\frac{2h}{2h+1}}a_{22} & 0\end{pmatrix}, \\ K(X,X)^{-1/2}XAX^*K(X,X)^{-1/2} &= \begin{pmatrix}\frac{2h}{2h+1}a_{22} & 0 \\ 0 & 0\end{pmatrix}.\end{aligned} \tag{4.4}$$

Consequently,

$$
\begin{aligned}
\widetilde{T_\phi^{(h)}}(X) &= \begin{pmatrix} 0 & \phi_{12}(0)\sqrt{2h} \\ \phi_{21}(0)\sqrt{2h} & \phi_{22}(0) \end{pmatrix} + O(h) \\
&+ \begin{pmatrix} 0 & \frac{\sqrt{2h}}{2}\Delta[Y^*\phi(Y)]_{22}(0) \\ 0 & 0 \end{pmatrix} + O(h) \\
&+ \begin{pmatrix} 0 & 0 \\ \frac{\sqrt{2h}}{2}\Delta[\phi(Y)Y]_{22}(0) & 0 \end{pmatrix} - O(h) \\
&+ \begin{pmatrix} \frac{1}{2}\Delta[Y^*\phi(Y)Y]_{22}(0) & 0 \\ 0 & 0 \end{pmatrix} + O(h).
\end{aligned}
$$

Note that, by a simple calculation,

$$
\begin{aligned}
\Delta[Y^*\phi(Y)]_{ab}(0) &= \sum_{k=1}^{2} \frac{\partial \phi_{kb}}{\partial y_{ka}}(0), \qquad (4.5) \\
\Delta[\phi(Y)Y]_{ab}(0) &= \sum_{k=1}^{2} \frac{\partial \phi_{ak}}{\partial \overline{y}_{kb}}(0), \\
\Delta[Y^*\phi(Y)Y]_{ab}(0) &= \delta_{ab} \cdot \operatorname{Tr}\phi(0).
\end{aligned}
$$

Letting $h \searrow 0$, (i) therefore follows immediately. For $\phi(Y) = y_{22}\sqrt{2}I$, only the second term in the last formula for $\widetilde{T_\phi^{(h)}}(X)$ gives a nonzero contribution, equal to

$$
\widetilde{T_\phi^{(h)}}(X) = \begin{pmatrix} 0 & \sqrt{h}\,\Delta[y_{22}Y^*]_{22}(0) \\ 0 & 0 \end{pmatrix} + O(h) = \sqrt{h}\,X + O(h),
$$

as $\Delta[y_{22}Y^*]_{22}(0) = 1$ by (4.5). This settles (ii). ■

Remark. We pause to observe that applying (4.3) to $f(Y) = Y^{*k}Y^k$ and comparing with (2.1) shows that

$$
\Delta^k(Y^{*k}Y^k)(0) = k!c_k I. \qquad (4.6)
$$

This gives, conceivably, a way of evaluating the numbers c_k without recourse to random matrix theory, and can also be used to show that the c_k have an interesting combinatorial meaning. Namely, expanding Δ^k and $Y^{*k}Y^k$ yields

$$\Delta^k(Y^{*k}Y^k) = \sum_{i_1,j_1,\dots,i_k,j_k=1}^{N} \partial_{i_1j_1}\overline{\partial}_{i_1j_1}\dots\partial_{i_kj_k}\overline{\partial}_{i_kj_k} \sum_{a_1,\dots,a_{2k-1}=1}^{N} \overline{y}_{a_1a}\overline{y}_{a_2a_1}\dots\overline{y}_{a_ka_{k-1}}y_{a_ka_{k+1}}y_{a_{k+1}a_{k+2}}\dots y_{a_{2k-1}b},$$

where, for the sake of brevity, we temporarily write ∂_{ij} for $\partial^2/\partial y_{ij}$, and similarly for $\overline{\partial}_{ij}$. Clearly a nonzero contribution only occurs if to each y there is applied precisely one ∂, and to each $\overline{y}$ precisely one $\overline{\partial}$. (We also see that the result will be independent of Y, i.e. a constant.) Thus

$$\Delta^k(Y^{*k}Y^k) = \sum_{i_1,j_1,\dots,i_k,j_k=1}^{N}\ \sum_{a_1,\dots,a_{2k-1}=1}^{N}\ \sum_{\sigma,\tau\in\mathfrak{S}_k} \overline{\partial}_{i_{\tau(1)}j_{\tau(1)}}\overline{y}_{a_1a}\dots\overline{\partial}_{i_{\tau(k)}j_{\tau(k)}}\overline{y}_{a_ka_{k-1}} \cdot \partial_{i_{\sigma(1)}j_{\sigma(1)}}y_{a_ka_{k+1}}\dots\partial_{i_{\sigma(k)}j_{\sigma(k)}}y_{a_{2k-1}b}.$$

Changing the order of summations and using the fact that $\partial_{ij}y_{kl} = \delta_{ik}\delta_{jl}$, this becomes

$$\Delta^k(Y^{*k}Y^k) = \sum_{\sigma,\tau\in\mathfrak{S}_k}\ \sum_{i_1,j_1,\dots,i_k,j_k=1}^{N}\ \sum_{a_1,\dots,a_{2k-1}=1}^{N} \delta_{i_{\tau(1)}a_1}\delta_{j_{\tau(1)}a}\cdot\dots\delta_{i_{\tau(k)}a_k}\delta_{j_{\tau(k)}a_{k-1}} \cdot\delta_{i_{\sigma(1)}a_k}\delta_{j_{\sigma(1)}a_{k+1}}\cdot\dots\delta_{i_{\sigma(k)}a_{2k-1}}\delta_{j_{\sigma(k)}b}.$$

Replacing $\mathbf{i} = (i_1,\dots,i_k)$, $\mathbf{j} = (j_1,\dots,j_k)$ by $\mathbf{i}\circ\tau^{-1}$, $\mathbf{j}\circ\tau^{-1}$ and setting $\mu = \tau^{-1}\sigma$, we thus get

$$\Delta^k(Y^{*k}Y^k) = k!\sum_{\mu\in\mathfrak{S}_k}\sum_{\mathbf{i},\mathbf{j}}\delta_{j_1a}\delta_{j_2i_1}\dots\delta_{j_ki_{k-1}} \cdot\delta_{i_{\mu(1)}i_k}\delta_{i_{\mu(2)}j_{\mu(1)}}\dots\delta_{i_{\mu(k)}j_{\mu(k-1)}}\delta_{j_{\mu(k)}b}.$$

In other words, $[\Delta^k(Y^{*k}Y^k)]_{ab}/k! = c_k\delta_{ab}$ is the constant equal to the number of triples $(\mu,\mathbf{i},\mathbf{j})$, where $\mu\in\mathfrak{S}_k$ and $\mathbf{i},\mathbf{j}\in\{1,\dots,N\}^k$, such that

$$\mathbf{j} = (a,i_1,\dots,i_{k-1}),\quad \mathbf{j}\circ\mu = (i_{\mu(2)},\dots,i_{\mu(k)},b),\quad\text{and}\quad i_k = i_{\mu(1)}.$$

It is evident that, indeed, this number is zero for $a\neq b$ (since the sequences $(a,i_1,\dots,i_{k-1},i_k)$ and $(i_{\mu(1)},i_{\mu(2)},\dots,i_{\mu(k)},b)$ must be permutations of each other), while for $a = b$ it is independent of a. It is also clear that c_k is always an integer, a fact definitely not apparent from (2.2). □

We conclude by giving a formula analogous to part (i) of the last theorem also for $\widetilde{T_\phi^{(h)} T_\psi^{(h)}}(X)$. It shows, in particular, that $\widetilde{T_\phi^{(h)} T_\psi^{(h)}}(X) - \widetilde{T_{\phi\psi}^{(h)}}(X)$ need not tend to 0 in general as $h \to 0$, but does so for scalar-valued ϕ and ψ.

Theorem 4.2. *Under the hypotheses of Theorem 4.1,*

$$\lim_{h\to 0} \widetilde{T_\phi^{(h)} T_\psi^{(h)}}(X) = \begin{pmatrix} \frac{\operatorname{Tr}\phi(0)}{2} \cdot \frac{\operatorname{Tr}\psi(0)}{2} & 0 \\ 0 & (\phi\psi)_{22}(0) \end{pmatrix}.$$

Proof. Using the formula for $\widetilde{T_\phi^{(h)} T_\psi^{(h)}}$ from Proposition 3.2 and (4.1), we have

$$\widetilde{\widetilde{T_\phi^{(h)} T_\psi^{(h)}}}(X) = \int_\Omega \int_\Omega \Big(I + \frac{XY^*}{2h}\Big)\phi(Y) K_h(Y,Z)\psi(Z)\Big(I + \frac{ZX^*}{2h}\Big)\, d\mu_h(Y)\, d\mu_h(Z).$$

Making again the change of variable $Y \mapsto Y\sqrt{h}$, $Z \mapsto Z\sqrt{h}$, the double integral becomes

$$\int_\Omega \int_\Omega \Big(I + \frac{XY^*\sqrt{h}}{2h}\Big)\phi(\sqrt{h}Y) K_1(Y,Z)\psi(\sqrt{h}Z)\Big(I + \frac{Z\sqrt{h}X^*}{2h}\Big)\, d\mu_1(Y)\, d\mu_1(Z).$$

The rest of the proof proceeds in the same way as in Theorem 4.1, using instead of (4.3) the expansion from the following lemma (applied also to $Y^*\phi(Y)$ and $\psi(Z)Z$ in place of $\phi(Y)$ and $\psi(Z)$, respectively), the fact that

$$\{\!\{Y^*\phi(Y), \psi(Z)Z\}\!\}_{22} = \frac{1}{4} \operatorname{Tr}\phi(0) \operatorname{Tr}\psi(0),$$

and the relations (4.4). We leave the details to the reader. ■

Lemma 4.3. *For $\phi, \psi \in C^\infty_{\mathbf{C}^{N\times N}}(\Omega)$,*

$$\int_\Omega \int_\Omega \phi(\sqrt{h}Y) K_1(Y,Z)\psi(\sqrt{h}Z)\, d\mu_1(Y)\, d\mu_1(Z)$$
$$= \phi(0)\psi(0) + h\big[\Delta\phi(0)\cdot\psi(0) + \phi(0)\cdot\Delta\psi(0) + \{\!\{\phi,\psi\}\!\}(0)\big] + O(h^2),$$

as $h \to 0$, where

$$\{\!\{\phi,\psi\}\!\} := \frac{1}{2} \sum_{i,j,k} \frac{\partial\phi}{\partial\overline{y}_{ij}} E_{ik} \frac{\partial\psi}{\partial z_{kj}},$$

where E_{ik} is the matrix $[E_{ik}]_{ab} = \delta_{ia}\delta_{kb}$.

Proof. Using the Taylor expansions for ϕ and ψ, we see that the integral asymptotically equals

$$\sum_{\alpha,\beta,\gamma,\delta} h^{\frac{|\alpha|+|\beta|+|\gamma|+|\delta|}{2}} \frac{\partial^\alpha \overline{\partial}^\beta \phi(0)}{\alpha!\beta!} \int_\Omega \int_\Omega y^\alpha \overline{y}^\beta K_1(Y,Z) z^\gamma \overline{z}^\delta \, d\mu_1(Y)\, d\mu_1(Z) \frac{\partial^\gamma \overline{\partial}^\delta \psi(0)}{\gamma!\delta!}.$$

(The summation extends over all multiindices $\alpha, \beta, \gamma, \delta$.) Note that the kernel satisfies $K_1(Y,Z) = K_1(\epsilon Y, \epsilon Z)$ for any $\epsilon \in \mathbf{C}$ of modulus one; hence the last integral vanishes unless $|\alpha| + |\gamma| = |\beta| + |\delta|$. Thus the coefficients at half-integer powers of h in fact vanish. The coefficient at h^0 is clearly $\phi(0)\psi(0)$, since

$$\int_\Omega \int_\Omega K_1(Y,Z)\, d\mu_1(Y)\, d\mu_1(Z) = \int_\Omega I \, d\mu_1(Y) = I$$

by the reproducing property of K_1 and (2.1). For the coefficient at h^1, the only nonzero contributions, by virtue of the last observation, come from $|\alpha| = |\beta| = 1$, or $|\gamma| = |\delta| = 1$, or $|\alpha| = |\delta| = 1$, or $|\beta| = |\gamma| = 1$ (i.e. from $y\overline{y}$, $z\overline{z}$, $y\overline{z}$, or $\overline{y}z$). Since

$$\int_\Omega \int_\Omega y_{ij}\overline{y}_{kl} K_1(Y,Z)\, d\mu_1(Y)\, d\mu_1(Z) = \int_\Omega y_{ij}\overline{y}_{kl} I \, d\mu_1(Y) = \delta_{ik}\delta_{jl} I$$

(and similarly for $z_{ij}\overline{z}_{kl}$), the first two possibilities contribute

$$\sum_{i,j,k,l} \frac{\partial^2 \phi(0)}{\partial y_{ij} \partial \overline{y}_{kl}} \delta_{ik}\delta_{jl} I \;\psi(0) = \Delta\phi(0) \cdot \psi(0)$$

and $\phi(0) \cdot \Delta\psi(0)$, respectively. For the $y\overline{z}$ possibility, the corresponding integral vanishes, since the integrand is a function holomorphic in the entries of Y and Z^* and vanishing at the origin. Finally, for the last possibility $\overline{y}z$ we use the series (2.4) to split the integral as

$$\begin{aligned} &\int_\Omega \int_\Omega \overline{y}_{ij} z_{kl} K_1(Y,Z)\, d\mu_1(Y)\, d\mu_1(Z) \\ &\qquad = \sum_{m=0}^{\infty} \frac{1}{c_m} \Big(\int_\Omega \overline{y}_{ij} Y^m \, d\mu_1(Y) \Big) \Big(\int_\Omega z_{kl} Z^{*m} \, d\mu_1(Z) \Big). \end{aligned} \tag{4.7}$$

Using again the invariance of $d\mu_1$ under the change of variable $Y \mapsto \epsilon Y$, we see that we only get nonzero contribution for $m = 1$. In that case,

$$\Big[\int_\Omega \overline{y}_{ij} Y \, d\mu_1(Y) \Big]_{ab} = \int_\Omega \overline{y}_{ij} y_{ab} \, d\mu_1(Y) = \delta_{ia}\delta_{jb} = [E_{ij}]_{ab},$$

and similarly for the Z integral. Thus the integral (4.7) equals

$$\frac{1}{c_1} E_{ij} E_{lk} = \frac{1}{2} \delta_{jl} E_{ik},$$

and the total contribution from the $\overline{y}z$ possibility is

$$\sum_{i,j,k,l} \frac{\partial \phi}{\partial \overline{y}_{ij}}(0) \frac{\delta_{jl}}{2} E_{ik} \frac{\partial \psi}{\partial z_{kl}}(0) = \{\{\phi, \psi\}\}(0),$$

which concludes the proof of the lemma. ■

5. Bad Behaviour: Normal Matrices

The matrix $X = \begin{pmatrix} 0 & 1 \\ 0 & 0 \end{pmatrix}$ featuring in the last section was not normal; this tempts one to hope that things might perhaps still work out fine for the domain Ω_{norm} of all normal matrices. We show that even in this case, unfortunately, the non-local behaviour described above still persists.

Theorem 5.1. *Consider the domain Ω_{norm} of all normal $N \times N$ matrices, with $N = 2$. Let $\mathcal{H}_h$ be the spaces from Section 2, with reproducing kernels K_h given by* (2.7). *Let X be the matrix*

$$X = \begin{pmatrix} 1 & 0 \\ 0 & 0 \end{pmatrix}$$

of the projection onto the first coordinate. Then

$$[\widetilde{T_\phi^{(h)}}(X)]_{11} = [\widetilde{T_\phi^{(h)}}(I)]_{11},$$
$$[\widetilde{T_\phi^{(h)}}(X)]_{22} = [\widetilde{T_\phi^{(h)}}(0)]_{22}.$$

Consequently, the asymptotic expansion (1.11) *cannot hold.*

Proof. As X is a projection, we have $X^j = X \; \forall j \geq 1$; thus

$$K_h(X, Y) = \sum_{j=0}^{\infty} \frac{X^j Y^{*j}}{j! h^j} = I + X \sum_{j=1}^{\infty} \frac{Y^{*j}}{j! h^j} = (I - X) + X K_h(I, Y).$$

Thus

$$\begin{aligned}\widetilde{\widetilde{T_\phi^{(h)}}}(X) &= \int_{\Omega_{\text{norm}}} K_h(X,Y)\phi(Y)K_h(Y,X)\,d\mu_h(Y)\\ &= X\cdot\int_{\Omega_{\text{norm}}} K_h(I,Y)\phi(Y)K_h(Y,I)\,d\mu_h(Y)\cdot X\\ &\quad + X\cdot\int_{\Omega_{\text{norm}}} K_h(I,Y)\phi(Y)\,d\mu_h(Y)\cdot(I-X)\\ &\quad + (I-X)\cdot\int_{\Omega_{\text{norm}}} \phi(Y)K_h(Y,I)\,d\mu_h(Y)\cdot X\\ &\quad + (I-X)\cdot\int_{\Omega_{\text{norm}}} \phi(Y)\,d\mu_h(Y)\cdot(I-X).\end{aligned}$$

But for any matrix $A = \begin{pmatrix} a_{11} & a_{12} \\ a_{21} & a_{22}\end{pmatrix}$, we have $XAX = \begin{pmatrix} a_{11} & 0 \\ 0 & 0\end{pmatrix}$, $XA(I-X) = \begin{pmatrix} 0 & a_{12} \\ 0 & 0\end{pmatrix}$, etc.; hence

$$[\widetilde{\widetilde{T_\phi^{(h)}}}(X)]_{11} = \left[\int_{\Omega_{\text{norm}}} K_h(I,Y)\phi(Y)K_h(Y,I)\,d\mu_h(Y)\right]_{11} = [\widetilde{\widetilde{T_\phi^{(h)}}}(I)]_{11},$$

and similarly, since $K_h(0,Y) = I$,

$$[\widetilde{\widetilde{T_\phi^{(h)}}}(X)]_{22} = \left[\int_{\Omega_{\text{norm}}} \phi(Y)\,d\mu_h(Y)\right]_{22} = [\widetilde{\widetilde{T_\phi^{(h)}}}(0)]_{22}.$$

Finally, since $K_h(X,X) = \begin{pmatrix} e^{1/h} & 0 \\ 0 & 1\end{pmatrix}$, $K_h(I,I) = e^{1/h}I$, and $K_h(0,0) = I$, the assertion about $\widetilde{T_\phi^{(h)}}(X) = K(X,X)^{-1/2}\widetilde{\widetilde{T_\phi^{(h)}}}(X)K(X,X)^{-1/2}$ follows. ∎

For completeness, we also state the analog of Theorem 4.2, which shows, among others, that the expansion (1.12) cannot hold. Its proof is the same as for Theorem 5.1.

Theorem 5.2. *In the situation of the preceding theorem,*

$$\begin{aligned}[\widetilde{T_\phi^{(h)}T_\psi^{(h)}}(X)]_{11} &= [\widetilde{T_\phi^{(h)}T_\psi^{(h)}}(I)]_{11},\\ [\widetilde{T_\phi^{(h)}T_\psi^{(h)}}(X)]_{22} &= [\widetilde{T_\phi^{(h)}T_\psi^{(h)}}(0)]_{22}.\end{aligned}$$

6. An Application of Stationary Phase

In this section we finally start exhibiting also results in the positive direction, namely, by using the stationary phase method we establish the existence of an (albeit non-local) semiclassical asymptotic expansion for $\widetilde{T_\phi^{(h)}}$ for the case of the normal matrices.

Recall that the stationary phase (WJKB) method tells us that if S, ϕ are smooth complex-valued functions on some domain in $\mathbf{C}^n$, such that S has a unique critical point x_0 (i.e. $S'(x_0) = 0$), which is nondegenerate (i.e. $\det S''(x_0)) \neq 0$) and is a global maximum for $\operatorname{Re} S$, and ϕ is compactly supported, then the integral

$$h^{-n} \int \phi(x)\, e^{S(x)/h}\, dx \tag{6.1}$$

has an asymptotic expansion

$$e^{S(x_0)/h} \sum_{j=0}^{\infty} h^j\, \mathcal{L}_j \phi(x_0) \qquad \text{as } h \to 0, \tag{6.2}$$

with some differential operators $\mathcal{L}_j$ whose coefficients are given by universal expressions in S and its partial derivatives. See e.g. [Hrm], Section 7.7. The hypothesis of the compact support of ϕ can be replaced by the requirement that the integral (6.1) exist for some $h = h_0 > 0$, and that the maximum of $\operatorname{Re} S$ at x_0 strictly dominate also the values of $\operatorname{Re} S$ at the boundary or at infinity, in the sense that $\operatorname{Re} S(x_n) \to \operatorname{Re} S(x_0) \implies x_n \to x_0$.

On the other hand, if the global maximum of $\operatorname{Re} S$ is not a critical point, then (6.1) decays faster than any power of h as $h \searrow 0$.

The formulas for the operators $\mathcal{L}_j$ are fairly complicated in general, but fortunately become quite explicit if the phase function S is quadratic (which will be the only case we will need). Namely, assume that

$$S(x) = -\langle Q(x - x_0), x - x_0 \rangle_{\mathbf{C}^n}$$

for some matrix Q with positive real part. Then x_0 is a unique critical point of S, is nondegenerate, and

$$\mathcal{L}_j = \frac{1}{j!} \mathcal{Q}^j, \qquad \text{where} \quad \mathcal{Q} = -\langle Q^{-1} \partial, \partial \rangle. \tag{6.3}$$

Let us now apply this to the integral defining $\widetilde{T_\phi^{(h)}}(X)$ in the case of the domain of normal matrices, viz.

$$\widetilde{T_\phi^{(h)}}(X) = \int_{\Omega_{\text{norm}}} K_h(X, X)^{-1/2} K_h(X, Y) \phi(Y) K_h(Y, X) K_h(X, X)^{-1/2}\, d\mu_h(Y).$$

Let

$$Y = UDU^*, \qquad X = VCV^* \tag{6.4}$$

be the spectral decompositions of Y and X, respectively. Observe that owing to the invariance of the kernels K_h and the measures μ_h under unitary transformations, we have $\widetilde{T_\phi^{(h)}}(X) = V\widetilde{T_{\phi^V}^{(h)}}(C)V^*$, where $\phi^V(Y) := V^*\phi(VYV^*)V$; thus it suffices to deal with the case of $V = I$, i.e. when $X = C = \mathrm{diag}(c_1, \ldots, c_N)$ is a diagonal matrix. From (6.4), we then have

$$\begin{aligned}
[K_h(X,Y)]_{ij} &= \sum_{k=0}^{\infty} \frac{[C^k U D^{*k} U^*]_{ij}}{k!h^k} \\
&= \sum_{k=0}^{\infty}\sum_{l=1}^{N} \frac{c_i^k u_{il} \overline{d}_l^k u_{lj}^*}{k!h^k} \\
&= \sum_{l=1}^{N} e^{c_i \overline{d}_l/h} u_{il}\overline{u}_{jl},
\end{aligned}$$

and similarly for $K_h(X,X)$. Thus the matrix entries of $\widetilde{T_\phi^{(h)}}(X)$ are given by

$$\begin{aligned}
[\widetilde{T_\phi^{(h)}}(X)]_{ab} = (\pi h)^{-N} \int_{U(N)}\int_{\mathbf{C}^N} \sum_{j,k,l,m=1}^{N} & e^{-|c_a|^2/2h} e^{c_a\overline{d}_l/h} u_{al}\overline{u}_{jl}\phi_{jk}(UDU^*) \\
& \cdot e^{\overline{c}_b d_m/h}\overline{u}_{bm}u_{km}e^{-|c_b|^2/2h}e^{-\|d\|^2/h}\, dU\, dD.
\end{aligned}$$

For simplicity, we will write $\phi(U; d_1, \ldots, d_N)$ instead of $\phi(UDU^*)$. The last integral over D is precisely of the form (6.1), with phase function given by

$$S(d_1, \ldots, d_N) = c_a\overline{d}_l + \overline{c}_b d_m - \|d\|^2 - \frac{|c_a|^2 + |c_b|^2}{2}.$$

The critical point condition $S' = 0$ amounts to

$$c_a\delta_{li} = d_i, \quad \overline{c}_b\delta_{mi} = \overline{d}_i, \qquad \forall i = 1, \ldots, N.$$

It follows that there is no critical point if $c_a \neq c_b$, or if $c_a = c_b \neq 0$ and $l \neq m$; while for $c_a = c_b \neq 0$ and $l = m$, or $c_a = c_b = 0$ and l, m arbitrary, there is a unique critical point

$$d = (0, \ldots, 0, \overbrace{c_a}^{l\text{-th slot}}, 0, \ldots, 0) \equiv c_a\chi_l,$$

which satisfies the assumptions for the application of the stationary phase method. The critical value is

$$S(c_a\chi_l) = |c_a|^2 + |c_a|^2 - |c_a|^2 - \frac{|c_a|^2 + |c_a|^2}{2} = 0,$$

and the operators $\mathcal{L}_j$ are equal to $\frac{1}{j!}\Delta^j$, by (6.3). By (6.2), it therefore follows that $[\widetilde{T^{(h)}_\phi}(X)]_{ab} = O(h^\infty)$ for $c_a \neq c_b$, while

$$[\widetilde{T^{(h)}_\phi}(X)]_{ab} \approx \sum_{j,k,l=1}^{N} \sum_{r=0}^{\infty} \frac{h^r}{r!} \int_{U(N)} u_{al}\overline{u}_{jl}u_{kl}\overline{u}_{bl}\, (\Delta^r_{(d)}\phi_{jk})(U; c_a\chi_l)\, dU \tag{6.5}$$

as $h \to 0$ if $c_a = c_b \neq 0$, and

$$\begin{aligned}[\widetilde{T^{(h)}_\phi}(X)]_{ab} &\approx \sum_{j,k,l,m,=1}^{N} \sum_{r=0}^{\infty} \frac{h^r}{r!} \int_{U(N)} u_{al}\overline{u}_{jl}u_{km}\overline{u}_{bm}\, (\Delta^r_{(d)}\phi_{jk})(U; 0)\, dU \\ &= \sum_{r=0}^{\infty} \frac{h^r}{r!} \int_{U(N)} (\Delta^r_{(d)}\phi_{ab})(U; 0)\, dU \\ &= \sum_{r=0}^{\infty} \frac{h^r}{r!} (\Delta^r_{(d)}\phi_{ab})(0) \qquad (\text{as } \phi(U;0) = \phi(0) \text{ is independent of } U)\end{aligned} \tag{6.6}$$

as $h \to 0$ if $c_a = c_b = 0$. Here the subscript at Δ indicates that it applies only to the d-variables in $\phi(U; d_1, \ldots, d_N)$.

Thus the coefficients at each h^r in the asymptotic expansion do not depend on the jet of ϕ at X, but rather on the behaviour of ϕ near the whole orbit $\{UP_aU^* : U \in U(N)\}$ of the spectral components $P_a := \text{diag}(0, \ldots, 0, c_a, 0, \ldots, 0)$ of X. Also, the off-diagonal entries asymptotically vanish (i.e. are $O(h^\infty)$) if $c_a \neq c_b$, which is quite unexpected.

Observe that setting $c_a = 0$ in (6.5) gives

$$[\widetilde{T^{(h)}_\phi}]_{ab} \approx \sum_{j,k=1}^{N} \sum_{r=0}^{\infty} \frac{h^r}{r!}\, \kappa_{jk}\, \Delta^r_{(d)}\phi_{jk}(0)$$

where

$$\kappa_{jk} := \sum_{l=1}^{N} \int_{U(N)} u_{al}\overline{u}_{jl}u_{kl}\overline{u}_{bl}\, dU.$$

It can be shown that

$$\kappa_{jk} = \frac{\delta_{aj}\delta_{kb} + \delta_{ab}\delta_{kj}}{N+1}, \tag{6.7}$$

and in fact,

$$\int_{U(N)} u_{al}\overline{u}_{jl}u_{kl}\overline{u}_{bl}\, dU = \frac{\delta_{aj}\delta_{kb} + \delta_{ab}\delta_{kj}}{N(N+1)}. \tag{6.8}$$

(The above relation can be obtained by an application of standard orthogonality relations for the matrix elements of irreducible representations of compact groups — in this case applied to the irreducible subrepresentation of $U(N)$, carried by second order symmetric tensors, in the decomposition of the natural representation of $U(N) \otimes U(N)$.) Thus we have

$$[\widetilde{T_\phi^{(h)}}]_{ab} \approx \sum_{r=0}^{\infty} \frac{h^r}{r!} \frac{\Delta^r_{(d)}(\phi_{ab} + \delta_{ab}\operatorname{Tr}\phi)(0)}{N+1},$$

which is different from (6.6). Thus we see that, in general, it is not possible to use the formula (6.5) in both cases $c_a = c_b \neq 0$ and $c_a = c_b = 0$.

For scalar-valued ϕ, (6.5) simplifies to

$$[\widetilde{T_\phi^{(h)}}(X)]_{ab} \approx \sum_{r=0}^{\infty}\sum_{l=1}^{N} \frac{h^r}{r!} \int_{U(N)} u_{al}\overline{u}_{bl}\ (\Delta^r_{(d)}\phi)(U; c_a\chi_l)\, dU;$$

and if in addition ϕ is independent of U, i.e. $\phi(UYU^*) = \phi(Y)\ \forall U \in U(N)$, then the last integral can be evaluated by Schur's orthogonality relations, yielding

$$[\widetilde{T_\phi^{(h)}}(X)]_{ab} \approx \delta_{ab} \cdot \sum_{r=0}^{\infty} \frac{h^r}{r!} (\Delta^r_{(d)}\phi)(c_a\chi_a). \tag{6.9}$$

In the general case, however, it does not seem that (6.5) can be simplified in any way.

In the same manner, one can also prove the following formula for the asymptotics of $\widetilde{T_\phi^{(h)}T_\psi^{(h)}}$, which of course reduces to (6.5) upon taking for ψ the constant function equal to I. The strange-looking operators $\mathfrak{M}_{mq}$ originate from the formula (6.3).

Theorem 6.1. *For any functions* $\phi, \psi \in C^\infty_{\mathbf{C}^{N\times N}}(\Omega_{\text{norm}})$ *and a diagonal matrix* $X = \operatorname{diag}(c_1, \ldots, c_N)$,

$$[\widetilde{T_\phi^{(h)}T_\psi^{(h)}}(X)]_{ab} = O(h^\infty) \qquad \text{as } h \to 0$$

if $c_a \neq c_b$*;*

$$[\widetilde{T_\phi^{(h)} T_\psi^{(h)}}(X)]_{ab} \approx \sum_{\substack{i,j,k,l,\\ m,p,q=1}}^{N} \sum_{r=0}^{\infty} \frac{h^r}{r!} \int_{U(N)} \int_{U(N)} u_{am} \overline{u}_{im} u_{jm} \overline{u}_{pm} w_{pq} \overline{w}_{kq} \cdot w_{lq} \overline{w}_{bq} \, (\mathfrak{M}^r_{mq} \phi_{ij} \psi_{kl})(U, W; c_a \chi_m, c_a \chi_q) \, dU \, dW \tag{6.10}$$

as $h \to 0$ *if* $c_a = c_b \neq 0$*, where*

$$(\mathfrak{M}^r_{mq} \phi_{ij} \psi_{kl})(U, W; d, e) := \left[\left(\Delta_{(d)} + \Delta_{(e)} + \frac{\partial^2}{\partial d_m \partial \overline{e}_q} \right)^r \phi_{ij}(U; d) \, \psi_{kl}(W; e) \right];$$

and

$$[\widetilde{T_\phi^{(h)} T_\psi^{(h)}}(X)]_{ab} \approx \sum_{\substack{i,j,k,l,m,\\ p,q,L,M=1}}^{N} \sum_{r=0}^{\infty} \frac{h^r}{r!} \int_{U(N)} \int_{U(N)} u_{aL} \overline{u}_{iL} u_{jm} \overline{u}_{pm} \cdot w_{pq} \overline{w}_{kq} w_{lM} \overline{w}_{bM} \, (\mathfrak{M}^r_{mq} \phi_{ij} \psi_{kl})(U, W; 0, 0) \, dU \, dW \tag{6.11}$$

as $h \to 0$ *if* $c_a = c_b = 0$.

The formula (6.11) can clearly be simplified upon carrying out the summations over L and M and performing the two integrations (which can be done since $\phi(U;0) = \phi(0)$ and $\psi(W;0) = \psi(0)$ are independent of U and W) via Schur's orthogonality relations; the result is

$$[\widetilde{T_\phi^{(h)} T_\psi^{(h)}}(X)]_{ab} \approx \sum_{m,q=1}^{N} \sum_{r=0}^{\infty} \frac{h^r}{r!} \, \mathfrak{M}^r_{mq} (\phi\psi)_{ab}(0, 0).$$

Similarly, as with (6.5) and (6.6), using (6.7) it can be shown that for $c_a = 0$ the formula (6.10) reduces to

$$[\widetilde{T_\phi^{(h)} T_\psi^{(h)}}(X)]_{ab} \approx \sum_{m,q=1}^{N} \sum_{r=0}^{\infty} \frac{h^r}{r!} \frac{\mathfrak{M}^r_{mq}[(\phi + I \operatorname{Tr} \phi)(\psi + I \operatorname{Tr} \psi)]_{ab}(0, 0)}{(N + 1)^2},$$

which is different from (6.11). We refrain from going into these details because they are not needed anywhere in the sequel.

We conclude this section by observing that (1.6) also fails in general.

Proposition 6.2. *Let ϕ be the $\mathbf{C}^{N\times N}$-valued function on Ω_{norm}, $N \geq 2$, defined by $\phi(Z) = |\det Z|^2 I$. Then*

$$\lim_{h\to 0} \widetilde{T_\phi^{(h)}}(X) = 0 \qquad \forall X \in \Omega_{\text{norm}}.$$

Proof. Since $\phi(VZV^*) = V\phi(Z)V^* \; \forall V \in U(N)$, it is enough to check the assertion for diagonal X, so let $X = \operatorname{diag}(c_1, \dots, c_N)$. As $\phi_{jk}(U;d) = \delta_{jk}|d_1 \dots d_N|^2$, we have

$$\begin{aligned}
(\Delta_{(d)}\phi_{jk})(U;d) &= \delta_{jk} \sum_m |d_1 \dots \hat{d}_m \dots d_N|^2, \\
(\Delta^2_{(d)}\phi_{jk})(U;d) &= \delta_{jk} \sum_{m\neq n} |d_1 \dots \hat{d}_m \dots \hat{d}_n \dots d_N|^2 \\
&= \delta_{jk}\, 2 \sum_{m_1<m_2} |d_1 \dots \hat{d}_{m_1} \dots \hat{d}_{m_2} \dots d_N|^2, \\
&\;\;\vdots \\
(\Delta^r_{(d)}\phi_{jk})(U;d) &= \delta_{jk}\, r! \sum_{m_1<m_2<\dots<m_r} |d_1 \dots \hat{d}_{m_1} \dots \hat{d}_{m_2} \dots \hat{d}_{m_r} \dots d_N|^2, \\
(\Delta^r_{(d)}\phi_{jk})(U;d) &= 0 \qquad \text{for } r > N.
\end{aligned}$$

(Here the hat $\widehat{}$ indicates that the corresponding variable is omitted.) Thus by (6.9) and (6.6)

$$[\widetilde{T_\phi^{(h)}}]_{ab} = \delta_{ab}(|c_a|^2 h^{N-1} + h^N) + O(h^\infty),$$

that is,

$$\widetilde{T_\phi^{(h)}}(X) = h^{N-1} X^* X + h^N + O(h^\infty)$$

as $h \to 0$, and the assertion follows. ∎

Similarly, it can be shown that (1.1) breaks down too: for instance, for $\phi(Z) = |\det Z|^2 e^{-\operatorname{Tr}(Z^*Z)} I$, one has

$$\|T_\phi^{(h)}\| \approx \|\phi\|_\infty^{1/N} h^{N-1} \quad \text{as } h \to 0$$

(where $\|\phi\|_\infty := \sup_{Z\in\Omega_{\text{norm}}} \|\phi(Z)\|_{\mathbf{C}^N\to\mathbf{C}^N}$). This can be proved by observing that the operator T_ϕ is diagonal with respect to the basis (2.3), with eigenvalues

$$\frac{(k+1)h^N}{(h+1)^{2N+k}};$$

and $\sup_k (k+1)/(h+1)^k \approx 1/(eh) = \|\phi\|_\infty^{1/N}/h$. We omit the details.

We now turn to classes of observables ϕ which are more manageable than the general case.

7. Spectral and U-Invariant Functions

A function $\phi(Z)$ of $Z \in \Omega_{\text{norm}}$ will be called *spectral* if it is a function of Z in the sense of the Spectral Theorem: that is, if there exists a function $f : \mathbf{C} \to \mathbf{C}$ such that $\phi = f^\#$, where

$$f^\#(Z) := U \cdot \operatorname{diag}_j(f(d_j)) \cdot U^* \qquad \text{if} \quad Z = U \cdot \operatorname{diag}_j(d_j) \cdot U^*. \tag{7.1}$$

Our first observation is that for spectral functions, all goes fine with the Berezin-Toeplitz quantization.

Theorem 7.1. *If $\phi = f^\#$ and $\psi = g^\#$ are two smooth spectral functions, then there exist unique spectral functions ρ_r, $r = 0, 1, 2, \dots,$ such that*

$$T_\phi^{(h)} T_\psi^{(h)} \approx \sum_{r=0}^{\infty} h^r\, T_{\rho_r}^{(h)} \qquad \text{as } h \to 0$$

in the sense of operator norms (i.e. as in (1.4)). In fact,

$$\rho_r = C_r(f, g)^\#,$$

where

$$C_r(f, g) = \frac{1}{r!}\,\partial^r f \cdot \bar{\partial}^r g \tag{7.2}$$

are the operators (1.10) for $n = 1$.

Proof. Recall that the monomials z^k, $k = 0, 1, 2, \dots,$ are orthogonal in the Segal-Bargmann space (1.7) for $n = 1$:

$$\langle z^k, z^l\rangle_{L^2_{\text{hol}}(\mathbf{C}, d\mu_h)} = \delta_{kl} k! h^k.$$

Comparing this with (2.6), we see that the mapping

$$\iota : Z^k \chi_j \longmapsto z^k \otimes \chi_j \tag{7.3}$$

is a unitary isomorphism of our space $\mathcal{H}_h$ onto the tensor product $L^2_{\mathrm{hol}}(\mathbf{C}, d\mu_h)\otimes\mathbf{C}^N$. Now if $\phi = f^\#$ is a spectral function and $\chi, \eta \in \mathbf{C}^N$, then

$$\begin{aligned}\langle T_\phi^{(h)} Z^k\chi, Z^l\eta\rangle &= \langle \phi Z^k\chi, Z^l\eta\rangle_{L^2_{\mathbf{C}^N}(\Omega_{\mathrm{norm}}, d\mu_h)}\\ &= \int_{\Omega_{\mathrm{norm}}} \eta^* Z^{*l}\phi(Z)Z^k\chi\, d\mu_h(Z)\\ &= \int_{\mathbf{C}^N}\int_{U(N)} \eta^* U D^{*l}\phi(D)D^k U^*\chi\, dU\, e^{-\mathrm{Tr}(D^*D)/h}\frac{dD}{(\pi h)^N}.\end{aligned}$$

However, for any matrix X,

$$\int_{U(N)} UXU^*\, dU = \frac{\mathrm{Tr}(X)}{N} I. \tag{7.4}$$

(Indeed, performing the change of variable $U \mapsto U_1U$ and using the invariance of the Haar measure, it transpires that the left-hand side commutes with any $U_1 \in U(N)$. Thus it must be a multiple of the identity. Taking traces and using the cyclicity of the trace, (7.4) follows.) Thus we can continue the above calculation with

$$\begin{aligned}&= \langle\chi,\eta\rangle\,\frac{1}{N}\int_{\mathbf{C}^N} \mathrm{Tr}(D^{*l}\phi(D)D^k)e^{-\mathrm{Tr}(D^*D)/h}\frac{dD}{(\pi h)^N}\\ &= \langle\chi,\eta\rangle\,\frac{1}{N}\sum_{j=1}^N\int_{\mathbf{C}^N} \overline{d}_j^l d_j^k f(d_j)\, e^{-\|d\|^2/h}\frac{dD}{(\pi h)^N}\\ &= \langle\chi,\eta\rangle\,\frac{1}{N}\sum_{j=1}^N \langle z^k f, z^l\rangle_{L^2(\mathbf{C},d\mu_h)}\\ &= \langle\chi,\eta\rangle\,\langle T_f^{(h)} z^k, z^l\rangle_{L^2_{\mathrm{hol}}(\mathbf{C},d\mu_h)}\\ &= \langle (T_f^{(h)}\otimes I)(z^k\otimes\chi), z^l\otimes\eta\rangle_{L^2_{\mathrm{hol}}(\mathbf{C},d\mu_h)\otimes\mathbf{C}^N}.\end{aligned}$$

Consequently, under the isomorphism $\imath$, the operator $T_\phi^{(h)}$ on $\mathcal{H}_h$ corresponds to the operator $T_f^{(h)} \otimes I$ on $L^2_{\mathrm{hol}}(\mathbf{C}, d\mu_h) \otimes \mathbf{C}^N$, and the desired assertions follow immediately from the ordinary Berezin-Toeplitz quantization on $\mathbf{C}$. ■

We list one more corollary of the above isomorphism $\imath$; it will not be needed in the sequel, but should be contrasted with Proposition 6.2 at the end of Section 6 and the example immediately thereafter. We omit the proof.

Proposition 7.2. *For any spectral function $\phi = f^\#$ and $x \in \mathbf{C}$,*

$$\widetilde{T_\phi^{(h)}}(xI) = \widetilde{T_f^{(h)}}(x)\cdot I \tag{7.5}$$

where the $\widetilde{T_f^{(h)}}$ *on the right-hand side is the ordinary scalar-valued Berezin transform of the operator* $T_f^{(h)}$ *on* $L^2_{\text{hol}}(\mathbf{C}, d\mu_h)$*. In particular,*

$$\lim_{h\to 0} \|\widetilde{T_\phi^{(h)}}\|_\infty = \lim_{h\to 0} \|T_\phi^{(h)}\| = \|\phi\|_\infty.$$

Remark. We pause to note that for the full matrix domain $\Omega = \mathbf{C}^{N\times N}$, the spaces $\mathcal{H}_h$ are not isomorphic to $L^2_{\text{hol}}(\mathbf{C}, d\nu)$ for any rotation invariant measure ν on $\mathbf{C}$. The reason is that the numbers c_k in (2.2), which take over the role of the $k!$, are not the moment sequence of any measure on $[0, \infty)$ if $N > 1$. This can be seen by checking that

$$\frac{1}{(k+1)(k+2)} \prod_{j=1}^{k+1} (N + j) = \int_{\mathbf{C}} |z|^{2k} \, d\nu_N(z)$$

where

$$d\nu_N(z) := \frac{1}{\pi} \sum_{j=0}^{N-1} \frac{(N-1)!(N-j)}{j!} \, |z|^{2j} \, e^{-|z|^2} \, dz;$$

thus for c_k to be a moment sequence (even of a measure which is not necessarily non-negative) it is necessary and sufficient that

$$\left\{ \frac{1}{(k+1)(k+2)} \prod_{j=1}^{k+1} (N - j) \right\}_{k=1}^{\infty} \tag{7.6}$$

be a moment sequence. However, the latter cannot be the case, since (7.6) has only a finite number of nonzero terms.

We restrict our attention exclusively to Ω_{norm} in the rest of this paper. □

Returning to the main line of discussion, we proceed to introduce another class of functions.

A $\mathbf{C}^{N\times N}$-valued function ϕ on Ω_{norm} will be called *U-invariant* if

$$\phi(UZU^*) = U\,\phi(Z)\,U^* \qquad \forall U \in U(N)\ \forall Z \in \Omega_{\text{norm}}. \tag{7.7}$$

Clearly, a spectral function is U-invariant, but not vice versa: an example is the function $\phi(Z) = |\det Z|^2 I$ from the end of Section 6. The relationship between spectral and U-invariant functions is clarified in the next proposition.

Proposition 7.3. *A function ϕ is U-invariant if and only if there exists a function $f(d_1; d_2, \ldots, d_N)$ from $\mathbf{C} \times \mathbf{C}^{N-1}$ into $\mathbf{C}$, symmetric in the $N-1$ variables $d_2, \ldots, d_N$, such that $\phi = f^\#$, where*

$$f^\#(UDU^*) := U \cdot \operatorname{diag}_j(f(d_j; d_1, \ldots, \hat{d}_j, \ldots, d_N)) \cdot U^*. \tag{7.8}$$

The function f is uniquely determined by ϕ.

Further, ϕ is spectral if and only if f depends only on the first variable, i.e. if and only if $f(d_1; d_2, \ldots, d_N) = f(d_1; 0, \ldots, 0)$.

Proof. For any complex numbers $\epsilon_1, \ldots, \epsilon_N$ of modulus one, consider the matrix $\epsilon = \operatorname{diag}(\epsilon_1, \ldots, \epsilon_N)$. Then $\epsilon \in U(N)$ and $\epsilon D \epsilon^* = D$ for any diagonal matrix D; thus by (7.7)

$$\phi(D) = \epsilon\, \phi(D)\, \epsilon^* \qquad \forall \epsilon_1, \ldots, \epsilon_N \in \mathbf{T}.$$

Consequently, $\phi(D)$ is also a diagonal matrix. Define the functions $f_1, \ldots, f_N$ on $\mathbf{C}^N$ by

$$f_j(d_1; d_2, \ldots, d_N) := \phi_{jj}(D) \qquad \text{where } D = \operatorname{diag}(d_1, \ldots, d_N). \tag{7.9}$$

For any permutation σ of the set $\{1, 2, \ldots, N\}$, let F_σ denote the permutation matrix $[F_\sigma]_{jk} = \delta_{\sigma(j),k}$. Then $F_\sigma \in U(N)$ and

$$F_\sigma D F_\sigma^* = \operatorname{diag}(d_{\sigma(1)}, \ldots, d_{\sigma(N)}) \qquad \text{if } D = \operatorname{diag}(d_1, \ldots, d_N).$$

Thus by (7.7) again

$$f_{\sigma(j)}(d_1; d_2, \ldots, d_N) = f_j(d_{\sigma(1)}; d_{\sigma(2)}, \ldots, d_{\sigma(N)}).$$

It follows that f_j is symmetric with respect to the $N-1$ variables $d_1, \ldots, \hat{d}_j, \ldots, d_N$ and $\phi = f^\#$ for $f = f_1$.

Conversely, it is easily seen that any function of the form (7.8) is U-invariant, and $f^\# = g^\# \iff f = g$.

Finally, the assertion concerning spectral functions is immediate upon comparing (7.8) and (7.1). ■

One consequence of the last proposition is that the mapping

$$f^\# \longmapsto (f^\flat)^\# \tag{7.10}$$

with $f^\flat : \mathbf{C} \to \mathbf{C}$ defined by

$$f^\flat(z) := f(z; 0, \ldots, 0)$$

is a projection from U-invariant functions onto spectral functions. (Here the first # in (7.10) is the one for U-invariant functions from (7.8), while the second is the one for spectral functions from (7.1); however, there is no danger of confusion in this abuse of notation.) In terms of $f^\# = \phi$, the function $f^\flat$ can be expressed directly by

$$f^\flat(z) = \phi_{11}(zE_{11}), \qquad z \in \mathbf{C},$$

where E_{11} is the matrix of projection onto the first coordinate, i.e. $[E_{11}]_{jk} = \delta_{1j}\delta_{1k}$. The projections $f \mapsto f^\flat$ and (7.10) will play a crucial role in the next section.

8. Quantization of U-Invariant Functions

We now proceed to establish our final result — a generalization of Theorem 7.1 to U-invariant functions. The key ingredient is played by the following specializations of the asymptotic expansions from Section 6.

Theorem 8.1. *For any smooth U-invariant functions $\phi = f^\#$ and $\psi = g^\#$ on Ω_{norm},*

$$\widetilde{T_\phi^{(h)}} \approx \sum_{r=0}^\infty h^r \, (l_r \phi)^\# \tag{8.1}$$

and

$$\widetilde{T_\phi^{(h)} T_\psi^{(h)}} \approx \sum_{r=0}^\infty h^r \, m_r(\phi, \psi)^\# \tag{8.2}$$

as $h \to 0$, where $l_r\phi$ and $m_r(\phi, \psi)$ are the functions on $\mathbf{C}$ defined by

$$l_r\phi(z) := \frac{1}{r!} (\Delta^r f)(z; 0, \ldots, 0) \tag{8.3}$$

(that is, $l_r\phi = \frac{1}{r!}(\Delta^r_{(d)}\phi)^\flat$) and

$$m_r(\phi, \psi)(z) := \left[\frac{1}{r!} \left(\Delta_{(d)} + \Delta_{(e)} + \frac{\partial^2}{\partial d_1 \partial \overline{e}_1} \right)^r f(d) g(e) \right] \Bigg|_{\substack{d=(z;0,\ldots,0) \\ e=(z;0,\ldots,0)}}. \tag{8.4}$$

Proof. In principle this could be gleaned from the formulas (6.5) and (6.6), but it is better to use directly the definitions: if $X = VCV^*$ with $V \in U(N)$ and $C = \operatorname{diag}(c_1, \dots, c_N)$, then by Proposition 3.2

$$\begin{aligned}
\widetilde{T_\phi^{(h)}}(X) &= \int_{\Omega_{\text{norm}}} K_h(X,X)^{-1/2} K_h(X,Y)\phi(Y)K_h(Y,X)K_h(X,X)^{-1/2}\, d\mu_h(Y) \\
&= \int_{\mathbf{C}^N}\int_{U(N)} VK_h(C,C)^{-1/2}V^* \sum_{k=0}^\infty \frac{VC^kV^*UD^{*k}U^*}{k!h^k}\phi(UDU^*) \\
&\qquad \cdot \sum_{l=0}^\infty \frac{UD^lU^*VC^{*l}V^*}{l!h^l} VK_h(C,C)V^*\, dU\, e^{-\operatorname{Tr}(D^*D)/h}\frac{dD}{(\pi h)^N} \\
&= \int_{\mathbf{C}^N}\int_{U(N)} Ve^{-CC^*/2h}\sum_k \frac{C^kV^*UD^{*k}}{k!h^k}\phi(D) \\
&\qquad \cdot \sum_l \frac{D^lU^*VC^{*l}}{l!h^l} e^{-CC^*/2h}V^*\, dU\, e^{-\operatorname{Tr}(D^*D)/h}\frac{dD}{(\pi h)^N} \\
&= \frac{1}{N}\int_{\mathbf{C}^N} Ve^{-CC^*/2h}\sum_{k,l}\frac{C^kC^{*l}}{k!l!h^{k+l}}\operatorname{Tr}(D^{*k}\phi(D)D^l) \\
&\qquad \cdot e^{-CC^*/2h}V^*\, e^{-\operatorname{Tr}(D^*D)/h}\frac{dD}{(\pi h)^N} \\
&= \frac{1}{N}\sum_{j=1}^N\int_{\mathbf{C}^N} Ve^{-CC^*/h}e^{(\overline{d}_jC+d_jC^*)/h}\phi_{jj}(D)\,V^*\,e^{-\|d\|^2/h}\frac{dD}{(\pi h)^N} \\
&= V\cdot\operatorname{diag}_k\left(\frac{1}{N}\sum_{j=1}^N\int_{\mathbf{C}^N} f(d_j;d_1,\dots,\hat{d}_j,\dots,d_N)\,e^{-\|c_k\chi_j-d\|^2/h}\frac{dD}{(\pi h)^N}\right)\cdot V^* \\
&= V\cdot\operatorname{diag}_k\left(\int_{\mathbf{C}^N} f(d_1;d_2,\dots,d_N)\,e^{-\|c_k\chi_1-d\|^2/h}\frac{dD}{(\pi h)^N}\right)\cdot V^* \\
&= V\cdot\operatorname{diag}_k\left(\sum_{r=0}^\infty \frac{h^r}{r!}(\Delta^r f)(c_k;0,\dots,0)\right)\cdot V^* \\
&= \sum_{r=0}^\infty h^r\,(l_r\phi)^\#(X).
\end{aligned}$$

Here we have used, in turn, the formula (2.7) for $K_h(X,Y)$; the U-invariance of ϕ; the formula (7.4) for the integral over $U(N)$; the fact that $\operatorname{Tr}(D^{*k}\phi(D)D^l) = \sum_{j=1}^N \overline{d}_j^k d_j^l \phi_{jj}(D)$, combined with the summation of the exponential series and the commutativity of C with C^*; the fact that $\phi = f^\#$; the independence of the integral on j; the stationary phase expansion; and (7.1) and the definition of $l_r\phi$.

The proof of (8.2) is similar:[3]

$$\widetilde{T_\phi^{(h)}T_\psi^{(h)}}(X) = \int_{\Omega_{\text{norm}}}\int_{\Omega_{\text{norm}}} K_h(X,X)^{-1/2}K_h(X,Y)\phi(Y)K_h(Y,Z)\psi(Z)$$

$$\cdot\, K_h(Z,X)K_h(X,X)^{-1/2}\,d\mu_h(Y)\,d\mu_h(Z)$$

$$= \int_{\mathbf{C}^N}\int_{\mathbf{C}^N}\int_{U(N)}\int_{U(N)} Ve^{-CC^*/2h}\sum_k \frac{C^kV^*UD^{*k}}{k!h^k}\phi(D)\sum_l \frac{D^lU^*WE^{*l}}{l!h^l}$$

$$\cdot\, \psi(E)\sum_m \frac{E^mW^*VC^{*m}}{m!h^m}e^{-CC^*/2h}V^*\,dU\,dW\,d\mu_h(D)\,d\mu_h(E)$$

$$= \frac{1}{N^2}\int_{\mathbf{C}^N}\int_{\mathbf{C}^N} Ve^{-CC^*/2h}\sum_{k,l,m}\frac{C^kC^{*m}}{l!k!m!h^{k+l+m}}\operatorname{Tr}(D^{*k}\phi(D)D^l)$$

$$\cdot\, \operatorname{Tr}(E^{*l}\psi(E)E^m)\,d\mu_h(D)\,d\mu_h(E)$$

$$= \frac{1}{N^2}\int_{\mathbf{C}^N}\int_{\mathbf{C}^N}\sum_{i,j=1}^N Ve^{-CC^*/h}e^{(\bar d_iC+d_i\bar e_j+e_jC^*)/h}\phi_{ii}(D)\psi_{jj}(E)\,d\mu_h(D)\,d\mu_h(E)$$

$$= V\cdot\operatorname{diag}_k\Bigg(\frac{1}{N^2}\sum_{i,j}\int_{\mathbf{C}^N}\int_{\mathbf{C}^N} e^{-|c_k|^2/h}e^{(\bar d_ic_k+d_i\bar e_j+e_j\bar c_k)/h}$$

$$\cdot\, e^{-(\|d\|^2+\|e\|^2)/h}\phi_{ii}(D)\psi_{jj}(E)\frac{dD}{(\pi h)^N}\frac{dE}{(\pi h)^N}\Bigg)\cdot V^*$$

$$= V\cdot\operatorname{diag}_k\Bigg(\int_{\mathbf{C}^N}\int_{\mathbf{C}^N} e^{-|c_k|^2/h}e^{(\bar d_1c_k+d_1\bar e_1+e_1\bar c_k)/h}e^{-(\|d\|^2+\|e\|^2)/h}$$

$$\cdot\, f(d_1;d_2,\ldots,d_N)g(e_1;e_2,\ldots,e_N)\frac{dD}{(\pi h)^N}\frac{dE}{(\pi h)^N}\Bigg)\cdot V^*$$

$$= V\cdot\operatorname{diag}_k\Bigg(\sum_{r=0}^\infty\frac{h^r}{r!}\left[\Delta_{(d)}+\Delta_{(e)}+\frac{\partial^2}{\partial d_1\partial\bar e_1}\right]^r f(d)g(e)\Big|_{\substack{d=(c_k;0,\ldots,0)\\ e=(c_k;0,\ldots,0)}}\Bigg)\cdot V^*$$

$$= \sum_{r=0}^\infty h^r\,(m_r(\phi,\psi))^\#(X).$$

Here the penultimate line comes from the formula (6.3). ■

[3]Of course, (8.1) can also be obtained from (8.2) upon setting $\psi\equiv I$; but it is more instructive to give a separate proof.

Corollary 8.2. *If g is a smooth function on $\mathbf{C}$ and $g^\#$ the corresponding spectral function on Ω_{norm}, then*

$$\widetilde{T^{(h)}_{g^\#}} \approx \sum_{r=0}^{\infty} \frac{h^r}{r!} (\Delta^r g)^\# \qquad \text{as } h \to 0. \tag{8.5}$$

In particular,

$$\lim_{h\to 0} \widetilde{T^{(h)}_{g^\#}}(X) = 0 \,\forall X \iff g \equiv 0. \tag{8.6}$$

Proof. Combine (8.3) with the last part of Proposition 7.3. ■

The following theorem is the main result of this section and indeed, of this paper.

Theorem 8.3. *For any smooth U-invariant functions ϕ, ψ on Ω_{norm}, there exist uniquely determined functions $g_0, g_1, \ldots,$ on $\mathbf{C}$ such that*

$$\widetilde{T^{(h)}_\phi T^{(h)}_\psi} \approx \sum_{m=0}^{\infty} h^m \, \widetilde{T^{(h)}_{g^\#_m}} \qquad \text{as } h \to 0. \tag{8.7}$$

Moreover, if $\phi = f^\#$ and $\psi = g^\#$, then the functions g_m are given by

$$g_m = G_m(f, g)^\flat \tag{8.8}$$

for some bidifferential operators G_m on $\mathbf{C}^N$ (independent of f and g). In particular,

$$\begin{aligned} G_0(f, g)^\flat &= f^\flat g^\flat, \qquad \text{and} \\ G_1(f, g)^\flat - G_1(g, f)^\flat &= \frac{i}{2\pi}\{f^\flat, g^\flat\}, \end{aligned} \tag{8.9}$$

the Poisson bracket of $f^\flat$ and $g^\flat$ on $\mathbf{C}$.

Proof. The uniqueness is immediate from (8.6). The existence is, by virtue of (8.2) and (8.5), equivalent to

$$\sum_{r=0}^{\infty} h^r \, m_r(\phi, \psi)^\# \approx \sum_{m,n=0}^{\infty} h^{m+n} \frac{(\Delta^n g_m)^\#}{n!}.$$

Comparing the expressions at like powers of h on both sides, this becomes

$$m_r(\phi, \psi) = \sum_{n=0}^{r} \frac{\Delta^n g_{r-n}}{n!},$$

which is solved by the recursive recipe

$$g_r = m_r(\phi, \psi) - \sum_{n=1}^{r} \frac{1}{n!} \Delta^n g_{r-n}. \tag{8.10}$$

From (8.4) it is also clear that g_m are of the form (8.8) with appropriate bidifferential operators G_m. Finally, a short computation using the special instances $r = 0, 1$ of (8.4),

$$m_0(\phi, \psi) = f^\flat g^\flat, \qquad m_1(\phi, \psi) = \Big(g\Delta f + f\Delta g + \frac{\partial f}{\partial d_1}\frac{\partial g}{\partial \bar{e}_1}\Big)^\flat,$$

gives (8.9). ■

Remark. Note that the quantities $G_m(\phi, \psi)^\flat$ do *not* depend only on $f^\flat$ and $g^\flat$: the bidifferential operators G_m involve derivatives also in other variables than d_1, e_1, and only after these are applied one takes the restriction to $d_2 = \cdots = d_N = e_2 = \cdots = e_N = 0$. It is therefore quite remarkable that $G_1(\phi, \psi)^\flat - G_1(\psi, \phi)^\flat$ depends only on $f^\flat$ and $g^\flat$ — the derivatives with respect to the other variables having cancelled out. □

We indicate another proof of the last theorem, based on the isomorphism (7.3). (We gave the proof above first since the isomorphism (7.3) is probably something peculiar to the domain of normal matrices, while the stationary phase method should work also in other situations. The proof below also requires a slightly stronger hypothesis on the functions ϕ and ψ.)

For a function f on $\mathbf{C}^N$ and $h > 0$, let $P_h f$ be the function on $\mathbf{C}$ defined by

$$P_h f(z_1) := \int_{\mathbf{C}^{N-1}} f(z_1, z_2, \ldots, z_N)\, e^{-(|z_2|^2+\cdots+|z_N|^2)/h} \frac{dz_2 \ldots dz_N}{(\pi h)^{N-1}}.$$

Theorem 8.4. *Let $\phi = f^\#$, $\psi = g^\#$ be smooth U-invariant functions on Ω_{norm} such that the partial derivatives of f and g of all orders are bounded, and let C_r be the bidifferential operators* (7.2). *Then*

$$T_\phi^{(h)} T_\psi^{(h)} \approx \sum_{r=0}^{\infty} h^r\, T^{(h)}_{C_r(P_h f, P_h g)^\#}$$

in the sense of operator norms. Consequently, (8.7) *holds for*

$$g_m = \sum_{\substack{j,k,r\geq 0,\\ j+k+r=m}} \frac{1}{j!k!r!} \partial^r (\Delta'^j f)^\flat \cdot \bar{\partial}^r (\Delta'^k g)^\flat,$$

where Δ' denotes the Laplacian with respect to the last $N-1$ variables $z_2, \ldots, z_N$.

Proof. By a computation similar to the one in the proof of Theorem 7.1, for any $\chi, \eta \in \mathbf{C}^N$,

$$
\begin{aligned}
\langle T_\phi^{(h)} Z^k \chi, Z^l \eta \rangle &= \int_{\Omega_{\mathrm{norm}}} \eta^* Z^{*l} \phi(Z) Z^k \chi \, d\mu_h(Z) \\
&= \int_{\mathbf{C}^N} \int_{U(N)} \eta^* U D^{*l} U^* \phi(U D U^*) U D^k U^* \chi \, dU \, e^{-\operatorname{Tr}(D^* D)/h} \frac{dD}{(\pi h)^N} \\
&= \int_{\mathbf{C}^N} \int_{U(N)} \eta^* U D^{*l} \phi(D) D^k U^* \chi \, dU \, e^{-\operatorname{Tr}(D^* D)/h} \frac{dD}{(\pi h)^N} \\
&\qquad \text{(by the } U\text{-invariance of } \phi) \\
&= \langle \chi, \eta \rangle \frac{1}{N} \int_{\mathbf{C}^N} \operatorname{Tr}(D^{*l} \phi(D) D^k) e^{-\operatorname{Tr}(D^* D)/h} \frac{dD}{(\pi h)^N} \\
&= \langle \chi, \eta \rangle \frac{1}{N} \sum_{j=1}^N \int_{\mathbf{C}^N} \overline{d}_j^l d_j^k f(d_j; d_1, \dots, \hat{d}_j, \dots, d_N) \, e^{-\|d\|^2/h} \frac{dD}{(\pi h)^N} \\
&= \langle \chi, \eta \rangle \int_{\mathbf{C}^N} \overline{d}_1^l d_1^k f(d_1; d_2, \dots, \dots, d_N) \, e^{-\|d\|^2/h} \frac{dD}{(\pi h)^N} \\
&= \langle \chi, \eta \rangle \int_{\mathbf{C}} \overline{d}_1^l d_1^k P_h f(d_1) \, e^{-|d_1|^2/h} \frac{dd_1}{\pi h} \\
&= \langle \chi, \eta \rangle \, \langle z^k P_h f, z^l \rangle_{L^2(\mathbf{C}, d\mu_h)} \\
&= \langle \chi, \eta \rangle \, \langle T_{P_h f}^{(h)} z^k, z^l \rangle_{L^2_{\mathrm{hol}}(\mathbf{C}, d\mu_h)} \\
&= \langle (T_{P_h f}^{(h)} \otimes I)(z^k \otimes \chi), z^l \otimes \eta \rangle_{L^2_{\mathrm{hol}}(\mathbf{C}, d\mu_h) \otimes \mathbf{C}^N}.
\end{aligned}
$$

Consequently, under the isomorphism (7.3), the operator $T_\phi^{(h)}$ on $\mathcal{H}_h$ corresponds to the operator $T_{P_h f}^{(h)} \otimes I$ on $L^2_{\mathrm{hol}}(\mathbf{C}, d\mu_h) \otimes \mathbf{C}^N$. Thus by the ordinary Berezin-Toeplitz quantization on $\mathbf{C}$,

$$
\begin{aligned}
T_\phi^{(h)} T_\psi^{(h)} &\cong T_{P_h f}^{(h)} T_{P_h g}^{(h)} \otimes I \\
&\approx \sum_{r=0}^\infty h^r \, T_{C_r(P_h f, P_h g)}^{(h)} \otimes I \\
&\cong \sum_{r=0}^\infty h^r \, T_{C_r(P_h f, P_h g)^\#}^{(h)}
\end{aligned}
$$

(the last isomorphism is the one from the proof of Theorem 7.1). This proves the first claim. The second part of the theorem follows upon inserting the expansion

$$
P_h f = \sum_{j=0}^\infty \frac{h^j}{j!} (\Delta'^j f)^\flat,
$$

which follows from the Taylor formula (or stationary phase), and taking Berezin transforms on both sides. (The hypothesis of boundedness of the derivatives of f and g is needed in order that the resulting expansion for $C_r(P_h f, P_h g)$ converge uniformly on $\mathbf{C}^N$, and thus imply the convergence of the corresponding expansion for $T^{(h)}_{C_r(P_h f, P_h g)^\#}$ by the inequality $\|T_\phi\| \leq \|\phi\|_\infty$.) ■

For two U-invariant functions $\phi = f^\#$, $\psi = g^\#$, define their "star product" $\phi * \psi$ as the formal power series

$$\phi * \psi := \sum_{r=0}^{\infty} h^r \, G_r(f, g)^{\flat\#}.$$

As usual, this product can be extended by $\mathbf{C}[[h]]$-linearity to all $\phi, \psi \in \mathcal{U}[[h]]$, the ring of all power series in h with coefficients in the algebra $\mathcal{U}$ of all U-invariant functions on Ω_{norm}. Alternatively, upon identifying $\phi = f^\# \in \mathcal{U}$ with f, we may view this as the star product

$$f * g := \sum_{r=0}^{\infty} h^r \, G_r(f, g)^{\flat}$$

on the algebra $\mathcal{S}$ of all functions $f(d_1; d_2, \dots, d_N)$ on $\mathbf{C} \times \mathbf{C}^{N-1}$ symmetric in the last $N-1$ variables, which again can be extended by $\mathbf{C}[[h]]$-linearity to all $f, g \in \mathcal{S}[[h]]$, the ring of formal power series with coefficients in $\mathcal{S}$. If we extend to $\mathcal{S}[[h]]$ by $\mathbf{C}[[h]]$-linearity also the operators G_r, then the extended star-product will still satisfy the relations (8.9). Further, $*$ is clearly associative, since the multiplication of operators is associative — both $(\phi * \psi) * \eta$ and $\phi * (\psi * \eta)$ originate from the asymptotic expansion as $h \to 0$ of $[T^{(h)}_\phi T^{(h)}_\psi T^{(h)}_\eta]^\sim$. (However, in contrast to a genuine star-product, the function constant one is <u>not</u> the unit element for $*$.)

The appearance of $f^\flat$ and $g^\flat$, and not f and g, in (8.9) means that the $\mathbf{C}^{N-1}$ part of f disappears in the semiclassical limit $h \to 0$, and only the projection $f^\flat$, which lives on $\mathbf{C}$, survives. As mentioned before, we are dealing here with a quantum system which has N internal degrees of freedom. This is made clear by the isomorphism (7.3), since the tensor product space $L^2_{\text{hol}}(\mathbf{C}, d\mu_h) \otimes \mathbf{C}^N$ is exactly the Hilbert space of a single quantum particle, moving on the phase space $\mathbf{C}$ and having N internal degrees of freedom. The full set of quantum observables of this system include those which do not have classical counterparts. The interesting fact that emerges from our analysis is that, it is exactly those observables which are Berezin quantized versions of U-invariant functions, that have classical counterparts. Since the internal degrees of freedom are purely quantum in this case, they do not survive in the semi-classical limit.

Acknowledgement

Part of this work was done while the second author was visiting the first; the support of the Mathematics Department of Concordia University is gratefully acknowledged.

References

[AAG] S.-T. Ali, J.-P. Antoine, J.-P. Gazeau: *Coherent States, Wavelets and Their Generalizations,* Springer-Verlag, New York, 2000.

[AE] S.-T. Ali, M. Engliš: *Quantization methods: a guide for physicists and analysts,* Rev. Math. Phys. **17** (2005), 391–490.

[AEG] S.-T. Ali, M. Engliš, J.-P. Gazeau: *Vector Coherent States from Plancherel's Theorem, Clifford Algebras and Matrix Domains,* J. Phys. A: Math. Gen. **37** (2004), 6067–6089.

[BMS] M. Bordemann, E. Meinrenken, M. Schlichenmaier: *Toeplitz quantization of Kähler manifolds and $gl(n)$, $n \to \infty$ limits,* Comm. Math. Phys. **165** (1994), 281–296.

[Gin] J. Ginibre: *Statistical ensembles of complex, quaternion, and real matrices,* J. Math. Phys. **6** (1965), 440–449.

[Hrm] L. Hörmander: *The analysis of linear partial differential operators, vol. I,* Grundlehren der mathematischen Wissenschaften, vol. 256, Springer-Verlag, Berlin - Heidelberg - New York - Tokyo, 1985.

[KS] A.V. Karabegov, M. Schlichenmaier: *Identification of Berezin-Toeplitz deformation quantization,* J. reine angew. Math. **540** (2001), 49–76.

[Kri] C. Kristjansen, J. Plefka, G.W. Semenoff, M. Staudacher: *A new double-scaling limit of $N = 4$ super-Yang-Mills theory and pp-wave strings,* Nuclear Physics B **643** (2002), 3–30.

[Me] E. Merzbacher, *Quantum Mechanics*, Wiley, New York, 1970.

Department of Mathematics and Statistics, Concordia University, Montréal, Québec, Canada H3G 1M8
E-mail: stali@mathstat.concordia.ca

Mathematics Institute, Žitná 25, 11567 Prague 1, Czech Republic *E-mail:* englis@math.cas.cz

Partial *-Algebras, A Tool for the Mathematical Description of Physical Systems

Jean-Pierre Antoine

We review the main steps in the development of partial *-algebras. First we discuss the algebraic structure stemming from the partial multiplication. Then we study in some detail the locally convex partial *-algebras, in particular, the Banach partial *-algebras, and we describe a number of concrete examples. Next we consider the partial *-algebras of closable operators in Hilbert spaces (partial O*-algebras), with a special emphasis on their *-automorphisms. Finally we sketch the representation theory of abstract partial *-algebras and give some instances of possible physical applications.

To Gérard Emch, colleague, friend, editor,
on the occasion of his 70th birthday

1. Introduction

Forty years ago, Haag and Kastler [27] introduced the algebraic approach to Quantum Field Theory, and soon their method was extended to statistical mechanics. The rationale was to free the theory from particular realizations, linked to specific models, and to focus on the basic structure, namely, the algebra of observables and states on it. This more abstract language, exploiting the deep mathematical theories of C*-algebras and von Neumann algebras, soon became the standard approach to the mathematically rigorous description of physical systems with infinitely many degrees of freedom. Textbooks flourished and, among many authors, Gérard Emch argued forcefully for the new approach [6].

Still there is a problem. Basically, the operator algebras that appear in the theory are algebras of *bounded* operators. Yet we know that most observables of a quantum system should be in fact represented by *unbounded* operators. This essentially follows from symmetry considerations. Indeed, if the system is invariant under a Lie group of transformations, the symmetry generators represent the corresponding Lie algebra, and as such they are realized by self-adjoint, mostly unbounded operators in the corresponding Hilbert space. Of course, unbounded operators may always be taken as *affiliated* to some von Neumann algebra, but this solution somehow masks the simplicity of the situation. Think of the position, the momentum or the energy observables.

Now unbounded operators in a Hilbert space are notoriously difficult to multiply, because of domain problems, so that the nice algebraic structure gets lost. A possible way out is to consider only as observables operators with a common, dense, *invariant* domain. In other words, one starts from an *algebra of unbounded operators*. This concept, originally introduced by Powers [31], was developed into a rich theory, mostly by the Leipzig group around Lassner and Uhlmann (see the monograph [32] for a review).

However, the problem is not yet completely solved, since in many cases it is difficult or unnatural to find an invariant domain common to all the observables of a given system. This suggests to look at the algebraic structure, if any, that would arise if one tries to multiply unbounded operators without the domain restriction. Indeed there is a rich structure behind, and not only at the algebraic level, as was found by W. Karwowski (Wrocław) and myself [5, 6]. Since then, several mathematicians have joined this circle of ideas and a full-fledged theory has emerged, thanks to the work of F. Mathot and J. Shabani (Louvain-la-Neuve), G. and G.A. Lassner (Leipzig), G. Epifanio, C. Trapani, F. Bagarello, and F. Tschinke (Palermo), A. Inoue, H. Ogi, I. Ikeda, and M. Takakura (Fukuoka). We will present here a quick overview of the theory, following essentially [11], [12], and [14], where the proofs and the original references may be found.

As will be clear in the sequel, the theory has developed mostly along mathematical lines, according to its own momentum. Yet physics is never far away. More precisely, one of the motivations was, and still is, to understand physical systems with infinitely many degrees of freedom, such as met in statistical mechanics or quantum field theory. In both cases, the system exhibits a singular behavior. For instance, the thermodynamical limit may fail to exist in the usual approaches in the presence of long range interactions. Also unsmeared quantum fields are very singular objects (operator-valued distributions). Another example is given by Hamiltonians corresponding to very singular interactions (δ or δ', pointlike or on a surface). All these applications will be described briefly in Section 6. Before that, we will examine successively the algebraic structure stemming from the partial multiplication (Section 2), the locally convex partial *-algebras, in particular, the Banach partial *-algebras, including several types of examples (Section 3), the partial *-algebras of closable operators in Hilbert spaces (also called partial O*-algebras) and their *-automorphisms and automorphism groups, (Section 4) and, finally the representation theory of abstract partial *-algebras (Section 5).

2. The Algebraic Structure

Definition 2.1. A *partial *-algebra* is a complex vector space $\mathfrak{A}$, endowed with an involution $x \mapsto x^*$ (that is, a bijection such that $x^{**} = x$) and a partial multiplication defined by a set $\Gamma \subset \mathfrak{A} \times \mathfrak{A}$ (a binary relation) such that:

(i) $(x, y) \in \Gamma$ implies $(y^*, x^*) \in \Gamma$;

(ii) $(x, y_1), (x, y_2) \in \Gamma$ implies $(x, \lambda y_1 + \mu y_2) \in \Gamma, \ \forall \lambda, \mu \in \mathbb{C}$;

(iii) for any $(x, y) \in \Gamma$, there is defined a product $xy \in \mathfrak{A}$, which is distributive with respect to the addition and satisfies the relation $(xy)^* = y^* x^*$.

Notice that the partial multiplication is *not* required to be associative (and often it is not). We shall assume the partial *-algebra $\mathfrak{A}$ contains a unit e, i.e., $e^* = e$, $(e, x) \in \Gamma$, $\forall x \in \mathfrak{A}$, and $ex = xe = x$, $\forall x \in \mathfrak{A}$. (If $\mathfrak{A}$ has no unit, it may always be embedded into a larger partial *-algebra with unit, in the standard fashion [1]).

Given the defining set Γ, spaces of *multipliers* are defined in the obvious way:

$$(x, y) \in \Gamma \quad \Leftrightarrow \quad x \in L(y) \text{ or } x \text{ is a left multiplier of } y$$
$$\Leftrightarrow \quad y \in R(x) \text{ or } y \text{ is a right multiplier of } x.$$

For any subset $\mathfrak{N} \subset \mathfrak{A}$, we write

$$L\mathfrak{N} = \bigcap_{x \in \mathfrak{N}} L(x), \qquad R\mathfrak{N} = \bigcap_{x \in \mathfrak{N}} R(x),$$

and, of course, the involution exchanges the two:

$$(L\mathfrak{N})^* = R\mathfrak{N}^*, \quad (R\mathfrak{N})^* = L\mathfrak{N}^*.$$

Clearly all these multiplier spaces are vector subspaces of $\mathfrak{A}$, containing e.

The partial *-algebra is *abelian* if $L(x) = R(x)$, $\forall x \in \mathfrak{A}$, and then $xy = yx$, $\forall y \in L(x)$. In that case, we write simply for the multiplier spaces $L(x) = R(x) := M(x)$, $L\mathfrak{N} = R\mathfrak{N} := M\mathfrak{N}$ ($\mathfrak{N} \subset \mathfrak{A}$).

Now the crucial fact is that the couple of maps (L, R) defines a *Galois connection* [4] on the complete lattice of all vector subspaces of $\mathfrak{A}$ (ordered by inclusion), which means that (i) both L and R reverse order; and (ii) both LR and RL are closures, that is,

$$\mathfrak{N} \subset LR\mathfrak{N} \text{ and } LRL = L, \qquad \mathfrak{N} \subset RL\mathfrak{N} \text{ and } RLR = R,$$

for any vector subspace $\mathfrak{N}$ of $\mathfrak{A}$. Let us denote by $\mathcal{F}^L$, resp. $\mathcal{F}^R$, the set of all LR-closed, resp. RL-closed, subspaces of $\mathfrak{A}$:

$$\mathcal{F}^L = \{\mathfrak{N} \subset \mathfrak{A} : \mathfrak{N} = LR\mathfrak{N}\}, \quad \mathcal{F}^R = \{\mathfrak{N} \subset \mathfrak{A} : \mathfrak{N} = RL\mathfrak{N}\},$$

both ordered by inclusion. Then, from standard results of universal algebra, one can deduce the following result.

Theorem 2.2. *(1) The set $\mathcal{F}^R$, ordered by inclusion, is a complete lattice with lattice operations*

$$\mathfrak{M} \wedge \mathfrak{N} = \mathfrak{M} \cap \mathfrak{N}, \quad \mathfrak{M} \vee \mathfrak{N} = RL(\mathfrak{M} + \mathfrak{N}).$$

The largest element is $\mathfrak{A}$, the smallest $R\mathfrak{A}$. A corresponding result holds for $\mathcal{F}^L$, exchanging L and R.

(2) Both $L : \mathcal{F}^R \to \mathcal{F}^L$ and $R : \mathcal{F}^L \to \mathcal{F}^R$ are lattice anti-isomorphisms: $L(\mathfrak{M} \wedge \mathfrak{N}) = L\mathfrak{M} \vee L\mathfrak{N}$, $L(\mathfrak{M} \vee \mathfrak{N}) = L\mathfrak{M} \wedge L\mathfrak{N}$, and similarly for R.

(3) The involution $\mathfrak{N} \leftrightarrow \mathfrak{N}^*$ is a lattice isomorphism between $\mathcal{F}^L$ and $\mathcal{F}^R$.

Actually, multiplier spaces determine completely the partial *-algebraic structure of $\mathfrak{A}$, as follows from the following statement:

$$(x, y) \in \Gamma \iff \exists (\mathfrak{N}, \mathfrak{M}) \in \mathcal{F}^L \times \mathcal{F}^R, \text{ such that } \mathfrak{N} = L\mathfrak{M},\ \mathfrak{M} = R\mathfrak{N}, \text{ and } x \in \mathfrak{N},\ y \in \mathfrak{M}.$$

(such pairs of subspaces are called *matching pairs*). We emphasize that the complete lattices $\mathcal{F}^R$, $\mathcal{F}^L$ are often difficult to describe explicitly, but much less is needed in practice, as we will see in the next section.

As examples of partial *-algebras, some of which we will encounter below, we may cite partial *-algebras of polynomials, of functions, of infinite matrices or kernels, topological quasi *-algebras, Banach partial *-algebras, CQ*-algebras, and partial *-algebras of closable operators in a Hilbert space (partial O*-algebras).

The last case is the most important in practice. It will also be needed to set up a representation theory, because a representation of a partial *-algebra $\mathfrak{A}$ is a homomorphism of $\mathfrak{A}$ into some partial O*-algebra (see Section 5). Here a **-homomorphism* of a partial *-algebra $\mathfrak{A}$ into another one $\mathfrak{B}$ is a linear map $\rho : \mathfrak{A} \to \mathfrak{B}$ such that (i) $\rho(x^*) = \rho(x)^*$ for every $x \in \mathfrak{A}$, and (ii) whenever $x \in L(y)$ in $\mathfrak{A}$, then $\rho(x) \in L(\rho(y))$ in $\mathfrak{B}$ and $\rho(x)\rho(y) = \rho(xy)$. The map ρ is a **-isomorphism* if it is a bijection and $\rho^{-1} : \mathfrak{B} \to \mathfrak{A}$ is also a *-homomorphism. In particular, for $\mathfrak{A} = \mathfrak{B}$, one speaks of a **-automorphism.*

3. Locally Convex Partial *-Algebras

3.1. Basic Definitions

Let $\mathfrak{A}$ be a partial *-algebra with unit and assume it carries a locally convex, Hausdorff, topology τ, which makes it into a locally convex topological vector space $\mathfrak{A}[\tau]$ (that is, the vector space operations are τ-continuous).

The partial *-algebraic structure of $\mathfrak{A}$ is completely characterized by its spaces of left, resp. right, multipliers. Thus, quite naturally, we describe the topological structure of $\mathfrak{A}[\tau]$ by providing all spaces of multipliers with appropriate topologies. Our goal is to make the algebraic and the topological structure coincide as much as possible.

We start with the following observation. Let $\mathfrak{M} \in \mathcal{F}^R$. To every $a \in L\mathfrak{M}$, one may associate a linear map L_a from $\mathfrak{M}$ into $\mathfrak{A}$:

$$L_a(x) = ax, \quad x \in \mathfrak{M},\ a \in L\mathfrak{M}.$$

Thus we may define the topology $\rho_{\mathfrak{M}}$ on $\mathfrak{M}$ as the weakest locally convex topology on $\mathfrak{M}$ such that all maps L_a, $a \in L\mathfrak{M}$, are continuous from $\mathfrak{M}$ into $\mathfrak{A}[\tau]$. This is, of course, a projective topology. In the same way, the topology $\lambda_{\mathfrak{N}}$ on $\mathfrak{N} \in \mathcal{F}^L$ is the weakest locally convex topology on $\mathfrak{N}$ such that all maps $R_b : x \mapsto xb$, $x \in \mathfrak{N}$, $b \in R\mathfrak{N}$, are continuous from $\mathfrak{N}$ into $\mathfrak{A}[\tau]$.

It follows immediately from the definition that, whenever $\mathfrak{M}_1, \mathfrak{M}_2 \in \mathcal{F}^R$ are such that $\mathfrak{M}_1 \subset \mathfrak{M}_2$, then the topology $\rho_{\mathfrak{M}_1}$ is finer than the topology $(\rho_{\mathfrak{M}_2} \lceil \mathfrak{M}_1)$ induced by $\mathfrak{M}_2$ on $\mathfrak{M}_1$. In other words, the embedding $\mathfrak{M}_1 \to \mathfrak{M}_2$ is a *continuous injection*.

Take now $\mathfrak{A}$ itself. It carries three topologies, τ, $\rho_{\mathfrak{A}}$ and $\lambda_{\mathfrak{A}}$, and it is easy to see that both $\rho_{\mathfrak{A}}$ and $\lambda_{\mathfrak{A}}$ are finer than τ. As a consequence, since τ was assumed to be Hausdorff, all topologies $\rho_{\mathfrak{M}}$, $\mathfrak{M} \in \mathcal{F}^R$, and $\lambda_{\mathfrak{N}}$, $\mathfrak{N} \in \mathcal{F}^L$, are Hausdorff.

Now, for reasons of coherence, it would be preferable that all three topologies on $\mathfrak{A}$, τ, $\rho_{\mathfrak{A}}$ and $\lambda_{\mathfrak{A}}$ be equivalent. Here is a handy criterion.

Lemma 3.1. *Let $\mathfrak{A}[\tau]$ be a partial *-algebra with locally convex topology τ. Then the projective topology $\rho_{\mathfrak{A}}$ on $\mathfrak{A}$ is equivalent to τ if and only if, for each $a \in L\mathfrak{A}$, the map $L_a : x \mapsto ax$ is continuous from $\mathfrak{A}[\tau]$ into itself. Similarly, the projective topology $\lambda_{\mathfrak{A}}$ on $\mathfrak{A}$ is equivalent to τ if and only if, for each $b \in R\mathfrak{A}$, the map $R_b : x \mapsto xb$ is continuous from $\mathfrak{A}[\tau]$ into itself.*

Moreover, if the involution $x \mapsto x^*$ is τ-continuous, then it is continuous from $\mathfrak{M}[\rho_{\mathfrak{M}}]$ into $\mathfrak{M}^*[\lambda_{\mathfrak{M}^*}] \in \mathcal{F}^L$, for every $\mathfrak{M} \in \mathcal{F}^R$.

According to our goal, we will naturally require that all three topologies $\rho_{\mathfrak{A}}$, $\lambda_{\mathfrak{A}}$ and τ on a topological partial *-algebra coincide and that the involution be continuous. Let us now look at multiplier spaces $\mathfrak{M} \in \mathcal{F}^R$. If $\mathfrak{M}_1 \subset \mathfrak{M}_2$, we have seen that the embedding is continuous. In order to make the structure tighter, we should also require that $\mathfrak{M}_1$ be *dense* in $\mathfrak{M}_2[\rho_{\mathfrak{M}_2}]$. This is true in many examples, typically the function spaces (see Section 3.3). Of course, it is enough to require that $R\mathfrak{A}$ be dense in each $\mathfrak{M}[\rho_{\mathfrak{M}}] \in \mathcal{F}^R$. Indeed, if $R\mathfrak{A} \subset \mathfrak{M}_1 \subset \mathfrak{M}_2$, and $R\mathfrak{A}$ is dense in $\mathfrak{M}_2$ for $\rho_{\mathfrak{M}_2}$, so is *a fortiori* $\mathfrak{M}_1$. But this condition is still too strong (and hardly verifiable in practice, because $\mathcal{F}^R$ is too large). Thus, we introduce the following notion, which is sufficient and much more manageable.

Definition 3.2. A subset $\mathcal{I}^R$ of $\mathcal{F}^R$ is called a *generating family* if

(i) $R\mathfrak{A} \in \mathcal{I}^R$ and $\mathfrak{A} \in \mathcal{I}^R$, and

(ii) $x \in L(y)$ if and only if $\exists\, \mathfrak{M} \in \mathcal{I}^R$ such that $y \in \mathfrak{M}$, $x \in L\mathfrak{M}$.

A generating family for $\mathcal{F}^L$ is defined in a similar way.

Thus a generating family determines completely the partial multiplication. Clearly, if $\mathcal{I}^R$ is a generating family for $\mathcal{F}^R$, $\mathcal{I}^L = L\mathcal{I}^R = \{L\mathfrak{M} : \mathfrak{M} \in \mathcal{I}^R\}$

is generating for $\mathcal{F}^L$, and similarly $\mathcal{I}^R_* = \{\mathfrak{M}^* : \mathfrak{M} \in \mathcal{I}^R\}$, but these two have *a priori* nothing in common. The following properties are obvious:

(i) if $\mathcal{I}^R$ is generating for $\mathcal{F}^R$, so is the sublattice $\mathcal{J}^R$ of $\mathcal{F}^R$ generated from $\mathcal{I}^R$ by *finite* lattice operations.

(ii) if $\mathcal{I}^R$ is generating, the *complete* lattice generated by $\mathcal{I}^R$ is $\mathcal{F}^R$ itself.

This last property allows us to weaken the density condition.

Proposition 3.3. – *Let $\mathfrak{A}[\tau]$ be a partial *-algebra with topology τ. Assume there exists a generating family $\mathcal{I}^R$ for $\mathcal{F}^R$ such that $R\mathfrak{A}$ is dense in $\mathfrak{M}[\rho_{\mathfrak{M}}]$ for every $\mathfrak{M} \in \mathcal{I}^R$. Then, for any pair $\mathfrak{M}_1, \mathfrak{M}_2 \in \mathcal{F}^R$ such that $\mathfrak{M}_1 \subset \mathfrak{M}_2$, $\mathfrak{M}_1$ is dense in $\mathfrak{M}_2[\rho_{\mathfrak{M}_2}]$.*

Summarizing, we may now state our definition of locally convex partial *-algebra.

Definition 3.4. – (a) Let $\mathfrak{A}[\tau]$ be a partial *-algebra, which is a topological vector space for the locally convex topology τ. Then $\mathfrak{A}[\tau]$ is called a *locally convex partial *-algebra* if the following two conditions are satisfied:

(i) the involution $x \mapsto x^*$ is τ-continuous;

(ii) the maps $x \mapsto ax$ and $x \mapsto xb$ are τ-continuous for all $a \in L\mathfrak{A}$ and $b \in R\mathfrak{A}$.

(b) The locally convex partial *-algebra $\mathfrak{A}[\tau]$ is said to be *tight*, if, in addition,

(iii) there is a generating family $\mathcal{J}^R$ for $\mathcal{F}^R$ such that $R\mathfrak{A}$ is dense in $\mathfrak{M}[\rho_{\mathfrak{M}}]$, $\forall\, \mathfrak{M} \in \mathcal{J}^R$.

This definition seems natural, since it forces the topological structure determined by τ to be consistent with the multiplier structure of $\mathfrak{A}$.

The simplest example is that of a *topological quasi *-algebra*, which is defined as follows. A topological quasi *-algebra is a pair $(\mathfrak{A}, \mathfrak{A}_o)$, where $\mathfrak{A}_o$ is a topological *-algebra such that the multiplication is separately, but not jointly, continuous, $\mathfrak{A}_o$ is not complete, and $\mathfrak{A}$ is the completion of $\mathfrak{A}_o$. Thus $\mathfrak{A}$ is only a partial *-algebra, since the product xy is defined only if either x or y belongs to $\mathfrak{A}_o$. Clearly, $(\mathfrak{A}, \mathfrak{A}_o)$ is a (trivial) partial *-algebra with $L\mathfrak{M} = R\mathfrak{M} = \mathfrak{A}_o$ and $\mathfrak{A}_o$ is dense in $\mathfrak{A}$. Thus every topological quasi *-algebra is a tight locally convex partial *-algebra.

Other examples are given by *Banach partial *-algebras* and by *Banach or locally convex partial *-algebras of functions*, that we will study in the next sections.

3.2. Banach Partial *-Algebras

Particularizing Definition 3.4, we obtain the following one [15].

Definition 3.5. (a) A partial *-algebra $\mathfrak{A}$ is said to be a *normed partial *-algebra* if it carries a norm $\|\cdot\|$ such that

(i) the involution $x \mapsto x^*$ is isometric : $\|x\| = \|x^*\|, \ \forall x \in \mathfrak{A}$;

(ii) for every $a \in L\mathfrak{A}$, there exists a constant $\gamma_a > 0$ such that $\|ax\| \leqslant \gamma_a \|x\|, \ \forall x \in \mathfrak{A}$.

(b) $\mathfrak{A}[\|\cdot\|]$ is called a *Banach partial *-algebra* if, in addition,

(iii) $\mathfrak{A}[\|\cdot\|]$ is a Banach space.

Then $\mathfrak{A}[\|\cdot\|]$ is a locally convex partial *-algebra and both $\rho_{\mathfrak{A}}$ and $\lambda_{\mathfrak{A}}$ are equivalent to the norm topology on $\mathfrak{A}$.

We describe now topologies on the spaces of multipliers of a Banach partial *-algebra. To begin with, let us consider the spaces of universal multipliers. The following sets of seminorms define, respectively, the topology $\rho_{R\mathfrak{A}}$ on $R\mathfrak{A}$ and $\lambda_{L\mathfrak{A}}$ on $L\mathfrak{A}$:

$$R\mathfrak{A} \ni b \mapsto \|xb\|, \ x \in \mathfrak{A}; \qquad L\mathfrak{A} \ni a \mapsto \|ax\|, \ x \in \mathfrak{A}.$$

In practice, it turns out that these topologies are often equivalent to norm topologies.

Take now an arbitrary matching pair $\mathfrak{M} \in \mathcal{F}^R, \ L\mathfrak{M} \in \mathcal{F}^L$. Apart from $\rho_{\mathfrak{M}}$ and $\lambda_{L\mathfrak{M}}$, other topologies can be defined on $\mathfrak{M}$ and $L\mathfrak{M}$, respectively, starting from the fact that, as above, $L\mathfrak{M}$ can be identified with a space of linear maps from $\mathfrak{M}$ into $\mathfrak{A}$. Let $\mathcal{G}$ be a bounded subset of $\mathfrak{M}[\rho_{\mathfrak{M}}]$ and $a \in L\mathfrak{M}$. We put

$$\|a\|_{\mathcal{G}} = \sup_{x \in \mathcal{G}} \|ax\|.$$

This family of seminorms endows $L\mathfrak{M}$ with a topology $\Lambda_{L\mathfrak{M}}$ finer than $\lambda_{L\mathfrak{M}}$. One defines in a similar way a topology $\mathrm{P}_{\mathfrak{M}}$ on $\mathfrak{M}$. In general, these topologies are neither normable, nor Fréchet.

At this point, it should be clear that there is a deep analogy between partial *-algebras and PIP-spaces [2, 3], the duality there being replaced here by the exchange under L or R. Thus we are led to consider Banach partial *-algebras for which "sufficiently many" multiplier spaces are themselves Banach spaces. In order to reach a proper definition, we study the relationship between the various topologies on a given matching pair $(\mathfrak{M}, L\mathfrak{M})$.

Let $\mathfrak{M} \in \mathcal{F}^R$ and let $\|\cdot\|_{\mathfrak{M}}$ be a norm on $\mathfrak{M}$ (for simplicity, we denote the corresponding norm topology by the same symbol). We say that $\|\cdot\|_{\mathfrak{M}}$ is *admissible* if one has $\rho_{\mathfrak{M}} \preceq \|\cdot\|_{\mathfrak{M}} \preceq \mathrm{P}_{\mathfrak{M}}$. Similarly, a norm $\|\cdot\|_{L\mathfrak{M}}$ on $L\mathfrak{M} \in \mathcal{F}^L$ is admissible whenever $\lambda_{L\mathfrak{M}} \preceq \|\cdot\|_{L\mathfrak{M}} \preceq \Lambda_{L\mathfrak{M}}$.

Let now $\|\cdot\|_{\mathfrak{M}}$ be an admissible norm on $\mathfrak{M} \in \mathcal{F}^R$. Then $L_a : \mathfrak{M}[\|\cdot\|_{\mathfrak{M}}] \to \mathfrak{A}[\|\cdot\|]$ is continuous, that is,

$$\|L_a x\| = \|ax\| \leqslant \gamma \, \|x\|_{\mathfrak{M}}, \quad x \in \mathfrak{M}.$$

Then we can define a norm $\| \cdot \|^{\circ}_{L\mathfrak{M}}$ on $L\mathfrak{M}$ by

$$\|a\|^{\circ}_{L\mathfrak{M}} = \sup_{\|x\|_{\mathfrak{M}} \leqslant 1} \|ax\|. \tag{3.1}$$

Since the unit ball of $\mathfrak{M}[\| \cdot \|_{\mathfrak{M}}]$ is bounded in $\mathfrak{M}[\rho_{\mathfrak{M}}]$, it follows that $\| \cdot \|^{\circ}_{L\mathfrak{M}}$ is admissible too. In the same way, we can define a new norm $\| \cdot \|^{\circ\circ}_{\mathfrak{M}}$ on $\mathfrak{M}$ by

$$\|x\|^{\circ\circ}_{\mathfrak{M}} = \sup_{\|a\|^{\circ}_{L\mathfrak{M}} \leqslant 1} \|ax\|. \tag{3.2}$$

It is easily seen that $\|x\|^{\circ\circ}_{\mathfrak{M}} \leqslant \|x\|_{\mathfrak{M}}$, for every $x \in \mathfrak{M}$, and that $\|x\|^{\circ\circ}_{\mathfrak{M}}$ is admissible.

Moreover,

$$\|ax\| \leqslant \|a\|^{\circ}_{L\mathfrak{M}} \|x\|^{\circ\circ}_{\mathfrak{M}}, \quad \forall a \in L\mathfrak{M},\ x \in \mathfrak{M},$$

which is closely reminiscent of the Hölder inequality.

If $\| \cdot \|^{\circ\circ}_{\mathfrak{M}}$ is strictly weaker than $\| \cdot \|_{\mathfrak{M}}$, then we can start the procedure again and define a new norm $\| \cdot \|^{\circ\circ\circ}_{L\mathfrak{M}}$ on $L\mathfrak{M}$. It is easily seen that $\| \cdot \|^{\circ\circ\circ}_{L\mathfrak{M}} \leqslant \| \cdot \|^{\circ}_{L\mathfrak{M}}$. But we cannot go further, the procedure stops there.

Proposition 3.6. *If $\| \cdot \|_{\mathfrak{M}}$ is admissible, then one has*

$$\| \cdot \|^{\circ\circ\circ}_{L\mathfrak{M}} = \| \cdot \|^{\circ}_{L\mathfrak{M}}.$$

This leads to a natural definition. The norm $\| \cdot \|_{\mathfrak{M}}$ on $\mathfrak{M}$ is said to be *reproducing* if $\| \cdot \|^{\circ\circ}_{\mathfrak{M}}$ is equivalent to $\| \cdot \|_{\mathfrak{M}}$. Then $\mathfrak{M}[\| \cdot \|_{\mathfrak{M}}]$ itself is said to be reproducing.

The central questions here are the completeness of $\mathfrak{M}$ in a suitable norm $\| \cdot \|_{\mathfrak{M}}$ and the possible equivalence between the three topologies $\rho_{\mathfrak{M}}$, $\| \cdot \|_{\mathfrak{M}}$ and $\mathrm{P}_{\mathfrak{M}}$. These properties rely on the sequential completeness of $\mathfrak{M}[\rho_{\mathfrak{M}}]$, but not on the completion of $\mathfrak{A}$. Indeed, one has:

Proposition 3.7. *Let $\mathfrak{A}$ be a normed partial *-algebra and let $\mathfrak{M}[\rho_{\mathfrak{M}}]$ be sequentially complete. Then $\mathfrak{M}[\| \cdot \|_{\mathfrak{M}}]$ is complete, i.e., a Banach space, for any reproducing norm $\| \cdot \|_{\mathfrak{M}}$.*

Note that, under these conditions, two comparable reproducing norms are necessarily equivalent.

A slightly stronger condition is that all the operators L_a be closed. Indeed:

Proposition 3.8. *Assume that $\mathfrak{A}$ is a Banach partial *-algebra and that, for every $a \in L\mathfrak{M}$, the map $L_a : x \in \mathfrak{M} \mapsto ax \in \mathfrak{A}$ is closed in $\mathfrak{A}$. Then $\mathfrak{M}[\rho_{\mathfrak{M}}]$ is sequentially complete.*

We come now to the central result of this section.

Theorem 3.9. *Let $\mathfrak{M}[\rho_{\mathfrak{M}}]$ be sequentially complete and let $\|\cdot\|_{\mathfrak{M}}$ be an admissible norm on $\mathfrak{M}$. Then the following statements are equivalent.*

(i) $\|\cdot\|_{\mathfrak{M}}$ *is reproducing;*

(ii) $\mathfrak{M}[\|\cdot\|_{\mathfrak{M}}]$ *is a Banach space;*

(iii) $\|\cdot\|_{\mathfrak{M}}$ *is the unique (up to equivalence) admissible Banach norm on $\mathfrak{M}$.*

From those results, we may deduce information about the equivalence of topologies on a given multiplier space $\mathfrak{M}$.

Proposition 3.10. *Given $\mathfrak{M} \in \mathcal{F}^R$, assume that $\mathfrak{M}[\|\cdot\|_{\mathfrak{M}}]$ is a Banach space. Then P_m is equivalent to $\|\cdot\|_{\mathfrak{M}}^{\diamond\diamond}$ and $\|\cdot\|_{\mathfrak{M}}$ is admissible if and only if it is reproducing. Similarly, if $L\mathfrak{M}[\|\cdot\|_{L\mathfrak{M}}^{\diamond}]$ is a Banach space, then $\Lambda_{L\mathfrak{M}}$ is equivalent to $\|\cdot\|_{L\mathfrak{M}}^{\diamond}$.*

Combining this proposition with the previous results, we get:

Corollary 3.11. *If $\mathfrak{M}[\rho_{\mathfrak{M}}]$ is sequentially complete and $\|\cdot\|_{\mathfrak{M}}$ is reproducing, then $\mathfrak{M}[\|\cdot\|_{\mathfrak{M}}]$ is Banach and all three topologies $\|\cdot\|_{\mathfrak{M}}$, $\|\cdot\|_{\mathfrak{M}}^{\diamond\diamond}$ and $\mathrm{P}_{\mathfrak{M}}$ are equivalent.*

After this discussion, we are now in a position to define what seems to be a natural class of Banach partial *-algebras. Following the analogy with PIP-spaces, we are led to impose a perfect symmetry between left and right multipliers, and thus to require that the two spaces of a pair of matching subspaces $(\mathfrak{M}, L\mathfrak{M})$ be *both* Banach spaces for an admissible norm. These norms are then automatically reproducing and coincide with $\|\cdot\|_{\mathfrak{M}} \sim \|\cdot\|_{\mathfrak{M}}^{\diamond\diamond}$ and $\|\cdot\|_{L\mathfrak{M}}^{\diamond}$, respectively.

Definition 3.12. A normed partial *-algebra or a Banach partial *-algebra $\mathfrak{A}[\|\cdot\|]$ is said to be *of type (B)* if there exists a generating family $\mathcal{I}^R$ such that, for each pair of matching subspaces $\mathfrak{M} \in \mathcal{I}^R$, $L\mathfrak{M} \in \mathcal{I}^L$, both spaces are Banach spaces for a reproducing norm.

We emphasize that completeness of $\mathfrak{A}[\|\cdot\|]$ may be relaxed (which leads to the definition of a normed partial *-algebra), but *not* that of the multiplier spaces $\mathfrak{M}[\|\cdot\|_{\mathfrak{M}}] \in \mathcal{I}^R$, where $\mathcal{I}^R$ is a generating family. Indeed, these spaces are completely determined by the partial multiplication (i.e., the set Γ). If one of them, say $\mathfrak{M}$, would be noncomplete, it could be embedded into its completion $\widetilde{\mathfrak{M}}$ with respect to $\|\cdot\|_{\mathfrak{M}}$, but nothing guarantees that the latter is still contained in $\mathfrak{A}$, and thus there is *a priori* no way of extending the partial multiplication to $\widetilde{\mathfrak{M}}$! We refer to [14] or [15] for further details (notice that the definitions of the latter paper are different from, and supersede, those of the former).

3.3. Examples of Banach or Locally Convex Partial *-Algebras

(i) L^p Spaces on a Finite Interval

A standard example of an abelian partial *-algebra is the space $L^1([0,1], dx)$, equipped with the partial multiplication [9]:

$$f \in M(g) \Leftrightarrow \exists q \in [1, \infty] \text{ such that } f \in L^q,\ g \in L^{\overline{q}},\ 1/q + 1/\overline{q} = 1. \quad (3.3)$$

A similar structure may be given for every L^p.

Thus we consider as generating family the scale of Banach spaces $\mathcal{I} = \{L^p([0,1], dx), 1 \leqslant p \leqslant \infty\}$, with $L^p \subset L^q$, $p > q$. For $1 < p < \infty$, every space L^p is a reflexive Banach space with dual $(L^p)' = L^{\overline{p}}$ $(1/p + 1/\overline{p} = 1)$.

Now, being a scale, $\mathcal{I}$ is of course a lattice, albeit not a complete one. The lattice completion of $\mathcal{I}$, denoted $\mathcal{F}$, is obtained by adding the so-called *nonstandard* spaces:

$$L^{p-} = \bigcap_{1 \leqslant q < p} L^q, \quad L^{p+} = \bigcup_{p < q \leqslant \infty} L^q.$$

Then, for $1 < p \leqslant \infty$, L^{p-}, with the projective topology, is a non-normable reflexive Fréchet space, with dual $L^{\overline{p}+}$. And for $1 \leqslant p < \infty$, L^{p+}, with the inductive topology, is a nonmetrizable complete DF-space, with dual $L^{\overline{p}-}$. Finally the following inclusions are strict:

$$L^{p+} \subset L^p \subset L^{p-} \subset L^{q+} \quad (1 < q < p < \infty), \quad (3.4)$$

all embeddings in (3.4) are continuous and have dense range. Then the complete lattice $\mathcal{F}$ generated by $\mathcal{I}$ is also a chain, obtained by replacing each L^p $(1 < p < \infty)$ by the corresponding triplet $L^{p+} \subset L^p \subset L^{p-}$ and adding the two spaces $L^{\infty-}$ and L^{1+}:

$$L^\infty \subset L^{\infty-} \subset \ldots \subset L^{p+} \subset L^p \subset L^{p-} \subset \ldots \subset L^{1+} \subset L^1.$$

Now we turn to the structure of partial *-algebra. The commutative partial multiplication on the space $L^1([0,1], dx)$ being defined as in (3.3), it is easy to see that

$$ML^p = L^{\overline{p}}, \quad ML^{p-} = L^{\overline{p}+}, \quad ML^{p+} = L^{\overline{p}-}.$$

As for topologies, take first the spaces L^p, $1 \leqslant p < \infty$. The following result is standard ([3] or [35, Chap.15]):

$$\|f\|_{\overline{p}}^{\diamond} = \sup_{g \in L^p, \|g\|_p \leqslant 1} \int_0^1 |fg| dx = \sup_{g \in L^p, \|g\|_p \leqslant 1} \left| \int_0^1 fg dx \right| = \|f\|_{\overline{p}}, \quad 1 \leqslant p < \infty.$$

By the same argument, $\|f\|_p^{\circ\circ} = \|f\|_p$. Combining this result with Proposition 3.10, we obtain

$$\rho_{L^p} \preceq \|\cdot\|_p^{\circ\circ} = \|\cdot\|_p \sim \mathrm{P}_{L^p}, \quad 1 \leqslant p < \infty. \tag{3.5}$$

As for the topology ρ_{L^p}, it is known that every L^p, $1 \leqslant p < \infty$ satisfies the closure condition of Proposition 3.8 [17]. Thus, every L^p, $1 < p \leqslant \infty$, is sequentially complete for ρ_{L^p}. For $p = 1$, one can prove directly that ρ_{L^1} coincides with the usual norm topology (as it should!), using the fact that the function $f_0(x) \equiv 1$ belongs to L^1 with $\|f_0\|_1 = 1$.[1]

The other multiplier spaces $L^{p\pm}$ do not belong to the generating family, so we don't have to take them into consideration (actually we don't have precise results on them, contrary to what is stated in [12] and in [14]).

In conclusion, the topological structure and the multiplier structure of $\mathcal{I}$ coincide and $\mathfrak{A} = L^1[(0,1), dx]$ is an abelian Banach partial *-algebra of type (B). It is tight, which means that the smallest space $M\mathfrak{A} = L^\infty[(0,1), dx] = ML^1$ is dense in every multiplier space L^p. The involution $f \mapsto \overline{f}$ is of course L^1-continuous. The multiplication is continuous from $L^\infty \times L^1$ into L^1. Actually, it is not only separately, but even jointly continuous, and similarly from $L^p \times L^{\overline{p}}$ and from $L^{p-} \times L^{\overline{p}+}$ into L^1, thanks to Hölder's inequality and the fact that all topologies are either Fréchet or DF [12].

Similarly, one may consider the scale : $\mathcal{I}_o = \{L^p([0,1], dx), 1 < p < \infty\}$ with largest space $L^{1+} = \bigcup_q L^q$. For the same reasons as before, $L^{1+}[0,1]$ is a tight abelian locally convex partial *-algebra. In addition, the chains $\mathcal{I}$ and $\mathcal{I}_o$ are partial inner product spaces [2, 3] and the latter structure coincides with the other two.

(ii) The Spaces $L^p(\mathbb{R}, dx)$

We turn now to the spaces $L^p(\mathbb{R}, dx)$ on the whole line. The difference with the previous case is that these no longer form a chain, no two of them being comparable. We have only

$$L^p \cap L^q \subset L^s, \quad \text{for every } s \text{ such that } p < s < q.$$

Hence we take the lattice generated by $\mathcal{I} = \{L^p(\mathbb{R}, dx), 1 \leqslant p \leqslant \infty\}$, that we call $\mathcal{J}$ (Figure 1). The extreme spaces of the lattice are, respectively:

$$V_J^\# = \bigcap_{1 \leqslant q \leqslant \infty} L^q, \quad \text{and} \quad V_J = \bigcup_{1 \leqslant q \leqslant \infty} L^q = \sum_{1 \leqslant q \leqslant \infty} L^q.$$

Here too, the lattice structure allows to give to V_J a structure of abelian Banach partial *-algebra. This partial *-algebra does have a unit, as we have assumed in

[1] This corrects some inexact statements made in [12] and in [14]: the topology ρ_{L^p} does not coincide with the $\|\cdot\|_p$-norm topology for $p > 1$.

general, namely the function $f_0(x) \equiv 1$, which belongs to L^∞, but not, of course, to any space $L^p(\mathbb{R}, dx),\ p < \infty$.

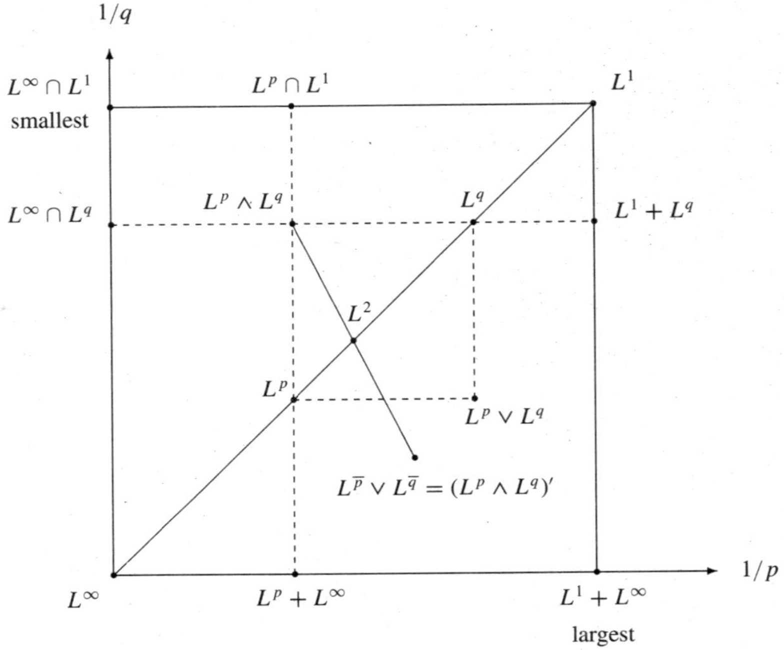

Figure 1. The lattice $\mathcal{J}$.

The lattice operations on $\mathcal{J}$ are those familiar in interpolation theory [18]:

- $L^p \wedge L^q = L^p \cap L^q$ is a Banach space, with the projective norm $\|f\|_{p\wedge q} = \|f\|_p + \|f\|_q$.
- $L^p \vee L^q = L^p + L^q$ is a Banach space, with the inductive norm $\|f\|_{p\vee q} = \inf\left(\|g\|_p + \|h\|_q\right),\ f = g + h,\ g \in L^p,\ h \in L^q$.
- For $1 < p, q < \infty$, both spaces $L^p \wedge L^q$ and $L^p \vee L^q$ are reflexive and $(L^p \wedge L^q)' = L^{\overline{p}} \vee L^{\overline{q}}$.

Notice that the lattice $\mathcal{J}$ is already obtained at the first generation. One has, for example, $L^{(r,s)} \wedge L^{(a,b)} = L^{(r\vee a, s\wedge b)}$, where $L^{(r,s)} = L^r \wedge L^s$, if $r > s$ and $L^{(r,s)} = L^r \vee L^s$, if $r < s$. Furthermore, in the lattice $\mathcal{J}$, inclusion means continuous embedding with dense range. Now we endow V_J with the natural partial multiplication

$$f \in M(g) \Leftrightarrow fg \in V_J, \text{ i.e., } fg \in L^s, \text{ for some } s,\ 1 \leqslant s \leqslant \infty.$$

Then the multipliers of the basic spaces are simple, namely, for $p > q$,[2]

$$M(L^p \wedge L^q) = L^{\overline{p}} + L^\infty := L^{(\overline{p},\infty)}, \quad M(L^p \vee L^q) = L^{\overline{q}} + L^\infty := L^{(\overline{q},\infty)},$$

and thus

$$MM(L^p \wedge L^q) = L^p + L^\infty := L^{(p,\infty)}, \quad MM(L^p \vee L^q) = L^q + L^\infty := L^{(q,\infty)}.$$

Thus matching pairs are of the form $(L^{(p,\infty)}, L^{(\overline{p},\infty)})$. Since $L^{(\overline{q},\infty)} \subset L^{(\overline{p},\infty)}$ for $q < p$, these multiplier spaces form a scale of Banach spaces.

In conclusion, we obtain another abelian Banach partial *-algebra of type (B), very similar to the previous one.

Two remarks are in order. First, here too, the lattice completion $\mathcal{F}$ of $\mathcal{J}$ and the multiplier spaces may be characterized explicitly [21]. Second, another structure of locally convex partial *-algebra may be given to the family $\mathcal{J}$ of spaces, simply replacing multiplication by convolution, with similar results [12, 14].

We note finally that the only difference between the two cases $\{L^p([0,1])\}$ and $\{L^p(\mathbb{R})\}$ lies in the type of order obtained: a chain $\mathcal{I}$ (total order) or a partially ordered lattice $\mathcal{J}$.

(iii) Amalgam Spaces

The lesson of the previous example is that an involutive lattice of (preferably reflexive) Banach spaces turns quite naturally into a (tight) locally convex partial *-algebra if it possesses a partial multiplication that satisfies a (generalized) Hölder inequality. A whole class of examples is given by the so-called *amalgam spaces* first introduced by N. Wiener (see [25] for a review). The simplest ones are the spaces (L^p, ℓ^q), consisting of functions on $\mathbb{R}$ which are locally in L^p and have ℓ^q behavior at infinity, in the sense that the L^p norms over the intervals $(n, n+1)$ form an ℓ^q sequence. For $1 < p, q < \infty$, the corresponding norm

$$\|f\|_{p,q} = \left\{ \sum_{n=-\infty}^{\infty} \left[\int_n^{n+1} |f(x)|^p dx \right]^{q/p} \right\}^{1/q}$$

turns the space (L^p, ℓ^q) into a reflexive Banach space. The same is true for the obvious extensions to p and/or q equal to 1 or ∞. Notice that $(L^p, \ell^p) = L^p$. Once again, the set of all spaces (L^p, ℓ^q) may be partially ordered by inclusion, and it turns out to be a complete lattice, with exactly the same order structure as that of the spaces $L^p(\mathbb{R}, dx)$. Thus one gets another abelian Banach partial *-algebra of type (B) [12, 14].

[2] The formula given in [12] and in [14] for $M(L^p \vee L^q)$ is incorrect, the result should read $L^{(\overline{p},\infty)}$ instead of $L^{(\overline{q},\infty)}$.

*(iv) Locally Convex Partial *-Algebras of Operators*

Let us consider the partial *-algebra of operators on a scale of Hilbert spaces, with the usual operator multiplication [3, 15]. There are several cases to consider here, so we start with the simplest one, namely, a single triplet of Hilbert spaces, associated to a positive self-adjoint operator $S \geqslant 1$ with domain $D(S)$,

$$\mathcal{H}_1 \subset \mathcal{H} \subset \mathcal{H}_{-1}, \tag{3.6}$$

in the sense that $\mathcal{H}_1 = D(S)$, equipped with the graph norm $\|f\|_1 = \|Sf\|$ and the corresponding inner product $\langle f|g\rangle_1 = \langle Sf|Sg\rangle$, and $\mathcal{H}_{-1} = \mathcal{H}_1^{\times}$, the conjugate dual of $\mathcal{H}_1$, itself a Hilbert space. Now, for every $\alpha > 0$, S^α is also a positive self-adjoint operator and $S^\alpha \geqslant 1$. So we obtain a continuous family of Hilbert spaces $\{\mathcal{H}_\alpha, -1 \leqslant \alpha \leqslant 1\}$, where, for $\alpha \geqslant 0$, $\mathcal{H}_\alpha = D(S^\alpha)$, with the graph norm, and $\mathcal{H}_{-\alpha} = \mathcal{H}_\alpha^{\times}$. Actually, this family is a chain, since, for $0 < \alpha < \beta < 1$, one has

$$\mathcal{H}_1 \subset \mathcal{H}_\beta \subset \mathcal{H}_\alpha \subset \mathcal{H} \subset \mathcal{H}_{-\alpha} \subset \mathcal{H}_{-\beta} \subset \mathcal{H}_{-1}, \tag{3.7}$$

and each embedding is continuous with dense range.

Now we consider the operators on the chain (3.7). Let $\mathcal{B}(\mathcal{H}_1, \mathcal{H}_{-1})$ be the Banach space of bounded operators from $\mathcal{H}_1$ into $\mathcal{H}_{-1}$ with its natural norm $\|\cdot\|_{1,-1}$. In $\mathcal{B}(\mathcal{H}_1, \mathcal{H}_{-1})$, we define an involution $A \mapsto A^*$ by the relation

$$< A^* f, g > = \overline{< Ag, f >}, \quad \forall f, g \in \mathcal{H}_1,$$

where $< \cdot, \cdot >$ is the form that puts $\mathcal{H}_1$ and $\mathcal{H}_{-1}$ in conjugate duality.

If $\alpha, \beta \in (-1, 1)$, we can also consider the Banach space $\mathcal{B}(\mathcal{H}_\alpha, \mathcal{H}_\beta)$ of bounded operators from $\mathcal{H}_\alpha$ into $\mathcal{H}_\beta$ with its natural norm $\|\cdot\|_{\alpha,\beta}$. Because of (3.7), the restriction to $\mathcal{H}_1$ of an operator of $\mathcal{B}(\mathcal{H}_\alpha, \mathcal{H}_\beta)$ belongs to $\mathcal{B}(\mathcal{H}_1, \mathcal{H}_{-1})$. Therefore,

$$\mathcal{B}(\mathcal{H}_\alpha, \mathcal{H}_\beta) \subset \mathcal{B}(\mathcal{H}_1, \mathcal{H}_{-1}), \ \forall \alpha, \beta \in [-1, 1].$$

Moreover, $\mathcal{B}(\mathcal{H}_\alpha, \mathcal{H}_\beta)^* = \mathcal{B}(\mathcal{H}_{-\beta}, \mathcal{H}_{-\alpha})$ for every $\alpha, \beta \in [-1, 1]$.

Let us now define the partial multiplication in $\mathcal{B}(\mathcal{H}_1, \mathcal{H}_{-1})$. Let $X, Y \in \mathcal{B}(\mathcal{H}_1, \mathcal{H}_{-1})$. We say that $X \in L(Y)$ if there exist $\alpha, \beta, \gamma \in [-1, 1]$ such that $Y \in \mathcal{B}(\mathcal{H}_\alpha, \mathcal{H}_\beta)$ and $X \in \mathcal{B}(\mathcal{H}_\beta, \mathcal{H}_\gamma)$. In this case $X \cdot Y$, the usual composition of the maps X and Y, is well-defined and belongs to $\mathcal{B}(\mathcal{H}_\alpha, \mathcal{H}_\gamma) \subset \mathcal{B}(\mathcal{H}_1, \mathcal{H}_{-1})$. It easily seen that, if $X \cdot Y$ is well-defined and belongs to $\mathcal{B}(\mathcal{H}_\alpha, \mathcal{H}_\gamma)$, then $Y^* \cdot X^*$ is also well-defined and belongs to $\mathcal{B}(\mathcal{H}_{-\gamma}, \mathcal{H}_{-\alpha})$. Moreover $(X \cdot Y)^* = Y^* \cdot X^*$. As a result, $\mathcal{B}(\mathcal{H}_1, \mathcal{H}_{-1})$ is a partial *-algebra with respect to the $\cdot$ multiplication.

The next step is the identification of the spaces of multipliers. By the definition of multiplication just given, it follows that the family of spaces $\{\mathcal{B}(\mathcal{H}_\alpha, \mathcal{H}_\beta), -1 \leqslant \alpha, \beta \leqslant 1\}$ is a generating sublattice for the lattice of left (or right) multipliers. A small calculation then shows that, for every $\alpha, \beta \in [-1, 1]$, $L\mathcal{B}(\mathcal{H}_\alpha, \mathcal{H}_\beta) = \mathcal{B}(\mathcal{H}_\beta, \mathcal{H}_{-1})$

and $R\mathcal{B}(\mathcal{H}_\alpha, \mathcal{H}_\beta) = \mathcal{B}(\mathcal{H}_1, \mathcal{H}_\alpha)$. Furthermore, it is easy to see that the topology $\rho_{\mathcal{B}(\mathcal{H}_\alpha, \mathcal{H}_\beta)}$ is equivalent to the norm topology defined by $\|\cdot\|_{\alpha,\beta}$. It follows that the norm of each space $\mathcal{B}(\mathcal{H}_\alpha, \mathcal{H}_\beta)$ is reproducing. In conclusion, $\mathcal{B}(\mathcal{H}_1, \mathcal{H}_{-1})$ is a Banach partial *-algebra of type (B) under the algebraic and topological structure defined above.

Moreover, we can here also 'enrich' the chain by introducing, for each $0 < \beta \leqslant 1$, the spaces

$$\mathcal{H}_{\beta^-} = \bigcap_{0<\alpha<\beta} \mathcal{H}_\alpha, \quad \mathcal{H}_{-\beta^+} = \bigcup_{0<\alpha<\beta} \mathcal{H}_{-\alpha}.$$

In general, $\mathcal{H}_{\beta^-}$ is a reflexive Fréchet space, whose dual is $\mathcal{H}_{-\beta^+}$, a reflexive DF-space. As in the example (i), the enriched chain is then the complete lattice generated by the chain $\{\mathcal{H}_\alpha\}$.

Starting from the construction just described, one can build several other examples. The first one consist in the discrete Hilbert scale built on the successive powers of the operator S,

$$\ldots \subset \mathcal{H}_2 \subset \mathcal{H}_1 \subset \mathcal{H} \subset \mathcal{H}_{-1} \subset \mathcal{H}_{-2} \subset \ldots, \tag{3.8}$$

where, for $n \in \mathbb{N}$, $\mathcal{H}_n = D(S^n)$, with the graph norm $\|f\|_n = \|S^n f\|$, and $\mathcal{H}_{-n} = \mathcal{H}_n^\times$, the conjugate dual of $\mathcal{H}_n$. From the discrete scale $\{\mathcal{H}_n, n \in \mathbb{Z}\}$, one can then build a continuous chain $\{\mathcal{H}_\alpha, \alpha \in \mathbb{R}\}$, by interpolation methods [18]. An example is the familiar chain of Sobolev spaces $W_s^2(\mathbb{R})$, $s \in \mathbb{R}$, where $f \in W_s^2(\mathbb{R})$ if its Fourier transform $\widehat{f}$ satisfies the condition $(1 + |.|^2)^{s/2}\, \widehat{f} \in L^2(\mathbb{R})$. Although the construction is similar, there is an essential difference, however. Whereas the operators on the bounded chain $\{\mathcal{H}_\alpha, -1 \leqslant \alpha \leqslant 1\}$ form a *Banach* partial *-algebra, in the case on an infinite scale one gets only a *locally convex* partial *-algebra. Actually the same statement is true for the 'open' chain $\{\mathcal{H}_\alpha, -1 < \alpha < 1\}$, exactly as with the L^p spaces discussed above: One gets a Banach partial *-algebra only if the extreme elements of the family are themselves Hilbert spaces. Interestingly enough, the same structure is obtained if one considers operators on a *lattice* of Hilbert spaces, instead of a scale, or for that matter, operators on a partial inner product space [2, 3].

*(v) Quasi *-Algebras of Operators in Rigged Hilbert Spaces*

Operators on an infinite scale of Hilbert spaces lead naturally to another type of partial *-algebras, namely operators on a rigged Hilbert space, and these turn out to be a quasi *-algebra.

First let us recall the definition of a rigged Hilbert space. Let $\mathcal{D}$ be a dense linear subspace of the Hilbert space $\mathcal{H}$ and t a locally convex topology on $\mathcal{D}$ finer than the topology induced by the Hilbert norm. Then the space $\mathcal{D}^\times$ of all continuous conjugate linear functionals on $\mathcal{D}[t]$, i.e., the conjugate dual of $\mathcal{D}[t]$, is a vector space that contains $\mathcal{H}$. The space $\mathcal{D}^\times$ will always be considered as endowed with the *strong dual topology* $t^\times = \beta(\mathcal{D}^\times, \mathcal{D})$ generated by the seminorms $\mathcal{D}^\times \ni \Theta \mapsto \sup_{\xi \in \mathcal{N}} | < \Theta, \xi > |$, where $< \cdot, \cdot >$ denotes the bilinear form that puts $\mathcal{D}$ and $\mathcal{D}^\times$ in duality and $\mathcal{N}$ runs over the family of all bounded sets of $\mathcal{D}[t]$. The Hilbert space $\mathcal{H}$ is dense in $\mathcal{D}^\times[t^\times]$. We get in this way a *Gel'fand triplet* or *rigged Hilbert space*

$$\mathcal{D}[t] \subset \mathcal{H} \subset \mathcal{D}^\times[t^\times], \tag{3.9}$$

where each embedding is continuous and has dense range. Clearly, this abstract set-up includes the familiar triplets of distribution spaces $\mathcal{D}(\Omega) \subset L^2(\Omega, d^n x) \subset \mathcal{D}'(\Omega)$, where Ω is a open domain in $\mathbb{R}^n$, or $\mathcal{S}(\mathbb{R}^n) \subset L^2(\mathbb{R}^n) \subset \mathcal{S}'(\mathbb{R}^n)$

In order to make contact with the scale (3.8) of Hilbert spaces built on the successive powers of the self-adjoint operator S, it suffices to define

$$\mathcal{D}_S := \mathcal{D}^\infty(S) = \bigcap_{n \in \mathbb{N}} D(S^n), \tag{3.10}$$

and to endow it with the projective topology defined by the family of Hilbert spaces $\{\mathcal{H}_n\}$, that is, the topology t_S of $\mathcal{D}^\infty(S)$ is defined by the set of seminorms $\xi \mapsto \|S^n \xi\|, \quad n = 0, 1, \ldots$. $\mathcal{D}_S[t_S]$ is a reflexive Fréchet space. The conjugate dual space of $\mathcal{D}_S$, denoted by $\mathcal{D}_S^\times$, contains each space $\mathcal{H}_{-n},\ n \geqslant 0$; more precisely,

$$\mathcal{D}_S^\times = \sum_{n \geqslant 0} \mathcal{H}_{-n} \,.$$

Given such a rigged Hilbert space $\mathcal{D}[t] \subset \mathcal{H} \subset \mathcal{D}^\times[t^\times]$, we consider the following spaces of continuous linear maps:

- $\mathfrak{L}(\mathcal{D}, \mathcal{D}^\times)$: the linear space of all continuous linear maps from $\mathcal{D}[t]$ into $\mathcal{D}^\times[t^\times]$;
- $\mathfrak{L}(\mathcal{D})$: the algebra of all continuous linear maps from $\mathcal{D}[t]$ into itself;
- $\mathfrak{L}(\mathcal{D}^\times)$: the algebra of all continuous linear maps from $\mathcal{D}^\times[t^\times]$ into itself.

Both $\mathfrak{L}(\mathcal{D})$ and $\mathfrak{L}(\mathcal{D}^\times)$ can be regarded as subspaces of $\mathfrak{L}(\mathcal{D}, \mathcal{D}^\times)$, in the sense that $\mathfrak{L}(\mathcal{D}, \mathcal{D}^\times)$ contains subspaces isomorphic to $\mathfrak{L}(\mathcal{D})$ and $\mathfrak{L}(\mathcal{D}^\times)$.

We remark that if $X \in \mathfrak{L}(\mathcal{D}, \mathcal{D}^\times)$ and $Y \in \mathfrak{L}(\mathcal{D})$, the product $X \cdot Y$, defined as the composition of X and Y makes sense and $XY \in \mathfrak{L}(\mathcal{D}, \mathcal{D}^\times)$. Similarly, if $Z \in \mathfrak{L}(\mathcal{D}^\times)$, then also $Z \cdot X$ is well-defined as an element of $\mathfrak{L}(\mathcal{D}, \mathcal{D}^\times)$. This defines a partial multiplication on $\mathfrak{L}(\mathcal{D}, \mathcal{D}^\times)$, and in fact gives the latter the structure of a quasi *-algebra. Indeed, define $\mathfrak{L}^\dagger(\mathcal{D}) = \mathfrak{L}(\mathcal{D}) \cap \mathfrak{L}(\mathcal{D}^\times)$. Then $\mathfrak{L}^\dagger(\mathcal{D})$ is a *-algebra and $\mathfrak{L}^\dagger(\mathcal{D}) \subset \mathfrak{L}(\mathcal{D}, \mathcal{D}^\times)$. Moreover, the map $X \mapsto X^\dagger$ is an involution of $\mathfrak{L}(\mathcal{D}, \mathcal{D}^\times)$, i.e., $X^{\dagger\dagger} = X$ for each $X \in \mathfrak{L}(\mathcal{D}, \mathcal{D}^\times)$, with all the required properties, so that we get:

Proposition 3.13. $(\mathfrak{L}(\mathcal{D}, \mathcal{D}^\times), \mathfrak{L}^\dagger(\mathcal{D}))$ *is a quasi *-algebra.*

Many more properties of $\mathfrak{L}(\mathcal{D}, \mathcal{D}^\times)$ as a partial *-algebra may be established, leading for instance to a new treatment of the multiplication of distributions. We refer to [14, Chap.10] for further details.

4. Partial *-Algebras of Closable Operators

A completely different type of example is given by partial *-algebras of closable operators in a Hilbert space. From now on, we will mostly concentrate on this class. We refer to [14] for a thorough analysis and the original references.

4.1. Basic Definitions and Properties

Let $\mathcal{H}$ be a complex Hilbert space and $\mathcal{D}$ a dense subspace of $\mathcal{H}$. We denote by $\mathcal{L}^\dagger(\mathcal{D}, \mathcal{H})$ the set of all (closable) linear operators X such that $\mathcal{D}(X) = \mathcal{D}$, $\mathcal{D}(X^*) \supseteq \mathcal{D}$. The set $\mathcal{L}^\dagger(\mathcal{D}, \mathcal{H})$ is a partial *-algebra with respect to the following operations: the usual sum $X_1 + X_2$, the scalar multiplication λX, the involution $X \mapsto X^\dagger = X^* \lceil \mathcal{D}$ and the *(weak)* partial multiplication $X_1 \,\square\, X_2 = {X_1}^{\dagger *} X_2$, defined whenever X_2 is a weak right multiplier of X_1 (equivalently, X_1 is a weak left multiplier of X_2), by which we mean $X_2\mathcal{D} \subset \mathcal{D}({X_1}^{\dagger *})$ and $X_1{}^*\mathcal{D} \subset \mathcal{D}({X_2}^*)$ (we write $X_2 \in R^{\rm w}(X_1)$ or $X_1 \in L^{\rm w}(X_2)$). When we regard $\mathcal{L}^\dagger(\mathcal{D}, \mathcal{H})$ as a partial *-algebra with those operations, we denote it by $\mathcal{L}^\dagger_{\rm w}(\mathcal{D}, \mathcal{H})$.

A *partial O*-algebra* on $\mathcal{D}$ is a *-subalgebra $\mathfrak{M}$ of $\mathcal{L}^\dagger_{\rm w}(\mathcal{D}, \mathcal{H})$, that is, $\mathfrak{M}$ is a subspace of $\mathcal{L}^\dagger_{\rm w}(\mathcal{D}, \mathcal{H})$, containing the identity and such that $X^\dagger \in \mathfrak{M}$ whenever $X \in \mathfrak{M}$, and $X_1 \,\square\, X_2 \in \mathfrak{M}$ for any $X_1, X_2 \in \mathfrak{M}$ such that $X_2 \in R^{\rm w}(X_1)$. Thus $\mathcal{L}^\dagger_{\rm w}(\mathcal{D}, \mathcal{H})$ itself is the largest partial O*-algebra on the domain $\mathcal{D}$.

On the space $\mathcal{L}^\dagger(\mathcal{D}, \mathcal{H})$ we will consider the *strong** topology τ_{s^*}, which is generated by the seminorms $p^*_\xi(X) = \|X\xi\| + \|X^\dagger\xi\|$, $\xi \in \mathcal{D}$. The space $\mathcal{L}^\dagger(\mathcal{D}, \mathcal{H})$ is complete for τ_{s^*}. For $\mathfrak{N} \subset \mathcal{L}^\dagger(\mathcal{D}, \mathcal{H})$, we denote by $[\mathfrak{N}]^{s^*}$ the τ_{s^*}-closure of $\mathfrak{N}$. We also meet the *weak* topology $\tau_{\rm w}$ on $\mathcal{L}^\dagger(\mathcal{D}, \mathcal{H})$, which is generated by the seminorms $p_{f,g}(X) = |\langle f|Xg\rangle|$, $f, g \in \mathcal{D}$, and the *quasi-uniform* topology, τ_*, defined by seminorms $p_{\mathcal{N}}(X) = \sup_{f\in\mathcal{N}}(\|Xf\| + \|X^\dagger f\|)$, where $\mathcal{N}$ is a bounded subset of $\mathcal{D}$, equipped with the projective topology determined by $\mathcal{L}^\dagger(\mathcal{D}, \mathcal{H})$.

If we restrict ourselves to those operators in $\mathcal{L}^\dagger(\mathcal{D}, \mathcal{H})$ that, together with their adjoint, leave the domain $\mathcal{D}$ invariant, we obtain a *-algebra, namely $\mathcal{L}^\dagger(\mathcal{D}) = \{A \in \mathcal{L}^\dagger(\mathcal{D}, \mathcal{H}) : A\mathcal{D} \subset \mathcal{D} \text{ and } A^*\mathcal{D} \subset \mathcal{D}\}$. Then an *O*-algebra* is defined as a *-subalgebra of $\mathcal{L}^\dagger(\mathcal{D})$; thus $\mathcal{L}^\dagger(\mathcal{D})$ is the maximal O*-algebra contained in $\mathcal{L}^\dagger(\mathcal{D}, \mathcal{H})$

and it is τ_{s^*}-dense in $\mathcal{L}^\dagger(\mathcal{D}, \mathcal{H})$, i.e., $\mathcal{L}^\dagger(\mathcal{D}, \mathcal{H}) = [\mathcal{L}^\dagger(\mathcal{D})]^{s^*}$. Clearly an O*-algebra is a particular case of a partial O*-algebra (see [32] for a comprehensive study of partial O*-algebras).

Given a partial O*-algebra $\mathfrak{M}$, we define *internal multipliers* as $R(X) = R^{\mathrm{w}}(X) \cap \mathfrak{M}$ and $L(X) = L^{\mathrm{w}}(X) \cap \mathfrak{M}$. Then the universal right multipliers of $\mathfrak{M}$ are the elements of the set:

$$R\mathfrak{M} = R^{\mathrm{w}}(\mathfrak{M}) \cap \mathfrak{M} = \{Y \in \mathfrak{M} : X \,\square\, Y \ \text{ is well-defined}, \ \forall X \in \mathfrak{M}\}.$$

A †-invariant subset $\mathfrak{N}$ of $\mathcal{L}^\dagger(\mathcal{D}, \mathcal{H})$ is called *fully closed* if $\mathcal{D} = \widehat{\mathcal{D}}(\mathfrak{N}) := \bigcap_{X \in \mathfrak{N}} \mathcal{D}(\overline{X})$. If $\mathfrak{N}$ is not fully closed, its full closure is the smallest fully closed set that contains it, that is, $\widehat{\mathfrak{N}} = \{\hat{\imath}(X) := \overline{X}\lceil\widehat{\mathcal{D}}(\mathfrak{N}),\ X \in \mathfrak{N}\}$. Let $\mathfrak{M}$ be a partial O*-algebra. If it is not fully closed, it may be embeddded into its full closure $\widehat{\mathfrak{M}} = \hat{\imath}(\mathfrak{M})$, which is a fully closed partial O*-algebra on the domain $\widehat{\mathcal{D}}(\mathfrak{M})$, isomorphic to $\mathfrak{M}$. Thus one may always restrict the analysis to fully closed partial O*-algebras without loss of generality. On the other hand, a partial O*-algebra $\mathfrak{M}$ is called *self-adjoint* if $\mathcal{D} = \mathcal{D}^*(\mathfrak{M}) := \bigcap_{X \in \mathfrak{M}} \mathcal{D}(X^*)$, and this *is* a strong restriction.

Given a †-invariant subset $\mathfrak{N}$ of $\mathcal{L}^\dagger(\mathcal{D}, \mathcal{H})$, we define, as usual, its *weak unbounded commutant*:

$$\mathfrak{N}'_\sigma = \{Y \in \mathcal{L}^\dagger(\mathcal{D}, \mathcal{H}) : (X\xi|Y\eta) = (Y^\dagger\xi|X^\dagger\eta),\ \forall \xi, \eta \in \mathcal{D},\ \forall X \in \mathfrak{N}\} \qquad (4.1)$$

and its *weak bounded commutant*:

$$\mathfrak{N}'_{\mathrm{w}} = \{C \in \mathcal{B}(\mathcal{H}) : (CX\xi|\eta) = (C\xi|X^\dagger\eta),\ \forall \xi, \eta \in \mathcal{D},\ \forall X \in \mathfrak{N}\}. \qquad (4.2)$$

The restriction to $\mathcal{D}$ of $\mathfrak{N}'_{\mathrm{w}}$ is the bounded part of $\mathfrak{N}'_\sigma$. Both $\mathfrak{N}'_\sigma$ and $\mathfrak{N}'_{\mathrm{w}}$ are weakly closed, †-invariant subspaces, but not necessarily algebras.

As for *bicommutant*, we consider here only the *weak unbounded* one, namely $\mathfrak{N}''_{\mathrm{w}\sigma} = (\mathfrak{N}'_{\mathrm{w}})'_\sigma$. Its bounded part is the (restriction to $\mathcal{D}$ of) $(\mathfrak{N}'_{\mathrm{w}})'$, where $\mathcal{B}'$ denotes the usual bounded commutant of a subset $\mathcal{B} \subset \mathcal{B}(\mathcal{H})$. We note the relation $(\mathfrak{N}''_{\mathrm{w}\sigma})''_{\mathrm{w}\sigma} = \mathfrak{N}''_{\mathrm{w}\sigma}$ and remark that $\mathfrak{N}''_{\mathrm{w}\sigma}$ is fully closed whenever $\mathfrak{N}$ is, because of the obvious inclusions $\mathcal{D} \subset \widehat{\mathcal{D}}(\mathfrak{N}''_{\mathrm{w}\sigma}) \subset \widehat{\mathcal{D}}(\mathfrak{N})$. The crucial fact is that, for any †-invariant subset $\mathfrak{N}$ of $\mathcal{L}^\dagger(\mathcal{D}, \mathcal{H})$, $\mathfrak{N}'_{\mathrm{w}}$ is a von Neumann algebra if, and only if, $\mathfrak{N}''_{\mathrm{w}\sigma} = \left[(\mathfrak{N}'_{\mathrm{w}})'\lceil\mathcal{D}\right]^{s^*}$.

A *partial GW*-algebra* $\mathfrak{M}$ on $\mathcal{D}$ is a fully closed partial O*-algebra which satisfies the two conditions $\mathfrak{M}'_{\mathrm{w}}\mathcal{D} = \mathcal{D}$ and $\mathfrak{M}''_{\mathrm{w}\sigma} = \mathfrak{M}$ (notice the analogy with the usual condition $\mathfrak{M}'' = \mathfrak{M}$ defining a von Neumann algebra). In that case, $\mathfrak{M}'_{\mathrm{w}}$ is a von Neumann algebra, the (closure of the) bounded part of $\mathfrak{M}$ is also a von Neumann algebra, namely $\mathfrak{M}_o := (\mathfrak{M}'_{\mathrm{w}})'$, and $\mathfrak{M} = \left[(\mathfrak{M}'_{\mathrm{w}})'\lceil\mathcal{D}\right]^{s^*}$. The good properties of partial GW*-algebras stem precisely from the fact that they contain a τ_{s^*}-dense subset of bounded operators.

The easiest way of constructing a partial GW*-algebra is to take a bicommutant. Indeed, if $\mathfrak{N}$ is a fully closed †-invariant subset of $\mathcal{L}^\dagger(\mathcal{D}, \mathcal{H})$, then $\mathfrak{N}''_{\mathrm{w}\sigma}$ is a partial GW*-algebra on $\mathcal{D}$ if and only if $\mathfrak{N}'_{\mathrm{w}}\mathcal{D} = \mathcal{D}$. On the other hand, if $\mathfrak{M}$ is a partial O*-algebra on $\mathcal{D}$ (not necessarily fully closed), such that $\mathfrak{M}'_{\mathrm{w}}\mathcal{D} = \mathcal{D}$ and $\mathfrak{M}''_{\mathrm{w}\sigma} = \mathfrak{M}$, then $\widehat{\mathfrak{M}}$ is a partial GW*-algebra on $\widehat{\mathcal{D}}(\mathfrak{M})$.

As a last point, we may ask the question whether a partial O*-algebra is a (tight) locally convex partial *-algebra. The answer, of course, depends on which topology τ one chooses, and many different ones are available, the strong* τ_{s^*}, the quasi-uniform τ_*, the weak τ_{w}, etc. We will not enter into the technical details, for lack of space, but only indicate a few general results. First, if $\mathcal{L}^\dagger_{\mathrm{w}}(\mathcal{D}, \mathcal{H})$ is self-adjoint, then it is a locally convex partial *-algebra for these three topologies, and it is complete for τ_{s^*} and τ_*. More generally, any self-adjoint partial O*-algebra $\mathfrak{M}$ is a locally convex partial *-algebra for the weak topology τ_{w}, and the same is true for τ_* if $R\mathfrak{M}$ contains only bounded operators. In all cases, tightness is open.

4.2. *-Automorphisms of Partial O*-Algebras

In the algebraic formulation of quantum theories [6], the observables of a physical system are represented by hermitian elements of a certain *-algebra $\mathfrak{A}$ and states by positive linear functionals on $\mathfrak{A}$. Then a symmetry of the system is realized by a *-automorphism σ of $\mathfrak{A}$, and a one parameter symmetry group by a *-automorphism group σ_t $(t \in \mathbb{R})$ of $\mathfrak{A}$. Given a state, the Gel'fand–Naĭmark–Segal (GNS) construction yields a representation π of $\mathfrak{A}$ by bounded operators in a Hilbert space $\mathcal{H}_\pi$ and a *-automorphism σ^π (resp. *-automorphism group σ_t^π) of $\pi(\mathfrak{A})$. Then the question is whether σ^π is *spatial*, that is, whether there exists a unitary operator U in $\mathcal{H}_\pi$ such that $\sigma^\pi(A) = U^* A\, U$, for every $A \in \pi(\mathfrak{A})$. Even more interesting is the case where U itself can be taken in $\pi(\mathfrak{A})$, i.e., the automorphism is *inner*. For a one parameter group σ_t, spatiality means that the automorphism group σ_t^π is unitarily implemented, i.e., $\sigma_t^\pi(A) = e^{-iHt} A e^{iHt}$, where H is a self-adjoint operator. In particular, if the automorphism is inner, this means that $H \in \pi(\mathfrak{A})''$ (or H is affiliated to $\pi(\mathfrak{A})''$), in other words that the operator H is an observable. For instance, if σ_t represents the time evolution of the system, then 'σ_t^π is inner' means that the Hamiltonian exists as an observable in the (GNS) representation at hand [19, 22, 29].

Now, if one decides to describe the set of observables of a given physical system by some partial *-algebra, in particular, a partial O*-algebra, one must generalize to that context the notion of *-automorphism, and of spatiality as well.

Let $\mathfrak{M}$ be a partial O*-algebra on $\mathcal{D}$, obtained, for instance, from the partial *-algebra of observables by a GNS construction (the latter indeed extends to partial *-algebras, as we shall see in Section 5.2). According to the general definition (see Section 2), a **-automorphism* of $\mathfrak{M}$ is a linear bijection $\sigma : \mathfrak{M} \to \mathfrak{M}$ such that (i) $\sigma(X^\dagger) = \sigma(X)^\dagger, \forall X \in \mathfrak{M}$; (ii) $\sigma(Y) \in R^{\mathrm{w}}(\sigma(X))$ if and only if $Y \in R^{\mathrm{w}}(X)$ and then $\sigma(X \square Y) = \sigma(X) \square \sigma(Y)$; and (iii) the same relations hold for σ^{-1}. It follows that $\sigma(\mathfrak{M}) = \mathfrak{M}$ and $\sigma(R\mathfrak{M}) = R\mathfrak{M}$. The *-automorphism σ is *spatial* if there exists a unitary operator $U \in \mathcal{H}$ such that $U_o := U\lceil\mathcal{D} \in R^{\mathrm{w}}(\mathfrak{M})$ and $\sigma(X) = U_o{}^*(X \square U_o), \forall X \in \mathfrak{M}$. It is *inner* if in addition $U_o \in \mathfrak{M}$, i.e., $U_o \in R\mathfrak{M}$. Notice that, if $\mathfrak{M}$ is self-adjoint and σ is spatial, then $U_o \in \mathcal{L}^\dagger(\mathcal{D})$ and $\sigma(X) = U_o{}^*XU_o$, for every $X \in \mathfrak{M}$.

Armed with these seemingly natural definitions, one may study under what conditions a *-automorphism is spatial or inner, and also the structure of the corresponding one-parameter groups of *-automorphisms, both globally and at the infinitesimal level (using the notion of derivation). As expected, one gets significant results only if one assumes that $\mathfrak{M}$ is a partial GW*-algebra, because then many known results on *-automorphisms of von Neumann algebras will be lifted from the bounded part of $\mathfrak{M}$ to $\mathfrak{M}$ itself [10]. Indeed, if $\mathfrak{M}$ is a partial GW*-algebra on the domain $\mathcal{D}$, its bounded part $\mathfrak{M}_b = \{X \in \mathfrak{M} : \overline{X} \in \mathcal{B}(\mathcal{H})\}$ is the restriction to $\mathcal{D}$ of the von Neumann algebra $\overline{\mathfrak{M}_b} = \{\overline{X} : X \in \mathfrak{M}_b\}$, with commutant $\overline{\mathfrak{M}_b}' = \mathfrak{M}'_{\mathrm{w}}$. The key observation is that every *-automorphism σ of $\mathfrak{M}$ induces a *-automorphism σ_b of the von Neumann algebra $\overline{\mathfrak{M}_b}$, by the simple relation $\sigma_b(\overline{A}) = \overline{\sigma(A)},\ A \in \mathfrak{M}_b$. Accordingly, one says that a *-automorphism σ of $\mathfrak{M}$ is *weakly spatial* (resp. *weakly inner*) if the corresponding *-automorphism σ_b of the von Neumann algebra $\overline{\mathfrak{M}_b}$ is spatial (resp. inner). This is the crucial link, which allows one to control in a rather complete way the properties of *-automorphisms and *-automorphism groups on partial O*-algebras.

It is clear that a weakly spatial *-automorphism will be spatial as soon as there is enough continuity for lifting it from $\mathfrak{M}_b$ to $\mathfrak{M}$. This is indeed the case, and of course the relevant topology is the strong* one. Indeed,

Theorem 4.1. *Let $\mathfrak{M}$ be a partial GW*-algebra on $\mathcal{D}$ and σ a *-automorphism of $\mathfrak{M}$. If σ is weakly spatial (resp. weakly inner) and τ_{s^*}-continuous, then σ is spatial (resp. inner).*

If $\mathfrak{M}$ is self-adjoint, the two conditions are in fact equivalent:

Corollary 4.2. *Let $\mathfrak{M}$ be a self-adjoint partial GW*-algebra on $\mathcal{D}$ and σ a *-automorphism of $\mathfrak{M}$. Then σ is spatial if and only if it is weakly spatial and τ_{s^*}-continuous.*

At this point, one may systematically list all the known results about spatial *-automorphisms of von Neumann algebras, as given in standard treatises such as [23] or [33], and try to lift them to partial GW*-algebras. The following one is very simple.

Corollary 4.3. *Let $\mathfrak{M}$ be a partial GW*-algebra with a cyclic and separating vector. Then every τ_{s^*}-continuous *-automorphism of $\mathfrak{M}$ is spatial.*

4.3. Automorphism Groups and *-Derivations of Partial O*-Algebras

From the point of view of physical applications, a crucial role is played by automorphism groups: They describe either the time evolution of the system or physical symmetries. This is the reason why it is worth considering them in the context of partial O*-algebras [10].

Let $\mathfrak{M}$ be a partial O*-algebra on $\mathcal{D}$ and let $\mathrm{Aut}_*(\mathfrak{M})$ denote the set of *-automorphisms of $\mathfrak{M}$. A one-parameter **-automorphism group* of $\mathfrak{M}$ is a map $\mathbb{R} \ni t \mapsto \alpha_t \in \mathrm{Aut}_*(\mathfrak{M})$ such that (i) $\alpha_0(X) = X$, $\forall X \in \mathfrak{M}$; and (ii) $\alpha_{s+t}(X) = \alpha_s(\alpha_t(X))$, $\forall X \in \mathfrak{M}$. If τ is any topology on $\mathcal{L}^\dagger_{\mathrm{w}}(\mathcal{D}, \mathcal{H})$, then the automorphism group α_t is called *τ-continuous* if $\mathbb{R} \ni t \mapsto \alpha_t(X)$ is continuous from $\mathbb{R}$ into $\mathcal{L}^\dagger_{\mathrm{w}}(\mathcal{D}, \mathcal{H})[\tau]$, $\forall X \in \mathfrak{M}$.

If $t \mapsto \alpha_t(X)$ is τ-continuous, we define its *infinitesimal generator* as

$$\delta_\alpha(X) = \tau - \lim_{t\to 0} t^{-1}(\alpha_t(X) - X),$$

on the domain $D_\tau(\delta_\alpha)$ consisting of all $X \in \mathfrak{M}$ for which the limit $\tau - \lim_{t\to 0} t^{-1}(\alpha_t(X) - X)$ exists in $\mathcal{L}^\dagger_{\mathrm{w}}(\mathcal{D}, \mathcal{H})$. If the involution $X \mapsto X^\dagger$ is τ-continuous, then $X \in D_\tau(\delta_\alpha)$ implies $X^\dagger \in D_\tau(\delta_\alpha)$ and $\delta_\alpha(X^\dagger) = \delta_\alpha(X)^\dagger$.

In analogy with the C*-situation, one would expect that $D_\tau(\delta_\alpha)$ is a partial O*-algebra and that δ_α is a *-derivation of it, but for this we need a suitable form of the Leibniz rule. Motivated by the properties of infinitesimal generators, we define a *weak *-derivation* of $\mathfrak{M}$ as a linear map $\delta : \mathfrak{M} \to \mathcal{L}^\dagger_{\mathrm{w}}(\mathcal{D}, \mathcal{H})$ satisfying the following conditions:

(i) $\delta(X)^\dagger = \delta(X^\dagger),\ \forall\, X \in \mathfrak{M}$,

(ii) $(\delta(X \,\square\, Y)\xi |\, \eta) = (Y\xi |\, \delta(X^\dagger)\,\eta) + (\delta(Y)\xi |\, X^\dagger \eta)$,
$\forall\, X, Y \in \mathfrak{M}$ such that $X \in L^{\mathrm{w}}(Y)$ and $\forall\, \xi, \eta \in \mathcal{D}$.

The weak *-derivation δ of $\mathfrak{M}$ is called *spatial*, resp. *inner*, if there exists an element $H_o = H_o^\dagger \in R^{\mathrm{w}}(\mathfrak{M})$, resp. $H \in R\mathfrak{M}$, such that

(i) H_o is the restriction to $\mathcal{D}$ of an operator $H \in \mathcal{L}(\mathfrak{M}\mathcal{D}, \mathcal{H})$ that satisfies the relation $(H X^\dagger \xi | Y\eta) = (X^\dagger \xi | H Y \eta),\ \forall\, X \in R^{\mathrm{w}}(\mathfrak{M}), \forall\, \xi, \eta \in \mathcal{D}$.

(ii) $\delta(X) = \delta_{H_o}(X) := i(H \,\square\, X - X \,\square\, H),\ X \in \mathfrak{M}$.

Clearly, the properties of these derivations will depend both on the continuity properties of the automorphism group and on the type of partial O*-algebra considered. As expected, partial GW*-algebras will again behave better. Indeed we have:

Proposition 4.4. *Let $\mathfrak{M}$ be a partial GW*-algebra. Let $t \mapsto \alpha_t$ be a strong*-continuous one parameter *-automorphism group of $\mathfrak{M}$ and δ_α the corresponding* *infinitesimal generator. Then $D(\delta_\alpha)$ is a partial O*-algebra and δ_α is a weak *-derivation of $D(\delta_\alpha)$ satisfying $\delta_\alpha\,(D(\delta_\alpha)) \subset \mathfrak{M}$.*

We consider first derivations which are generators of the automorphism group α_t^H generated by a self-adjoint operator H in $\mathcal{H}$, that is,

$$\alpha_t^H(X) = e^{itH} X e^{-itH}, \quad X \in \mathcal{L}_{\rm w}^{\dagger}(\mathcal{D}, \mathcal{H}).$$

Proposition 4.5. *Let $\mathfrak{M}$ be a partial GW*-algebra and H a self-adjoint operator in $\mathcal{H}$. Assume that the corresponding unitary group $\{e^{itH}, t \in \mathbb{R}\}$ satisfies the following three conditions:*

(i) $e^{itH}\mathcal{D} = \mathcal{D}, \forall t \in \mathbb{R}$;

(ii) $e^{itH}\mathfrak{M}\, e^{-itH} = \mathfrak{M}, \forall t \in \mathbb{R}$;

(iii) $t \mapsto e^{itH}\xi$ *is* $t_{\mathfrak{M}}$*-continuous,* $\forall \xi \in \mathcal{D}$.

Then,

(1) $\forall X \in \mathfrak{M}$, $\alpha_t^H(X)$ *is a strong*-continuous one parameter *-automorphism group of* $\mathfrak{M}$.

(2) $D_{\rm w}(\delta_\alpha)$ *is a partial O*-algebra and* δ_α *is a weak *-derivation of* $D_{\rm w}(\delta_\alpha)$.

(3) If either $D_{\rm w}(\delta_\alpha)\mathcal{D} \subset \mathcal{D}(H)$, *or* $\mathcal{D} \subset \mathcal{D}(H)$, *then* δ_α *is spatial, that is,* $\delta_\alpha(X) = \delta_{H_o}(X) := i(H \,\square\, X - X \,\square\, H)$, $X \in \mathfrak{M}$, *where* $H_o := H \restriction \mathcal{D}$.

Let now δ be an arbitrary weak *-derivation of $\mathfrak{M}$ such that $\delta(\mathfrak{M}) \subset \mathfrak{M}$. Then $\delta_b \equiv \delta \restriction \mathfrak{M}_b$ is a *-derivation of $\mathfrak{M}_b$ into $\mathfrak{M}$, which provides a simple tool for the study of δ.

Theorem 4.6. *Let $\mathfrak{M}$ be a partial GW*-algebra and δ a weak *-derivation of $\mathfrak{M}$ such that $\delta(\mathfrak{M}) \subset \mathfrak{M}$ and $\delta(\mathfrak{M}_b) \subset \mathfrak{M}_b$. If δ is weakly continuous, then there exists $H = H^\dagger \in \mathfrak{M}_b \cap R\mathfrak{M}$ such that*

$$\delta(X)\xi = i(H \,\square\, X - X \,\square\, H)\xi, \ \forall X \in \mathfrak{M}, \forall \xi \in \mathcal{D}.$$

The converse is also true if $\mathfrak{M}$ is self-adjoint.

4.4. Tomita–Takesaki Theory of Modular Automorphisms

In the standard approach to quantum statistical mechanics, the equilibrium states of a physical system (Gibbs states) are described by the so-called KMS states on the observable algebra $\mathfrak{A}$, and these states are derived with the celebrated Tomita–Takesaki theory of modular automorphisms [20]. Once again, the obvious question is, what happens if one starts from a partial O*-algebra $\mathfrak{M}$ as observable "algebra"? It turns out that the Tomita–Takesaki theory can be extended to that case, but the construction is rather involved. Hence we will give here only a brief, nontechnical summary. The full story may be found in the monograph [14].

The key notion (due to Tomita) for answering the question is that of a *generalized vector* on $\mathfrak{M}$. By this, one means a map $\lambda : \mathfrak{M} \to \mathcal{H}$, defined on a domain $\mathcal{D}(\lambda)$, such that there exists a subspace $B(\lambda)$ of $\mathfrak{M}$ with the following properties:

(i) $\mathcal{D}(\lambda) = \text{linear span}\,\{Y \,\square\, X : X \in B(\lambda), Y \in L(X)\}$;

(ii) λ is linear on $\mathcal{D}(\lambda)$;

(iii) $\lambda(B(\lambda)) \subset \mathcal{D}$;

(iv) $\lambda(Y \,\square\, X) = Y\lambda(X)$ for every $X \in B(\lambda)$ and $Y \in L(X)$.

Such a subspace $B(\lambda)$ is called a *core* for λ, and every core $B(\lambda)$ can be embedded in a maximal core $B_M(\lambda)$.

Then, under suitable conditions (coded into the expression '$(\mathfrak{M}, \lambda)$ is a cyclic system'), one defines the *commutant* $\lambda^c : K\lambda(X) = X\lambda^c(K)$, $K \in \mathfrak{M}'_{\mathrm{w}}$, and it turns out that λ^c is a generalized vector for the von Neumann algebra $\mathfrak{M}'_{\mathrm{w}}$. Under similar conditions (one says, '$(\mathfrak{M}, \lambda, \lambda^c)$ is a cyclic and separating system'), one defines the *bicommutant* $\lambda^{cc} : A\lambda^c(K) = K\lambda^{cc}(A)$, $A \in (\mathfrak{M}'_{\mathrm{w}})'$, and again λ^{cc} is a generalized vector for the von Neumann algebra $(\mathfrak{M}'_{\mathrm{w}})'$.

At this stage, one defines the modular involutions (on suitable domains):

$$S_\lambda : \lambda(X) \mapsto \lambda(X^\dagger), \qquad S_\lambda^{cc} : \lambda^{cc}(A) \mapsto \lambda^{cc}(A^*).$$

As in the usual von Neumann case, one shows that S_λ and $S_{\lambda^{cc}}$ are closable conjugate linear operators in $\mathcal{H}$. Consider then the polar decomposition of the respective closures (denoted by the same symbol, for simplicity), $S_\lambda = J_\lambda \Delta_\lambda^{1/2}$ and $S_{\lambda^{cc}} = J_{\lambda^{cc}} \Delta_{\lambda^{cc}}^{1/2}$. Using these and a number of technical conditions, one introduces the key notion of *standard* generalized vector, in terms of which the fundamental theorem may be stated as follows.

Theorem 4.7. *Let λ be a standard generalized vector for a partial O^*-algebra $\mathfrak{M}$. Then :*

(1) $S_\lambda = S_{\lambda^{cc}}$, and thus $J_\lambda = J_{\lambda^{cc}}$ and $\Delta_\lambda = \Delta_{\lambda^{cc}}$.

*(2) Define $\sigma_t^\lambda(X) = \Delta_\lambda^{it} X \Delta_\lambda^{-it}$, $X \in \mathfrak{M}$, $t \in \mathbb{R}$. Then $\{\sigma_t^\lambda\}_{t\in\mathbb{R}}$ is a one-parameter group of *-automorphisms of $\mathfrak{M}$.*

(3) λ satisfies the KMS condition with respect to $\{\sigma_t^\lambda\}_{t\in\mathbb{R}}$, i.e., $\forall\, X, Y \in B_M(\lambda)^\dagger \cap B_M(\lambda)$, there exists a function $f_{X,Y}$, bounded and continuous in the strip $-1 \leqslant \mathrm{Im} z \leqslant 0$ and analytic in the interior, such that, for all $t \in \mathbb{R}$,

$$f_{X,Y}(t) = (\lambda(\sigma_t^\lambda(X))|\lambda(Y)),$$

$$f_{X,Y}(t - i) = (\lambda(Y^\dagger)|\lambda(\sigma_t^\lambda(X^\dagger))).$$

The next step in the theory is to relax somewhat the assumptions, with help of the weaker notion of *modular* generalized vector, then to consider the particular case where the partial *-algebra is a partial GW*-algebra. Finally one can prove a generalized Connes cocycle theorem, that allows to compare two generalized vectors on the same partial GW*-algebra.

In summary, essentially the complete Tomita–Takesaki theory extends to partial *-algebras, provided the appropriate notions are identified.

5. Representation Theory

5.1. Generalities

A *-representation of a partial *-algebra $\mathfrak{A}$ is a *-homomorphism of $\mathfrak{A}$ into $\mathcal{L}^{\dagger}_{\mathrm{w}}(\mathcal{D}, \mathcal{H})$, for some pair $\mathcal{D} \subset \mathcal{H}$, that is, a linear map $\pi : \mathfrak{A} \to \mathcal{L}^{\dagger}_{\mathrm{w}}(\mathcal{D}, \mathcal{H})$ such that: (i) $\pi(x^*) = \pi(x)^{\dagger}$ for every $x \in \mathfrak{A}$; (ii) $x \in L(y)$ in $\mathfrak{A}$ implies $\pi(x) \in L^{\mathrm{w}}(\pi(y))$ and $\pi(x) \,\square\, \pi(y) = \pi(xy)$.

Let π be a *-representation of a partial *-algebra $\mathfrak{A}$. It is called *fully closed* if $\pi(\mathfrak{A})$ is fully closed. In any case, a *-representation π can always be extended to a fully closed *-representation $\widehat{\pi}(\mathfrak{A})$, namely $\widehat{\pi}(x) = \widehat{\pi(x)}$, $x \in \mathfrak{A}$, on the domain $\mathcal{D}(\widehat{\pi}) = \widehat{\mathcal{D}}(\pi(\mathfrak{A}))$.

Next we define the weak commutants of a *-representation of a partial *-algebra. Besides the usual weak bounded commutant $\pi(\mathfrak{A})'_{\mathrm{w}}$ of $\pi(\mathfrak{A})$, as defined in (4.2), we introduce a new one, called *quasi-weak*, which takes explicitly into account the possible lack of associativity:

$$\mathcal{C}_{\mathrm{qw}}(\pi) = \big\{C \in \pi(\mathfrak{A})'_{\mathrm{w}} : (C\pi(x_1{}^*)\xi|\pi(x_2)\eta) = (C\xi|\pi(x_1x_2)\eta), \\ \forall x_1, x_2 \in \mathfrak{A} \text{ such that } x_1 \in L(x_2) \text{ and } \forall \xi, \eta \in \mathcal{D}(\pi)\big\}. \tag{5.1}$$

$\mathcal{C}_{\mathrm{qw}}(\pi)$ is a weakly closed *-invariant subspace of $\mathcal{B}(\mathcal{H})$, contained in $\pi(\mathfrak{A})'_{\mathrm{w}}$, and, moreover, $\mathcal{C}_{\mathrm{qw}}(\widehat{\pi}) = \mathcal{C}_{\mathrm{qw}}(\pi)$. Thus we have now two possible definitions of irreducibility, namely:

- π is *irreducible* if $\mathcal{C}_{\mathrm{qw}}(\pi) = \pi(\mathfrak{A})'_{\mathrm{w}} = \mathbb{C}I$;
- π is *weakly irreducible* if $\mathcal{C}_{\mathrm{qw}}(\pi) = \mathbb{C}I$.

5.2. The GNS Construction

As always, the crucial question is how to build concrete representations. For *-algebras, the Gel'fand–Naĭmark–Segal (GNS) construction is usually the answer [20, 6]. In order to extend it to partial *-algebras, we must first have an appropriate notion of state. In the case of a *-algebra $\mathfrak{A}$, a state is a normalized positive linear form on $\mathfrak{A}$. But the positivity condition requires the existence of products a^*a, which need not exist in a partial *-algebra. The alternative, of course, is to use *sesquilinear* forms. However, for a *-algebra $\mathfrak{A}$, the GNS construction works only if the starting sesquilinear form ϕ on $\mathfrak{A} \times \mathfrak{A}$ is *invariant*, in the sense that $\phi(x^*y, z) = \phi(y, xz)$, for all $x, y, z \in \mathfrak{A}$. Clearly this definition is inapplicable for a partial *-algebra, since the products x^*y and xz need not exist. An obvious solution is to impose this relation for $y, z \in R\mathfrak{A}$ only, and this gives us a good hint.

Let $\mathfrak{A}$ be a partial *-algebra, and let φ be a sesquilinear form on $D(\varphi) \times D(\varphi)$, where $D(\varphi)$ is a subspace of $\mathfrak{A}$. The form φ is said to be *positive* if $\varphi(x, x) \geqslant 0, \forall x \in D(\varphi)$. Then we have:

$$\varphi(x, y) = \overline{\varphi(y, x))}, \ \forall x, y \in D(\varphi), \tag{5.2}$$

$$|\varphi(x, y)|^2 \leqslant \varphi(x, x)\varphi(y, y), \ \forall x, y \in D(\varphi), \tag{5.3}$$

and hence

$$\begin{aligned} N_\varphi &:= \{x \in D(\varphi) : \varphi(x, x) = 0\} \\ &= \{x \in D(\varphi) : \varphi(x, y) = 0 \text{ for all } y \in D(\varphi)\}, \end{aligned} \tag{5.4}$$

and so N_φ is a subspace of $\mathfrak{A}$. For each $x \in D(\varphi)$, we denote by $\lambda_\varphi(x)$ the element of $D(\varphi)/N_\varphi$ which contains x, and define an inner product $(.|.)$ on $\lambda_\varphi(D(\varphi)) = D(\varphi)/N_\varphi$ by

$$(\lambda_\varphi(x)|\lambda_\varphi(y)) = \varphi(x, y), \ x, y \in D(\varphi). \tag{5.5}$$

We denote by $\mathcal{H}_\varphi$ be the Hilbert space obtained by the completion of the pre-Hilbert space $\lambda_\varphi(D(\varphi))$. We are now ready to introduce the relevant class of sesquilinear forms.

Definition 5.1. (a) Let φ be a positive sesquilinear form on $D(\varphi) \times D(\varphi)$. A subspace $B(\varphi)$ of $D(\varphi)$ is said to be a *pre-core* for φ if

(i) $B(\varphi) \subset R\mathfrak{A}$;

(ii) $\{ax : a \in \mathfrak{A}, x \in B(\varphi)\} \subset \mathcal{D}(\varphi)$;

(iii) $\varphi(ax, y) = \varphi(x, a^*y), \ \forall a \in \mathfrak{A}, \forall x, y \in B(\varphi)$;

(iv) $\varphi(a^*x, by) = \varphi(x, (ab)y), \ \forall a \in L(b), \forall x, y \in B(\varphi)$;

(b) The subspace $B(\varphi)$ is called a *core* if, in addition,

(v) $\lambda_\varphi(B(\varphi))$ is dense in $\mathcal{H}_\varphi$.

Condition (iv) takes care of the possible non-associativity of $\mathfrak{A}$.

We denote by $\mathcal{B}_\varphi$ the set of all cores $B(\varphi)$ for φ and call *biweight* on $\mathfrak{A}$ a positive sesquilinear form φ on $D(\varphi) \times D(\varphi)$ such that $\mathcal{B}_\varphi$ is nonempty [13].

This notion is biweight is precisely the one that allows the GNS construction on a partial $*$-algebra, as shown in the next proposition.

Proposition 5.2. *Let φ be a biweight on $\mathfrak{A}$ with a core $B(\varphi)$. Put*

$$\pi_\varphi^\circ(a)\lambda_\varphi(x) = \lambda_\varphi(ax), \quad a \in \mathfrak{A}, x \in B(\varphi).$$

Then π_φ° is a $*$-representation of $\mathfrak{A}$ into $\mathcal{L}_{\mathrm{w}}^\dagger(\lambda_\varphi(B(\varphi)), \mathcal{H}_\varphi)$.

Denote by π_φ^B the closure of π_φ°. We call the triple $(\pi_\varphi^B, \lambda_\varphi, \mathcal{H}_\varphi)$ the *GNS construction* for the biweight φ on $\mathfrak{A}$ with core $B(\varphi)$.

In view of this result, it is natural to ask whether one can characterize the GNS construction in terms of a core. Some care is needed, since it might happen that $\pi_\varphi^{B_1} = \pi_\varphi^{B_2}$ with $B_1(\varphi) \neq B_2(\varphi)$. The key observation is that the set of all cores $B_1(\varphi) \in \mathcal{B}_\varphi$ giving the same GNS representation as $B(\varphi)$, i.e., $\pi_\varphi^{B_1} = \pi_\varphi^B$, has a maximal element, namely

$$B_L(\varphi) = \{x \in D(\varphi) \cap R\mathfrak{A} : \lambda_\varphi(x) \in D(\pi_\varphi^B), \text{ and } \\ ax \in D(\varphi) \text{ and } \lambda_\varphi(ax) = \pi_\varphi^B(a)\lambda_\varphi(x) \text{ for all } a \in \mathfrak{A}\}.$$

We say that $B(\varphi)$ is *GNS-maximal* whenever $B(\varphi) = B_L(\varphi)$. Then one gets a unique characterization:

Proposition 5.3. *Let φ be a biweight on $\mathfrak{A}$ and $B_1(\varphi)$, $B_2(\varphi)$ two GNS-maximal cores for φ. Then :*

- $\pi_\varphi^{B_1} \subset \pi_\varphi^{B_2} \Leftrightarrow B_1(\varphi) \subset B_2(\varphi)$
- $\pi_\varphi^{B_1} = \pi_\varphi^{B_2} \Leftrightarrow B_1(\varphi) = B_2(\varphi)$

We may now define the appropriate notion of pure state. Let $\mathfrak{A}$ be a partial $*$-algebra with unit e, φ a state or normalized biweight, i.e., $\varphi(e, e) = 1$. We say that the state φ is *pure* if it cannot be decomposed in a convex combination of two states φ_1, φ_2 *with the same core* $B(\varphi)$ as φ :

$$\varphi \neq \lambda\varphi_1 + (1-\lambda)\varphi_2, \ 0 < \lambda < 1, \ B(\varphi) \text{ is a core for } \varphi_1 \text{ and } \varphi_2.$$

The interest of this concept is that the equivalence between the purity of a state φ and the irreducibility of its GNS representation π_φ^B extends to partial $*$-algebras, essentially with the same proof:

Proposition 5.4. *Let $\mathfrak{A}$ be a partial $*$-algebra with unit, φ a state on $\mathfrak{A}$, with core $B(\varphi)$. Then the GNS representation π_φ^B is weakly irreducible, i.e., $C_{\mathrm{qw}}(\pi_\varphi^B) = \mathbb{C}I$, if and only if φ is a pure state.*

Actually, biweights seem a crucial concept, for they allow the extension to partial $*$-algebras of several classical results from von Neumann algebra theory.

For instance, the Radon–Nikodým theorem, concerning the relative domination of one biweight over another one, and the Lebesgue decomposition theorem of a biweight into an absolutely continuous biweight and a singular biweight. In addition, the definition of standard biweights on a partial GW*-algebra leads to a further generalization of the Tomita-Takesaki theory of modular automorphism groups.

5.3. Biweights vs. *-Homomorphisms

It turns out that biweights exhibit unfamiliar features (in particular when compared with the sesquilinear forms defined by positive linear functionals on a *-algebra $\mathfrak{A}$). For instance, the restriction of a biweight to a partial *-subalgebra is not necessarily a biweight. Also, if $\mathfrak{A}_1$ and $\mathfrak{A}_2$ are partial *-algebras, φ a biweight on $\mathfrak{A}_2$ and Φ a *-homomorphism from $\mathfrak{A}_1$ to $\mathfrak{A}_2$, then the natural *composition* $\varphi \circ \Phi$ of φ and Φ may fail to be a biweight. Clearly, it is the density condition (v) of Definition 5.1 that creates the problem. Thus one needs to give sufficient conditions for a biweight to pass safely through these two operations [16].

Let us look first at the problem of the restriction of a biweight to a partial *-subalgebra or, more generally, to a subspace. Let $\mathfrak{A}$ be a partial *-algebra and $\mathfrak{B}$ a *-invariant subspace of $\mathfrak{A}$. Given a biweight φ on $\mathfrak{A}$ with domain $D(\varphi)$ and core $B(\varphi)$, we consider its restriction $\varphi_{\mathfrak{B}}$ to $\mathfrak{B}$ and define

$$D(\varphi_{\mathfrak{B}}) = \mathfrak{B} \cap \mathcal{D}(\varphi), \quad B(\varphi_{\mathfrak{B}}) = \mathfrak{B} \cap B(\varphi) \quad \text{and} \quad N_{\varphi_{\mathfrak{B}}} = N_\varphi \cap \mathfrak{B}.$$

Then we perform the standard construction, which yields the Hilbert space $\mathcal{H}_{\varphi_{\mathfrak{B}}}$ as the completion of $\lambda_{\varphi_{\mathfrak{B}}}(D(\varphi_{\mathfrak{B}}))$ with respect to the inner product $\langle \lambda_{\varphi_{\mathfrak{B}}}(x) | \lambda_{\varphi_{\mathfrak{B}}}(y) \rangle = \varphi_{\mathfrak{B}}(x, y)$. Next, we define the map

$$I_{\mathfrak{B}} : \lambda_{\varphi_{\mathfrak{B}}}(x) \mapsto \lambda_\varphi(x), \quad \forall x \in D(\varphi_{\mathfrak{B}}),$$

which is well-defined, injective and isometric. Thus it extends to to an isometric operator $\widehat{I_{\mathfrak{B}}}$ from $\mathcal{H}_{\varphi_{\mathfrak{B}}}$ into $\mathcal{H}_\varphi$. It follows that $\widehat{I_{\mathfrak{B}}}\mathcal{H}_{\varphi_{\mathfrak{B}}}$ is a closed subspace of $\mathcal{H}_\varphi$. Let P be the projection from $\mathcal{H}_\varphi$ onto $\widehat{I_{\mathfrak{B}}}\mathcal{H}_{\varphi_{\mathfrak{B}}}$. Then the condition we are looking for is given by the following proposition.

Proposition 5.5. *The following statements are equivalent:*

(i) $\lambda_{\varphi_{\mathfrak{B}}}(B(\varphi_{\mathfrak{B}}))$ *is dense in* $\mathcal{H}_{\varphi_{\mathfrak{B}}}$;

(ii) $I_{\mathfrak{B}} \lambda_{\varphi_{\mathfrak{B}}}(B(\varphi_{\mathfrak{B}})) \subseteq P\lambda_\varphi(B(\varphi)) \subseteq \overline{I_{\mathfrak{B}} \lambda_{\varphi_{\mathfrak{B}}}(B(\varphi_{\mathfrak{B}}))}$.

Notice that the first inclusion in condition (ii) is automatic.

Now we turn to the heredity problem of biweights through a *-homomorphism. In general, if $\mathfrak{A}_1, \mathfrak{A}_2$ are partial *-algebras, φ a biweight on $\mathfrak{A}_2$ with domain $\mathcal{D}(\varphi)$

and core $B(\varphi)$, and Φ is a *-homomorphism of $\mathfrak{A}_1$ into $\mathfrak{A}_2$, then φ_Φ is not necessarily a biweight on $\mathfrak{A}_1$. Several examples of that situation may be found in [16]. However, the following proposition shows that, under certain conditions, biweights can be pulled back from one partial *-algebra to another one by a *-homomorphism.

Proposition 5.6. *Let $\mathfrak{A}_1, \mathfrak{A}_2$ be partial *-algebras, φ a positive sesquilinear form on $\mathfrak{A}_2$ defined on $D(\varphi) \times D(\varphi)$ and $B(\varphi)$ a pre-core for φ. Let Φ be a *-homomorphism of $\mathfrak{A}_1$ into $\mathfrak{A}_2$. Define*

$$D(\varphi_\Phi) = \{a \in \mathfrak{A}_1 : \Phi(a) \in \mathcal{D}(\varphi)\},$$
$$B(\varphi_\Phi) = \{x \in R\mathfrak{A}_1 : \Phi(x) \in B(\varphi)\},$$

and

$$\varphi_\Phi(a, b) = \varphi(\Phi(a), \Phi(b)).$$

If $\Phi(B(\varphi_\Phi))$ is a core for φ, then $B(\varphi_\Phi)$ is a core for φ_Φ and, therefore, φ_Φ is a biweight on $\mathfrak{A}_1$ with domain $D(\varphi_\Phi)$ and core $B(\varphi_\Phi)$. In particular, if Φ is surjective, then $\Phi(B(\varphi_\Phi))$ is a core for φ if, and only if, $B(\varphi_\Phi)$ is a core for φ_Φ.

Proposition 5.6 then suggests the following definition.

Definition 5.7. Let $\mathfrak{A}_1, \mathfrak{A}_2$ be partial *-algebras, $\Phi : \mathfrak{A}_1 \to \mathfrak{A}_2$ a *-homomorphism and φ a biweight on $\mathfrak{A}_2$ with domain $D(\varphi)$. We put $D(\varphi_\Phi) = \{a \in \mathfrak{A}_1 : \Phi(a) \in D(\varphi)\}$. If $B(\varphi)$ is a core for φ, then we put $B(\varphi_\Phi) = \{x \in R\mathfrak{A}_1 : \Phi(x) \in B(\varphi)\}$. Then we say that:

(a) Φ *preserves biweights* if, for any biweight φ on $\mathfrak{A}_2$ and for any core $B(\varphi)$ for φ, $\Phi(B(\varphi_\Phi))$ is a core for φ.

(b) Φ *strictly preserves biweights* if, for any biweight φ and for any core $B(\varphi)$, the equality $\Phi(B(\varphi_\Phi)) = B(\varphi)$ holds.

Then one can prove the following results.

Proposition 5.8. *(a) Every *-isomorphism strictly preserves biweights.*

*(b) Every *-homomorphism $\Phi : \mathfrak{A}_1 \to \mathfrak{A}_2$ such that $\Phi(R\mathfrak{A}_1) = R\mathfrak{A}_2$ strictly preserves biweights.*

Once again we verify the usual situation: Most of the time, results which are standard for *-algebras can be extended to partial *-algebras, provided some technical conditions are imposed. This is the price to pay for the far greater generality.

6. Physical Applications

Two important potential applications of partial O*-algebras in physics have been described in the previous sections, namely, the description of symmetries of quantum systems by automorphism groups (Section 4.3) and the Tomita–Takesaki theory of modular automorphisms, central in quantum statistical mechanics (Section 4.4).

Another instance where partial *-algebras are useful is Wightman's approach to (axiomatic) quantum field theory. In a concrete realization, the point-like (unsmeared) field $A(x)$ is represented as a sesquilinear form on a certain domain $\mathcal{D}$ in a Hilbert space $\mathcal{H}$ [28, 34] and one of the basic Wightman axioms is that

$$A(f) = \int_{\mathbb{R}^4} A(x) f(x) dx, \quad f \in C_0^\infty(\mathbb{R}^4), \tag{6.1}$$

is a well-defined operator in $\mathcal{D}$. The interesting point is that the field $A(x)$ may be defined as a $\mathfrak{L}(\mathcal{D}, \mathcal{D}^\times)$*-valued field*, in the sense of Section 3.3. By this we mean a map A from Minkowski space-time $\mathbb{M}$ into $\mathfrak{L}(\mathcal{D}, \mathcal{D}^\times)$, $A : x \in \mathbb{M} \to A(x) \in \mathfrak{L}(\mathcal{D}, \mathcal{D}^\times)$, satisfying the usual axioms of translation invariance, existence (and uniqueness) of a translation invariant vacuum and spectral condition. In practice, one chooses for $\mathcal{D}$ the domain $\mathcal{D} = \mathcal{D}^\infty(H)$ where $H = P^0$ is the energy operator. Then one can prove the following result.

Proposition 6.1. *Let $x \mapsto A(x)$ be an $\mathfrak{L}(\mathcal{D}, \mathcal{D}^\times)$-valued field with $\mathcal{D} = \mathcal{D}^\infty(H)$. Then the integral*

$$< A(f)\xi, \eta > := \int_{\mathbb{M}} f(x) < A(x)\xi, \eta > d^4x$$

converges for all $\xi, \eta \in \mathcal{D}$ and defines for each $f \in \mathcal{S}(\mathbb{M})$, the Schwartz space over $\mathbb{M}$, an operator of $\mathfrak{L}(\mathcal{D}, \mathcal{D}^\times)$. For every $\xi, \eta \in \mathcal{D}$, the map $f \mapsto < A(f)\xi, \eta >$ is a tempered distribution. Moreover $A(f) \in \mathcal{L}^\dagger(\mathcal{D})$, for all $f \in \mathcal{S}(\mathbb{M})$.

The Wightman field $A(f)$ so obtained satisfies all the required conditions. Additional boundedness properties of $A(f)$ may also derived, see [14, Chap.11] for further details.

Another field where partial *-algebras have had a impact is quantum statistical physics. There, the observable algebra $\mathfrak{A}$ (which is supposed to be a quasi-local C*-algebra) does not contain, in general, the thermodynamical limit of the local Heisenberg dynamics. Then the procedure to follow for circumventing this difficulty is to define in $\mathfrak{A}$ a new locally convex topology, τ, called, for obvious reasons, the *physical topology*, in such a way that the dynamics in the thermodynamical limit belongs to the completion of $\mathfrak{A}$ with respect to τ [30]. In spite of possible technical difficulties, the ideas are very simple.

In statistical physics, one has to deal with systems consisting of a very large number of particles, so large that one usually considers this number to be *infinite*.

One begins by considering systems living in a *local region* V and requires that the set of local regions be directed, i.e., if V_1, V_2 are two local regions, then there exists a third local region V_3 containing both V_1 and V_2. The thermodynamical problem consists in finding macroscopic properties of the system, starting from a given interaction between the particles. Thus one asks the question, how does the system behave when V becomes larger and larger ($|V| \to \infty$)? One starts with the local algebra $\mathfrak{A}_V$ of observables living in V, including the local Hamiltonian H_V. Then one considers the **-algebra of local observables* $\mathfrak{A}_o = \bigcup_V \mathfrak{A}_V$, which is well-defined as a *-algebra since the set of local regions is directed. The *observable algebra* $\mathfrak{A}$ is taken as the completion of $\mathfrak{A}_o$ in the physical topology (which is defined from a family of *-representations of $\mathfrak{A}_o$). Thus $(\mathfrak{A}, \mathfrak{A}_o)$ is a quasi *-algebra. This technique has been applied successfully to a number of spin systems with long range correlations, such as the BCS model and almost mean field models. It yields also a neat treatment of the method of effective Hamiltonians, and in another direction, a nice derivation of the Bogoliubov inequality for Bose gases. More details and the original references may be found in [14, Chap.11].

Another instance where partial *-algebras show up is the proper definition of Hamiltonians corresponding to very singular interactions (δ or δ', pointlike or on a surface) [1, 7, 26]. The point is that such a Hamiltonian has to be defined via von Neumann's theory of self-adjoint extensions of symmetric operators, using sesquilinear forms. These in turn are neatly formulated in terms of operators on a scale of Hilbert spaces, as discussed in Section 3.3 (iv), and we know that such operators do constitute a partial *-algebra.

7. Outcome

In conclusion of this rapid survey, we may say that the theory of partial *-algebras has reached a reasonable stage of maturity. Many nontrivial examples have been studied, both abelian and nonabelian, although no classification has been made so far. The representation theory is well under control. In particular, many standard results extend to partial *-algebras, such as the GNS construction or various structure properties. Two offshoots, in particular, have undergone a rapid development, namely CQ*-algebras and partial O*-algebras. The latter, and among them partial GW*-algebras, are a far reaching generalization of *-algebras of operators, both bounded and unbounded. Their structure is quite complex, yet a substantial body of information is available. Besides the representation theory associated to various notions of generalized vectors and weights, progress has been achieved also in the study of *-automorphisms and *-derivations, in particular the spatial theory.

These last results point toward a promising direction of research, namely the study of dynamical systems based on partial O*-algebras. In view of the results obtained so far, it is reasonable to expect progress for the case of partial GW*-algebras, since then the powerful theory of von Neumann algebras is available. In particular, the modular theory of Tomita-Takesaki extends, with suitable modifications, to partial GW*-algebras. Another active direction is a deeper analysis of particular types of partial *-algebras, for instance Banach partial *-algebras, for which some significant progress has been made [15, 16] since the publication of the monograph [14].

Where do we go from here? Admittedly, physical applications are scarce so far. However, the mathematical tool is there and continues to be developed for its own sake. Future research will decide which physical systems, if any, are complex enough to *require* the use of this approach.

References

[1] S. Albeverio, F. Gesztesy, R. Hoegh-Krohn, and, H. Holden, *Solvable Models in Quantum Mechanics*, Springer-Verlag, New York, Berlin, Heidelberg (1988).

[2] J-P. Antoine and A. Grossmann, Partial inner product spaces. I. General properties. II. Operators, *J. Funct. Anal.* **23**, 369–378, 379–391 (1976).

[3] J-P. Antoine, Partial inner product spaces. III. Compatibility relations revisited. IV. Topological considerations, *J. Math. Phys.* **21**, 268–279, 2067–2079 (1980).

[4] J-P. Antoine and K. Gustafson, Partial inner product spaces and semi-inner product spaces, *Adv. in Math.* **41**, 281–300 (1981).

[5] J-P. Antoine and W. Karwowski, Partial *-algebras of closed operators, in *Quantum Theory of Particles and Fields*, pp. 13–30; B. Jancewicz and J. Lukierski (eds.), World Scientific, Singapore (1983).

[6] J-P. Antoine and W. Karwowski, Partial *-algebras of closed linear operators in Hilbert space, *Publ. RIMS, Kyoto Univ.* **21**, 205–236 (1985); Add./Err. *ibid.* **22**, 507–511 (1986).

[7] J-P. Antoine, F. Gesztesy, and J. Shabani, Exactly solvable models of sphere interactions in Quantum Mechanics, *J. of Physics A : Math. Gen.* **20**, 3687–3712 (1987).

[8] J-P. Antoine and F. Mathot, Partial *-algebras of closed operators and their commutants. I. General structure, *Ann. Inst. H. Poincaré* **46**, 299–324 (1987).

[9] J-P. Antoine, A. Inoue, and C. Trapani, Partial *-algebras of closable operators I. The basic theory and the abelian case, *Publ. RIMS, Kyoto Univ.* **26**, 359–395 (1990).

[10] J-P. Antoine, A. Inoue, and C. Trapani, Spatial theory of *-automorphisms on partial O*-algebras, *J. Math. Phys.* **35**, 3059–3073 (1994).

[11] J-P. Antoine, A. Inoue, and C. Trapani, Partial *-algebras of closable operators: A review, *Reviews Math. Phys.* **8**, 1–42 (1996).

[12] J-P. Antoine, F. Bagarello and C. Trapani, Topological partial *-algebras: Basic properties and examples, *Reviews Math. Phys.* **11**, 267–302 (1999).

[13] J-P. Antoine, A. Inoue, and C. Trapani, Biweights on partial *-algebras, *J. Math. Anal. and Appl.* **242**, 164–190 (2000).

[14] J-P. Antoine, A. Inoue, and C. Trapani, *Partial *-Algebras and Their Operator Realizations*, Mathematics and Its Applications, vol. 553, Kluwer, Dordrecht, NL (2002).

[15] J-P. Antoine and C. Trapani, A note on Banach partial *-algebras, *Mediterr. j. math.* **3**, 67–86 (2006).

[16] J-P. Antoine, C. Trapani, and F. Tschinke, Biweights and *-homomorphisms of partial *-algebras, *Int. J. Math. and Math. Sci.* (2006) art. ID 54793, 1–20.

[17] F. Bagarello and C. Trapani, L^p-spaces as quasi *-algebras, J. Math. Anal. Appl., **197**, 810–824 (1996).

[18] J. Bergh and J. Löfström, *Interpolation Spaces*, Springer, Berlin (1976).

[19] H. J. Borchers, Energy and momentum as observables in quantum field theory, *Commun. Math. Phys.* **2**, 49–54 (1966).

[20] O. Bratteli and D. W. Robinson, *Operator Algebras and Quantum Statistical Mechanics I, II*, Springer-Verlag, Berlin (1979).

[21] H. W. Davis, F. J. Murray, and J. K. Weber, Families of L_p spaces with inductive and projective topologies, Pacific J. Math. **34**, 619–638 (1970); Inductive and projective limits of L_p spaces, *Portug. Math.* **31**, 21–29 (1972).

[22] G. F. dell'Antonio, On some groups of automorphisms of physical observables, *Commun. Math. Phys.* **2**, 384–397 (1966).

[23] J. Dixmier, *Les algèbres d'opérateurs dans l'espace hilbertien (Algèbres de von Neumann)*, Gauthier-Villars, Paris (1957, 1969)

[24] G.G. Emch, *Algebraic Methods in Statistical Mechanics and Quantum Field Theory*, Wiley-Interscience, New York (1972).

[25] J. J. F. Fournier and J. Stewart, Amalgams of L^p and ℓ^q, *Bull. Amer. Math. Soc.* **13**, 1–21 (1985).

[26] A. Grossmann, R. Hoegh-Krohn and M. Mebkhout, A class of explicity soluble, local, many-center Hamiltonians for one-particle quantum mechanics in two and three dimensions, *J. Math. Phys.* **21** (1980) 2376-2385

[27] R. Haag and D. Kastler, An algebraic approach to Quantum Field Theory, *J. Math. Phys.* **5**, 848–861 (1964).

[28] R. Jost, *The General Theory of Quantized Fields*, Amer. Math. Soc., Providence, RI (1965).

[29] R. V. Kadison and J. R. Ringrose, Derivations and automorphisms of operator algebras, *Commun. Math. Phys.* **4**, 32–63 (1967).

[30] G. Lassner, Topological algebras of operators, *Rep. Math. Phys.* **3**, 279–293 (1972).

[31] R. T. Powers, Self-adjoint algebras of unbounded operators. I,*Commun. Math. Phys.* **21**, 85–124 (1971).

[32] K. Schmüdgen, *Unbounded Operator Algebras and Representation Theory*, Akademie-Verlag, Berlin (1990).

[33] S. Stratila and L. Zsidó, *Lectures on von Neumann Algebras*, Abacus Press, Tunbridge Wells, UK (1979).

[34] R. F. Streater and A. S. Wightman, *PCT, Spin and Statistics, and All That*, Benjamin, New York (1964).

[35] A. C. Zaanen, *Integration*, North-Holland, Amsterdam, 1967.

Institut de Physique Théorique, Université catholique de Louvain B-1348 Louvain-la-Neuve, Belgium
E-mail: `Antoine@fyma.ucl.ac.be`

Some Variations on Maxwell's Equations

Giorgio A. Ascoli and Gerald A. Goldin

In the first sections of this article, we discuss two variations on Maxwell's equations that have been introduced in earlier work—a class of nonlinear Maxwell theories with well-defined Galilean limits (and correspondingly generalized Yang-Mills equations), and a linear modification motivated by the coupling of the electromagnetic potential with a certain nonlinear Schrödinger equation. In the final section, revisiting an old idea of Lorentz, we write Maxwell's equations for a theory in which the electrostatic force of repulsion between like charges differs fundamentally in magnitude from the electrostatic force of attraction between unlike charges. We elaborate on Lorentz' description by means of electric and magnetic field strengths, whose governing equations separate into two fully relativistic Maxwell systems—one describing ordinary electromagnetism, and the other describing a universally attractive or repulsive long-range force. If such a force cannot be ruled out *a priori* by known physical principles, its magnitude should be determined or bounded experimentally. Were it to exist, interesting possibilities go beyond Lorentz' early conjecture of a relation to (Newtonian) gravity.

It is a pleasure to dedicate this paper to Gérard Emch, whose skeptical perspective helps motivate those who know him to the pursuit of deeper scientific understandings.

1. Introduction

Maxwell's equations are among the most beautiful in physics, unifying the forces of electricity and magnetism in a classical field theory that explains electromagnetic waves [1]. Some well-known, profoundly-motivated variations on Maxwell's equations have included the Born-Infeld theory (a nonlinear but Lorentz-covariant modification, that introduces an effective upper bound to the electric field strength), and the Yang-Mills equations (introducing non-Abelian gauge potentials) [2, 3]. These ideas go back many decades, and have deeply influenced the development of theoretical physics. Indeed, there has been a recent resurgence of interest in non-Abelian Born-Infeld Lagrangians [4], which turn out to have important application in string theory and related subjects [5, 6, 7, 8]. More recently, variations of Maxwell's equations have been considered as "test theories," with respect to which observations in astrophysics can provide upper bounds to deviations from the usual equations or laws of physics [9, 10]. We nevertheless seek to approach the idea of modifying Maxwell's equations in new ways with appropriate humility. None of the variations considered in this article is *ad hoc*. Rather, each occurred in answer to a specific question in fundamental physics.

Sections 2 and 3 review two such modifications considered by the second author in recent years. The first of these, proposed in joint work with Vladimir Shtelen [11, 12], is a class of Galilean nonlinear Maxwell theories, together with non-Abelian versions that generalize the Yang-Mills equations and the non-Abelian Born-Infeld equations. These possibilities arise in answer to the question of whether and how Maxwell's equations for the four fields $\mathbf{E}$, $\mathbf{B}$, $\mathbf{D}$ and $\mathbf{H}$ can survive unchanged in the Galilean limit of $c \to \infty$, a feature that is present in neither the usual, linear Maxwell theory, nor the usual Born-Infeld theory. The second variation is a simple, linear modification, that can be associated with a change over time in some "constants" of electromagnetism. This possibility comes up in answer to the question of how to write gauge-invariant expressions for the electromagnetic field strengths $\mathbf{E}$ and $\mathbf{B}$, when Maxwell's equations are coupled with a natural and very general family of *nonlinear* Schrödinger time-evolutions in quantum mechanics [13]. Such a family of nonlinear Schrödinger equations was developed and studied in a series of articles, in joint work with Heinz-Dietrich Doebner and Peter Nattermann [14, 15, 16]. In the general nonlinear equation, a certain "frictional" term originally proposed by Kostin [17] is the one requiring a change in Maxwell's equations for the field strengths, after nonlinear gauge transformations are taken into account.

Finally, in Section 4, we take up a different question. In a 1900 article that seems to be not very well-known today, Lorentz explored the idea that (Newtonian) gravity could be explained if the electrostatic force of repulsion between like charges were smaller in absolute magnitude than the electrostatic force of attraction between unlike charges [18]. Setting aside Lorentz' conjectured relation to gravity, we want to reopen the possibility of a difference in magnitude between these forces. There is then a straightforward and elegant description of the situation by means of Maxwell's equations, that was partially written down in Lorentz' original article. Under the given hypothesis, one may introduce new electric and magnetic fields whose governing equations separate into two fully relativistic Maxwell systems—one describing ordinary electromagnetism, and the other describing an overall attractive or repulsive long-range force that couples to an "absolute charge." While the latter force might conceivably have something to do with gravity, it is more plausible to regard it as an extremely small, but theoretically possible, correction to ordinary electromagnetism. Such a correction could, in principle, be time-dependent, and serve as a further "test theory" for astrophysical measurements. Whether attractive or repulsive, a modified electromagnetism could be important in modeling the early universe. Unless a known physical principle rules out such a force *a priori,* its magnitude should be regarded as an experimental question. Well-known nonlinear and non-Abelian generalizations of Maxwell's equations, and their unification with weak interactions, could then equally well be constructed from the new equations, opening up interesting possibilities.

In the remainder of this section we establish notation, summarize some elementary, familiar background material, and make a few relevant remarks. All of our discussions pertain to $(3+1)$-dimensional space-time.

Let us write Maxwell's equations in SI units, as follows [1]:

$$\nabla \times \mathbf{E} = -\frac{\partial \mathbf{B}}{\partial t}, \quad \nabla \cdot \mathbf{B} = 0, \quad \nabla \times \mathbf{H} = \frac{\partial \mathbf{D}}{\partial t} + \mathbf{j}, \quad \nabla \cdot \mathbf{D} = \rho, \tag{1.1}$$

where $\mathbf{E}(\mathbf{x}, t)$ is the electric field, $\mathbf{D}(\mathbf{x}, t)$ the electric displacement, $\mathbf{B}(\mathbf{x}, t)$ the magnetic induction, and $\mathbf{H}(\mathbf{x}, t)$ the magnetic field; $\rho(\mathbf{x}, t)$ is the charge density, and $\mathbf{j}(\mathbf{x}, t)$ the electric current density.

The first pair of Eqs. (1.1) imply that we can write $\mathbf{E}$ and $\mathbf{B}$ in terms of potentials $(\Phi, \mathbf{A})$,

$$\mathbf{B} = \nabla \times \mathbf{A}, \quad \mathbf{E} = -\frac{\partial \mathbf{A}}{\partial t} - \nabla \Phi. \tag{1.2}$$

The choice of Φ and $\mathbf{A}$ is not unique. For an arbitrary smooth function $\Theta(\mathbf{x}, t)$, new potentials $\mathbf{A}'$ and Φ' that are obtained from the *gauge transformation*

$$\mathbf{A}' = \mathbf{A} + \nabla \Theta, \quad \Phi' = \Phi - \frac{\partial \Theta}{\partial t} \tag{1.3}$$

give just the same fields $\mathbf{E}$ and $\mathbf{B}$. Thus $\mathbf{E}$ and $\mathbf{B}$ are said to be gauge invariant. The condition $\nabla \cdot \mathbf{B} = 0$ expresses the nonexistence in nature of magnetic monopoles, which in this article we do not consider changing. Our choice of SI units here is motivated by the desire to avoid incorporating the speed of light c into the definitions of any of the fields, as we shall later be interested in considering the $c \to \infty$ limit.

From the second pair of Eqs. (1.1), there follows immediately the equation of continuity,

$$\frac{\partial \rho}{\partial t} + \nabla \cdot \mathbf{j} = 0, \tag{1.4}$$

expressing conservation of electric charge.

Thus far, the system is underdetermined. To complete Maxwell's equations in the presence of matter, one introduces *constitutive equations* relating the fields $\mathbf{D}$, $\mathbf{H}$ to the fields $\mathbf{E}$, $\mathbf{B}$. The usual system of Maxwell equations *in vacuo* is obtained using *linear* constitutive equations,

$$\mathbf{D} = \varepsilon_0 \mathbf{E}, \quad \mathbf{H} = \frac{1}{\mu_0} \mathbf{B}, \tag{1.5}$$

where $\varepsilon_0 \mu_0 = 1/c^2$. However, we shall shortly be considering a certain class of nonlinear constitutive equations.

One may take the point of view, given the absence of magnetic monopoles, that the only *physically detectable* fields are $\mathbf{E}$ and $\mathbf{B}$. These are defined operationally *via* the observed Lorentz force $\mathbf{F}$ on a small "test particle" with electric charge q moving with velocity $\mathbf{v}$:

$$\mathbf{F} = q\,\mathbf{E} + q\,\mathbf{v} \times \mathbf{B}\,. \tag{1.6}$$

The fields $\mathbf{H}$ and $\mathbf{D}$ can then be regarded as unobservable constructs used to describe, by way of the latter two Maxwell equations and the constitutive equations, how the observable fields are *produced* by charges and currents.

As noted in Ref. [11], we then actually have a more general class of linear constitutive equations,

$$\begin{bmatrix} \mathbf{D} \\ \mathbf{H} \end{bmatrix} = \begin{bmatrix} \varepsilon_0 & \lambda \\ -\lambda & 1/\mu_0 \end{bmatrix} \begin{bmatrix} \mathbf{E} \\ \mathbf{B} \end{bmatrix}, \tag{1.7}$$

which—when combined with Eqs. (1.1)—lead *for all* values of the real parameter λ to the *same* set of equations for the observable fields $\mathbf{E}$ and $\mathbf{B}$. Thus the choice $\lambda = 0$ resulting in Eqs. (1.5) is arbitrary.

Also relevant to the forthcoming discussion is the well-known *minimal coupling* of the electromagnetic potentials $(\Phi, \mathbf{A})$ with Schrödinger's equation for the (complex-valued) wave function $\psi(\mathbf{x}, t)$ of a single quantum-mechanical particle having charge q,

$$i\hbar \frac{\partial \psi}{\partial t} = \frac{1}{2m}(-i\hbar\nabla - q\mathbf{A})^2 \psi + q\Phi\psi\,. \tag{1.8}$$

Local $U(1)$ gauge transformations act on ψ according to the formula,

$$\psi'(\mathbf{x}, t) = e^{i\theta(\mathbf{x},t)}\psi(\mathbf{x}, t)\,, \tag{1.9}$$

and it is easily checked that if ψ obeys Eq. (1.8), then ψ' obeys a gauge-transformed equation of the same form, with new electromagnetic potentials given by Eqs. (1.3) in which $\Theta(\mathbf{x}, t) = (\hbar/q)\,\theta(\mathbf{x}, t)$.

The gauge-invariant fields $\mathbf{E}$ and $\mathbf{B}$, which exert the electric and magnetic forces on the charged quantum particle, are obtained from Φ and $\mathbf{A}$ using Eqs. (1.2), and satisfy Eqs. (1.1). The gauge-invariant probability and probability flux densities for the particle are given, respectively, by

$$\rho^{gi} = \bar{\psi}\psi, \quad \mathbf{J}^{gi} = \frac{\hbar}{2im}[\bar{\psi}\nabla\psi - (\nabla\bar{\psi})\psi] - \frac{q}{m}\bar{\psi}\psi\,\mathbf{A}\,; \tag{1.10}$$

these also obey an equation of continuity.

Let us remark on the fact that Eqs. (1.1) respect the Lorentz transformations of special relativity, while Schrödinger's equation respects Galilean transformations.

The minimal coupling of Eq. (1.8) is compatible with these facts because Eqs. (1.1) *also* respect Galilean transformations. It is the *linear constitutive equations* that impose Lorentz symmetry on the usual Maxwell equations, breaking the Galilean symmetry (see below).

We reproduce Lorentz transformations here in SI units for completeness. Let the subscript $\|$ indicate the component of a vector in the direction of the velocity $\mathbf{v}$ of an inertial frame of reference, let the subscript $\perp$ indicate the component perpendicular to $\mathbf{v}$, and let $v = |\mathbf{v}|$. Then with

$$\gamma = \frac{1}{\sqrt{1 - v^2/c^2}}, \tag{1.11}$$

the space-time transformation under the Lorentz boost is

$$\mathbf{x}'_{\|} = \gamma(\mathbf{x}_{\|} - \mathbf{v}t), \quad \mathbf{x}'_{\perp} = \mathbf{x}_{\perp}, \quad t' = \gamma(t - \frac{\mathbf{v}\cdot\mathbf{x}}{c^2}); \tag{1.12}$$

the field transformations are

$$\mathbf{B}'_{\|} = \mathbf{B}_{\|}, \quad \mathbf{B}'_{\perp} = \gamma(\mathbf{B} - \frac{1}{c^2}\mathbf{v}\times\mathbf{E})_{\perp}, \quad \mathbf{E}'_{\|} = \mathbf{E}_{\|}, \; \mathbf{E}'_{\perp} = \gamma(\mathbf{E} + \mathbf{v}\times\mathbf{B})_{\perp},$$

$$\mathbf{H}'_{\|} = \mathbf{H}_{\|}, \quad \mathbf{H}'_{\perp} = \gamma(\mathbf{H} - \mathbf{v}\times\mathbf{D})_{\perp}, \tag{1.13}$$

$$\mathbf{D}'_{\|} = \mathbf{D}_{\|}, \quad \mathbf{D}'_{\perp} = \gamma(\mathbf{D} + \frac{1}{c^2}\mathbf{v}\times\mathbf{H})_{\perp};$$

and the electric current and charge density transformations are

$$\mathbf{j}'_{\|} = \gamma(\mathbf{j}_{\|} - \rho v), \quad \mathbf{j}'_{\perp} = \mathbf{j}_{\perp}, \quad \rho' = \gamma(\rho - \frac{\mathbf{v}\cdot\mathbf{j}}{c^2}). \tag{1.14}$$

The corresponding electromagnetic potential transformations in SI units are

$$\Phi' = \gamma(\Phi - \mathbf{v}\cdot\mathbf{A}), \quad \mathbf{A}' = \gamma(\mathbf{A} - \frac{\mathbf{v}}{c^2}\Phi). \tag{1.15}$$

Under Lorentz transformation, the following combinations of the fields are then invariant:

$$I_1 = \mathbf{B}^2 - \frac{1}{c^2}\mathbf{E}^2, \quad I_2 = \mathbf{B}\cdot\mathbf{E};$$

$$I_3 = \mathbf{D}^2 - \frac{1}{c^2}\mathbf{H}^2, \quad I_4 = \mathbf{H}\cdot\mathbf{D}; \tag{1.16}$$

$$I_5 = \mathbf{B}\cdot\mathbf{H} - \mathbf{E}\cdot\mathbf{D}, \quad I_6 = \mathbf{B}\cdot\mathbf{D} + \frac{1}{c^2}\mathbf{E}\cdot\mathbf{H}.$$

The Born-Infeld Lagrangian as a function of these invariants is

$$\mathcal{L} = 1 - R, \quad R = \frac{b^2}{\mu_0 c^2}\sqrt{1 + \frac{c^2}{b^2} I_1 - \frac{c^2}{b^4} I_2^2}\,. \tag{1.17}$$

Next let us write the above in covariant notation. Define $x^\mu = (ct, \mathbf{x})$, $\mu = 0, 1, 2, 3$, and $x_\mu = g_{\mu\nu} x^\nu = (ct, -\mathbf{x})$, where the metric tensor $g_{\mu\nu} = \mathrm{diag}\,(1, -1, -1, -1)$, and where summation over repeated Lorentz indices is understood; for example, $x_\mu x^\mu = c^2 t^2 - \mathbf{x}^2$. We further define $\partial_\mu \equiv \partial/\partial x^\mu = [\,(1/c)\,\partial/\partial t,\, \nabla\,]$; and we shall use the antisymmetric Levi-Civita tensor $\varepsilon^{\alpha\beta\mu\nu}$, with $\varepsilon^{0123} = 1$. Then the usual relativistic tensor fields $F_{\alpha\beta}$ and $\mathcal{F}^{\alpha\beta}$, constructed from the fields $\mathbf{E}$ and $\mathbf{B}$, are

$$F_{\alpha\beta} = \begin{bmatrix} 0 & (1/c)E_1 & (1/c)E_2 & (1/c)E_3 \\ -(1/c)E_1 & 0 & -B_3 & B_2 \\ -(1/c)E_2 & B_3 & 0 & -B_1 \\ -(1/c)E_3 & -B_2 & B_1 & 0 \end{bmatrix}, \tag{1.18}$$

$$\mathcal{F}^{\alpha\beta} = \frac{1}{2}\,\varepsilon^{\alpha\beta\mu\nu} F_{\mu\nu} = \begin{bmatrix} 0 & -B_1 & -B_2 & -B_3 \\ B_1 & 0 & (1/c)E_3 & -(1/c)E_2 \\ B_2 & -(1/c)E_3 & 0 & (1/c)E_1 \\ B_3 & (1/c)E_2 & -(1/c)E_1 & 0 \end{bmatrix},$$

with

$$F^{\alpha\beta} = g^{\alpha\mu} g^{\beta\nu} F_{\mu\nu}\,, \quad \mathcal{F}_{\alpha\beta} = g_{\alpha\mu} g_{\beta\nu} \mathcal{F}^{\mu\nu}\,. \tag{1.19}$$

Likewise,

$$G^{\alpha\beta} = \begin{bmatrix} 0 & -cD_1 & -cD_2 & -cD_3 \\ cD_1 & 0 & -H_3 & H_2 \\ cD_2 & H_3 & 0 & -H_1 \\ cD_3 & -H_2 & H_1 & 0 \end{bmatrix}, \; G_{\alpha\beta} = g_{\alpha\mu} g_{\beta\nu} G^{\mu\nu}\,, \tag{1.20}$$

and so forth. Maxwell's equations (1.1) then become

$$\partial_\alpha \mathcal{F}^{\alpha\beta} = 0; \quad \partial_\alpha G^{\alpha\beta} = j^\beta, \tag{1.21}$$

with $j^\beta = (c\rho, \mathbf{j})$. With $A_\mu = (\mathbf{\Phi}, -\mathbf{A})$, we have from the first of Eqs. (1.21),

$$\mathcal{F}^{\alpha\beta} = \epsilon^{\alpha\beta\mu\nu} \partial_\mu A_\nu\,, \quad F_{\mu\nu} = \partial_\mu A_\nu - \partial_\nu A_\mu\,. \tag{1.22}$$

The first two invariants of Eqs. (1.16), that enter Eq. (1.17) for the Born-Infeld Lagrangian density, are now written

$$I_1 = \frac{1}{2} F_{\mu\nu} F^{\mu\nu}, \quad I_2 = -\frac{c}{4} F_{\mu\nu} \mathcal{F}^{\mu\nu}. \tag{1.23}$$

Note that our strategy, following Refs.[11] and [12], has been to postpone writing constitutive equations for as long as possible. These now relate **D**, **H** to **E**, **B**; or, equivalently, they relate $G^{\alpha\beta}$ to $\mathcal{F}^{\alpha\beta}$. The general form for Lorentz-invariant constitutive equations is given by [19]

$$\mathbf{D} = M\mathbf{B} + \frac{1}{c^2} N\mathbf{E}, \quad \mathbf{H} = N\mathbf{B} - M\mathbf{E}, \tag{1.24}$$

where M and N are functions of the Lorentz invariants in Eqs. (1.16), or

$$\mathbf{B} = R\mathbf{D} + \frac{1}{c^2} Q\mathbf{H}, \quad \mathbf{E} = Q\mathbf{D} - R\mathbf{H}, \tag{1.25}$$

where Q and R are likewise functions of the invariants. The linear constitutive equations (1.5) correspond to the choices $M = 0$, $N = 1/\mu_0$, in which case $\varepsilon_0 = 1/\mu_0 c^2$. Other choices lead to nonlinear relativistic field equations, such as Born-Infeld or Euler-Kockel electrodynamics. It is natural in Eqs. (1.24) to take M and N to be functions of just the first two invariants I_1 and I_2 (which depend only on **E** and **B**), or in the inverted Eqs. (1.25) to take R and Q to be functions of just the invariants I_3 and I_4 (which depend only on **D** and **H**).

The usual approach to writing nonlinear Maxwell theories is to begin with the Lagrangian $\mathcal{L}$, which can be written as a function of the invariants. The constitutive equations then follow from the Euler-Lagrange equations. But not all Lorentz-covariant theories are Lagrangian, and the approach *via* constitutive equations is more general.

In covariant form, Eqs. (1.24) become

$$G^{\mu\nu} = NF^{\mu\nu} + cM\mathcal{F}^{\mu\nu} \equiv M_1 \frac{\partial I_1}{\partial F^{\mu\nu}} + M_2 \frac{\partial I_2}{\partial F^{\mu\nu}}, \tag{1.26}$$

where M_1 and M_2 are likewise functions of the Lorentz invariants.

Now we are ready to discuss the three variations to which this article is devoted.

2. Maxwell Equations Having Galilean Limits

When Maxwell's equations are written for the four fields E, **B**, **D**, and **H** as in Eqs. (1.1), the system is underdetermined. In the SI units we are using, these equations are independent of the speed of light c. Furthermore, the corresponding equations for the primed fields defined by Eqs. (1.13) and the primed currents defined by

Eqs. (1.14), when written in the primed coordinates defined by Eqs. (1.12), are unchanged from Eqs. (1.1)—even though the Lorentz transformations given in Eqs. (1.12)-(1.14) *are* parameterized explicitly by c. Since the invariance of Eqs. (1.1) holds for every value of c, we should not be at all surprised that in the limit $c \to \infty$, these equations also respect the resulting Galilean transformations,

$$\begin{gathered}\mathbf{x}' = \mathbf{x} - \mathbf{v}t\,, \quad t' = t\,, \\ \mathbf{B}' = \mathbf{B}\,, \quad \mathbf{E}' = \mathbf{E} + \mathbf{v} \times \mathbf{B}\,, \\ \mathbf{H}' = \mathbf{H} - \mathbf{v} \times \mathbf{D}\,, \quad \mathbf{D}' = \mathbf{D}\,, \\ \mathbf{j}' = \mathbf{j} - \rho\,\mathbf{v}\,, \quad \rho' = \rho\,.\end{gathered} \tag{2.1}$$

Now, the value of c *does* appear in the constitutive equations. Thus the choice between Lorentz or Galilei symmetry, or the selection of a particular, finite value of c under which the Lorentz symmetry holds, resides entirely in the constitutive equations.

For the linear constitutive equations (1.5) or (1.7), the speed of light is specified by $c = (\varepsilon_0 \mu_0)^{-1/2}$. In this case, taking a Galilean limit requires that some aspect of Maxwell's equations be sacrificed, as discussed in detail by Le Bellac and Levy-Leblond [20].

The constitutive equations that select Galilean symmetry, when combined with Eqs. (1.1), are

$$\mathbf{D} = \hat{M}\mathbf{B}, \quad \mathbf{H} = \hat{N}\mathbf{B} - \hat{M}\mathbf{E}, \tag{2.2}$$

or, equivalently

$$\mathbf{B} = \hat{R}\mathbf{D}, \quad \mathbf{E} = \hat{Q}\mathbf{D} - \hat{R}\mathbf{H}, \tag{2.3}$$

where $\hat{M}$ and $\hat{N}$, $\hat{Q}$ and $\hat{R}$ are arbitrary functions of Galilean invariants,

$$\begin{gathered}\hat{I}_1 = \mathbf{B}^2, \quad \hat{I}_2 = \mathbf{B} \cdot \mathbf{E}; \\ \hat{I}_3 = \mathbf{D}^2, \quad \hat{I}_4 = \mathbf{H} \cdot \mathbf{D}; \\ \hat{I}_5 = \mathbf{B} \cdot \mathbf{H} - \mathbf{E} \cdot \mathbf{D}, \quad \hat{I}_6 = \mathbf{B} \cdot \mathbf{D}.\end{gathered} \tag{2.4}$$

These constitutive equations and field invariants are respectively the formal limits as $c \to \infty$ of their Lorentz invariant counterparts.

As discussed in Ref. [20], however, taking the mathematical step of letting $c \to \infty$ is not precisely the same thing as imposing the low velocity condition $v/c << 1$ on a class of physical systems governed by the dynamical equations with

Lorentz symmetry. For instance, when **E** and **B** are held fixed, the limit $c \to \infty$ does not allow the "electric limit" of Ref. [20], although $v/c << 1$ is compatible with it.

Letting $\hat{M}$ be a constant in Eqs. (2.2) requires (since $\nabla \cdot \mathbf{B} = 0$) that the charge density $\rho \equiv 0$. Hence the answer to the question of a consistent Galilean electrodynamics, retaining Maxwell's equations, the continuity equation, and the Lorentz force, is a class of *essentially nonlinear* theories, that can arise as the $c \to \infty$ limit of a class of essentially nonlinear Lorentz-covariant theories. Indeed, Le Bellac and Levy-LeBlond emphasize (always assuming linear constitutive equations) the mutual incompatibility of Galilean invariance, the continuity equation with non-zero values, and magnetic forces between electric currents.

With nonlinear constitutive equations, these features are no longer incompatible. Nontrivial choices of $\hat{M}$ and $\hat{N}$ in Eqs. (2.2), or $\hat{R}$ and $\hat{Q}$ in Eqs. (2.3), combined with Maxwell's equations, yield fully consistent Galilean versions of electrodynamics.

For example, Ref. [11] proposes to set

$$\hat{Q} = \frac{1}{\varepsilon} \qquad \hat{R} = \alpha + 2\alpha^2 \varepsilon \, \frac{\mathbf{H} \cdot \mathbf{D}}{|\mathbf{D}|^2}\,, \tag{2.5}$$

which are homogeneous functions of the field strengths. This can be shown to lead to an interesting, albeit non-Lagrangian, theory.

Writing the Lagrangian for a nonlinear relativistic theory as

$$\mathcal{L} = \mathcal{L}(I_1,\, I_2), \tag{2.6}$$

a short calculation in Ref. [12] demonstrates from the Euler-Lagrange equations that

$$N = 2\,\frac{\partial \mathcal{L}}{\partial I_1}\,, \qquad M = -\frac{\partial \mathcal{L}}{\partial I_2}\,. \tag{2.7}$$

Therefore the necessary compatibility condition for the constitutive equations to describe such a Lagrangian theory is given by,

$$2\,\frac{\partial M}{\partial I_1} + \frac{\partial N}{\partial I_2} = 0\,. \tag{2.8}$$

In the Galilean limit, we would take $\mathcal{L} = \mathcal{L}(\hat{I}_1,\, \hat{I}_2)$, and argue similarly.

The usual Born-Infeld theory does not have a nontrivial Galilean limit. The Lagrangian $\mathcal{L}(I_1,\, I_2)$ is given by

$$\mathcal{L} = 1 - \mathcal{R}\,, \qquad \mathcal{R} = \frac{b^2}{\mu_0 c^2} \sqrt{1 + \frac{c^2}{b^2} I_1 - \frac{c^2}{b^4} I_2^2}\,, \tag{2.9}$$

which leads to the constitutive equations (1.24) with

$$M(I_1,\, I_2) = \frac{I_2}{\mu_0 b^2 \mathcal{R}}\,, \qquad N = \frac{1}{\mu_0 \mathcal{R}}\,. \tag{2.10}$$

Taking $c \to \infty$, we have of course $I_1 \to \hat{I}_1 = \mathbf{B}^2$, and $I_2 \to \hat{I}_2 = \mathbf{B}\cdot\mathbf{E}$. But for large c, one has $\mathcal{R} \approx (c/b)[\hat{I}_1 - \hat{I}_2^2/b^2]^{1/2}$, whence the limits of M and N are both zero.

It is therefore suggested in Ref. [12] to modify the Born-Infeld Lagrangian, replacing $\mathcal{R}$ by

$$\tilde{\mathcal{R}} = \sqrt{1 + \frac{c^2}{b^2}\,[\,(1+\lambda_1 c^2)I_1 - \frac{1}{b^2}(1+\lambda_2 c^2)I_2^2\,]}\,, \tag{2.11}$$

where λ_1, λ_2 are new constants with the dimensions of $1/c^2$. Now, taking $c \to \infty$, one obtains the Galilean constitutive equations (2.2), with

$$\hat{M} = \frac{\lambda_2 \hat{I}_2}{\mu_0 b \sqrt{\lambda_1 \hat{I}_1 - \lambda_2 \hat{I}_2^2/b^2}}\,, \qquad \hat{N} = \frac{b\lambda_1}{\mu_0 \sqrt{\lambda_1 \hat{I}_1 - \lambda_2 \hat{I}_2^2/b^2}}\,. \tag{2.12}$$

Similarly, generalizations of classical (non-Abelian) Yang-Mills theory are written by means of Lorentz-covariant, nonlinear constitutive equations. Again, with appropriate choices for the dynamics, the new systems can have fully consistent Galilean-covariant limits as $c \to \infty$. In analogy with Eqs. (2.11)-(2.12), one obtains a class of generalizations of non-Abelian Born-Infeld theories that are of this type [12].

3. Modification From a Nonlinear Schrödinger Equation

Another variation on Maxwell's equations occurs as a result of considering the coupling of external electromagnetic fields with nonlinear Schrödinger time-evolutions [13, 16]. First we write the class of Schrödinger equations under consideration. Refs. [14] and [15] provide extensive motivation and development, that we omit here; we mainly follow the discussion in Ref. [13].

Letting $\psi(\mathbf{x}, \mathbf{t})$ be the quantum-mechanical wave function, and define

$$\hat{\rho}\,(\mathbf{x}, t) = \overline{\psi}\psi\,, \quad \hat{\mathbf{j}}\,(\mathbf{x}, t) = (1/2i)\,[\,\overline{\psi}\nabla\psi - (\nabla\overline{\psi})\psi\,]\,. \tag{3.1}$$

In this article we shall use the notation $\hat{\rho}\,(\mathbf{x}, t)$ to refer to the spatial probability density for the quantum-mechanical particle, to distinguish it from the net charge density $\rho\,(\mathbf{x}, t)$ that appears in Secs. 1 and 2. Thus $\hat{\rho}$ is here the expression we called ρ^{gi} in the first of Eqs. (1.10).

Define the real, homogeneous functionals $R_1[\psi], \ldots, R_5[\psi]$, by

$$R_1 = \frac{\nabla \cdot \hat{\mathbf{j}}}{\hat{\rho}}, \quad R_2 = \frac{\nabla^2 \hat{\rho}}{\hat{\rho}}, \quad R_3 = \frac{\hat{\mathbf{j}}^2}{\hat{\rho}^2}, \quad R_4 = \frac{\hat{\mathbf{j}} \cdot \nabla \hat{\rho}}{\hat{\rho}^2}, \quad R_5 = \frac{(\nabla \hat{\rho})^2}{\hat{\rho}^2}. \tag{3.2}$$

The Laplacian in the linear Schrödinger equation (1.8) can be expanded with respect to this basis of functionals,

$$\frac{\nabla^2 \psi}{\psi} = i R_1[\psi] + \frac{1}{2} R_2[\psi] - R_3[\psi] - \frac{1}{4} R_5[\psi], \tag{3.3}$$

so that it does not appear explicitly in the equation we shall next write down. The general family of nonlinear Schrödinger equations takes the form,

$$\begin{aligned} i \frac{\dot{\psi}}{\psi} = {} & i \left[\sum_{j=1}^{2} \nu_j R_j[\psi] + \frac{\nabla \cdot (\mathcal{A}(\mathbf{x}, t) \hat{\rho})}{\hat{\rho}} \right] + \sum_{j=1}^{5} \mu_j R_j[\psi] + \\ & + U(\mathbf{x}, t) + \frac{\nabla \cdot (\mathcal{A}_1(\mathbf{x}, t) \hat{\rho})}{\hat{\rho}} + \frac{\mathcal{A}_2(\mathbf{x}, t) \cdot \hat{\mathbf{j}}}{\hat{\rho}} + \alpha_1 \ln \hat{\rho} + \alpha_2 S, \end{aligned} \tag{3.4}$$

where $\dot{\psi}$ is $\partial \psi / \partial t$, the coefficients ν_j $(j = 1, 2)$, μ_j $(j = 1, \ldots, 5)$, and α_j $(j = 1, 2)$ are all continuously differentiable, real-valued functions of t; $S(\mathbf{x}, t)$ is $\arg[\psi(\mathbf{x}, t)]$; $U(\mathbf{x}, t)$ is a real-valued scalar function; and $\mathcal{A}$, $\mathcal{A}_1$, and $\mathcal{A}_2$ are distinct real-valued, time-dependent vector fields.

Writing a class of nonlinear Schrödinger equation by adding terms of the form (3.2) to the usual, linear Schrödinger equation as in Ref. [15], we have

$$i\hbar \frac{\partial \psi}{\partial t} = H_0 \psi + \frac{i}{2} \hbar D R_2[\psi] \psi + \hbar \sum_{j=1}^{5} D_j' R_j[\psi] \psi, \tag{3.5}$$

where $H_0 \psi$ is given by the right-hand side of Eq. (1.8), and D and the D_j' have the dimension of diffusion coefficients. Then Eq. (3.5) is obtained from Eq. (3.4) with the values,

$$\nu_1 = -\frac{\hbar}{2m}, \quad \nu_2 = \frac{1}{2} D, \quad \mathcal{A} = \frac{q}{2m} \mathbf{A},$$

$$\mu_1 = D_1', \quad \mu_2 = -\frac{\hbar}{4m} + D_2', \quad \mu_3 = \frac{\hbar}{2m} + D_3', \quad \mu_4 = D_4', \quad \mu_5 = \frac{\hbar}{8m} + D_5',$$

$$U(\mathbf{x}, t) = \frac{q}{\hbar} \Phi + \frac{q^2}{2m\hbar} \mathbf{A}^2, \quad \mathcal{A}_1 = 0, \quad \mathcal{A}_2 = -\frac{q}{m} \mathbf{A},$$

$$\alpha_1 = \alpha_2 = 0. \tag{3.6}$$

The motivation for the form adopted in writing Eq. (3.4), for the presence of the terms with α_1, α_2 and $\mathcal{A}_1 \neq 0$, and the time-dependence of the coefficients, is the possibility of introducing a group of *nonlinear gauge transformations* that leave this family of equations invariant (as a class). With $\psi = R \exp[iS]$, these take the form $\psi \mapsto \psi' = R' \exp[iS']$, with

$$R' = R, \quad S' = \Lambda S + \gamma \ln R + \theta; \tag{3.7}$$

where γ and Λ are continuously differentiable, real-valued functions of t, $\Lambda \neq 0$, and θ is a continuously differentiable, real-valued function of $\mathbf{x}$ and t. Then nonlinear gauge transformations obey the group law,

$$(\Lambda_1, \gamma_1, \theta_1)(\Lambda_2, \gamma_2, \theta_2) = (\Lambda_1\Lambda_2, \gamma_1 + \Lambda_1\gamma_2, \theta_1 + \Lambda_1\theta_2). \tag{3.8}$$

With quantum-mechanical measurements characterized as sequences of positional measurements (at distinct times), together with the application of external fields between positional measurements [21, 22], and maintaining the standard interpretation of $\hat{\rho} = |\psi|^2$ as a probability density for the outcomes of positional measurements, such transformations then leave the distribution of outcomes of all measurements invariant. They are also local in space-time, and respect a separation condition for multiparticle product wave functions [16, 23]. Notice that with $\gamma \equiv 0$ and $\Lambda \equiv 1$, we recover the usual local $U(1)$ gauge group of electromagnetism (acting linearly on ψ) as a subgroup of the larger group of nonlinear gauge transformations.

Under the nonlinear transformation in Eq. (3.7), we have

$$\hat{\rho}' = \overline{\psi'}\psi' = \hat{\rho},$$

$$\hat{\mathbf{j}}' = \frac{1}{2i}[\overline{\psi'}\nabla\psi' - (\nabla\overline{\psi'})\psi'] = \Lambda\hat{\mathbf{j}} + \frac{\gamma}{2}\nabla\hat{\rho} + \hat{\rho}\nabla\theta; \tag{3.9}$$

so that $\hat{\rho}$ (as desired) is gauge-invariant (for nonlinear as well as linear gauge transformations). Thus we shall also write $\hat{\rho} = \rho^{\text{gi}}$ when we want to emphasize this. But $\hat{\mathbf{j}}$ is not gauge-invariant—one must write a new gauge-invariant current (see below), to fully generalize Eqs. (1.10). Moreover, if ψ satisfies Eq. (3.4) then ψ' likewise satisfies an equation of this form, but with new (primed) coefficients and new external fields. The coefficients transformed under $(\gamma, \Lambda, \theta)$ are given by

$$\nu_1' = \frac{\nu_1}{\Lambda}, \quad \nu_2' = -\frac{\gamma}{2\Lambda}\nu_1 + \nu_2,$$

$$\mu_1' = -\frac{\gamma}{\Lambda}\nu_1 + \mu_1, \quad \mu_2' = \frac{\gamma^2}{2\Lambda}\nu_1 - \gamma\nu_2 - \frac{\gamma}{2}\mu_1 + \Lambda\mu_2,$$

$$\mu_3' = \frac{\mu_3}{\Lambda}, \quad \mu_4' = -\frac{\gamma}{\Lambda}\mu_3 + \mu_4, \quad \mu_5' = \frac{\gamma^2}{4\Lambda}\mu_3 - \frac{\gamma}{2}\mu_4 + \Lambda\mu_5,$$

$$\alpha_1' = \Lambda\alpha_1 - \frac{\gamma}{2}\alpha_2 + \frac{1}{2}\left(\frac{\dot{\Lambda}}{\Lambda}\gamma - \dot{\gamma}\right), \quad \alpha_2' = \alpha_2 - \frac{\dot{\Lambda}}{\Lambda}. \tag{3.10}$$

Observe that even if one begins with $\alpha_1 = \alpha_2 = 0$ and with time-independent coefficients ν_j and μ_j, the fact that γ and Λ can be time-dependent *requires* that in Eq. (3.4) the α_j be permitted to take nonzero values, and that all the ν_j, μ_j, and α_j be, in general, time-dependent. However, the nonlinear gauge transformations do not mix the coefficients α_j with the coefficients ν_j and μ_j.

The nonlinear term with coefficient α_1 was first proposed as a modification of linear quantum mechanics by Bialynicki-Birula and Micielski [24], and the term with coefficient α_2 was proposed still earlier by Kostin [17]. The term with coefficient $\nu_2 \neq 0$ was derived by Doebner and Goldin from considerations of local current algebra representations [14], which led to the generalized equation containing the nonlinear functionals R_j [15].

The external fields of Eq. (3.4) transformed under $(\gamma, \Lambda, \theta)$ are given in Ref. [13] by

$$\mathcal{A}' = \mathcal{A} - \frac{\nu_1}{\Lambda}\nabla\theta,$$

$$\mathcal{A}_1' = \Lambda\mathcal{A}_1 - \gamma\mathcal{A} - \frac{\gamma}{2}\mathcal{A}_2 + \left(\frac{\gamma}{\Lambda}\nu_1 - \mu_1 + \frac{\gamma}{\Lambda}\mu_3 - \mu_4\right)\nabla\theta,$$

$$\mathcal{A}_2' = \mathcal{A}_2 - \frac{2\mu_3}{\Lambda}\nabla\theta,$$

$$U' = \Lambda U - \dot{\theta} + \left(\frac{\dot{\Lambda}}{\Lambda} - \alpha_2\right)\theta + \frac{\mu_3}{\Lambda}[\nabla\theta]^2 +$$

$$\left(\mu_4 - \mu_3\frac{\gamma}{\Lambda}\right)\nabla^2\theta + \frac{\gamma}{2}\nabla\cdot\mathcal{A}_2 - \mathcal{A}_2\cdot\nabla\theta. \tag{3.11}$$

Observe that Eqs. (3.10)-(3.11) imply both the nonzero $\mathcal{A}_1$ and nontrivial $\mathcal{A}_2$ values are required in Eq. (3.4). Even if one begins with $\mathcal{A}_1 \equiv 0$ and $\mathcal{A}_2 \equiv -2\mathcal{A}$, as in the linear Schrödinger equation [*cf.* Eqs. (3.6)], the nonlinear gauge transformations compel one to introduce more general values for these fields. Nonlinear Schrödinger equations with arbitrary values of $\mathcal{A}_2$ were proposed by Haag and Bannier [25], while the interaction with a general external vector field $\mathcal{A}_1$ was considered in Ref. [26].

Next let us write the equations of motion described by this class of nonlinear Schrödinger equations entirely in terms of gauge-invariant quantities, as in Ref. [13]—where "gauge invariance" is interpreted with respect to the group of *nonlinear* gauge transformations. To start, a gauge-invariant current density $\mathbf{J}^{gi}$ may be written

$$\mathbf{J}^{gi} = -2\nu_1 \hat{\mathbf{j}} - 2\nu_2 \nabla\hat{\rho} - 2\hat{\rho}\mathcal{A}, \tag{3.12}$$

which evidently reduces to Eq. (1.10) for the linear Schrödinger equation [when $\nu_1 = -\hbar/2m,\ \nu_2 = 0,\ \mathcal{A} = (q/2m)\mathbf{A}$]. We thus have (again) a continuity equation for the probability density, $\partial\rho^{gi}/\partial t = -\nabla \cdot \mathbf{J}^{gi}$. Refs. [15] and [16] provide a set of gauge-invariant parameters necessary for the desired description,

$$\tau_1 = \nu_2 - \frac{1}{2}\mu_1, \quad \tau_2 = \nu_1\mu_2 - \nu_2\mu_1, \quad \tau_3 = \frac{\mu_3}{\nu_1}, \quad \tau_4 = \mu_4 - \mu_1\frac{\mu_3}{\nu_1},$$

$$\tau_5 = \nu_1\mu_5 - \nu_2\mu_4 + \nu_2^2\frac{\mu_3}{\nu_1},$$

$$\beta_1 = \nu_1\alpha_1 - \nu_2\alpha_2 + \nu_2\frac{\dot{\nu}_1}{\nu_1} - \dot{\nu}_2, \quad \beta_2 = \alpha_2 - \frac{\dot{\nu}_1}{\nu_1}. \tag{3.13}$$

When $\tau_1 \neq 0$, $\tau_4 \neq 0$, or $\beta_2 \neq 0$, time-reversal invariance is violated. When $\tau_3 \neq -1$ or $\tau_4 \neq 0$, Galilean invariance is violated.

Gauge-invariant external magnetic and electric fields are now given by the formulas [13]

$$\mathcal{B} = \nabla \times \mathcal{A} = \frac{q}{2m}\mathbf{B},$$

$$\mathcal{E} = -\nabla\hat{U} - \frac{\partial\mathcal{A}}{\partial t} - \beta_2\mathcal{A} = \frac{q}{2m}\mathbf{E}, \tag{3.14}$$

where

$$\hat{U} = -\nu_1 U - \tau_3\mathcal{A}^2 - (\tau_4 - 2\tau_1\tau_3)\nabla\cdot\mathcal{A} + \mathcal{A}\cdot\mathcal{A}_2 - \nu_2\nabla\cdot\mathcal{A}_2. \tag{3.15}$$

Thus $\hat{U}$ is to be identified with $(q/2m)\Phi$ (which may be directly checked for the linear Schrödinger equation); but the main point here is that the formula for $\mathbf{E}$ in terms of Φ and $\mathbf{A}$ has been modified from Eq. (1.2) to include an extra term, originating with Kostin's nonlinearity:

$$\mathbf{E} = -\nabla\Phi - \frac{\partial\mathbf{A}}{\partial t} - \beta_2\mathbf{A}. \tag{3.16}$$

The extra term is *necessary*—if we leave it out of Eq. (3.14), $\mathcal{E}$ fails to be gauge-invariant. We also have new gauge-invariant external vector fields,

$$\mathcal{A}_1^{gi} = \nu_1\mathcal{A}_1 + \left(\frac{2\nu_2\mu_3}{\nu_1} - \mu_1 - \mu_4\right)\mathcal{A} - \nu_2\mathcal{A}_2, \quad \mathcal{A}_2^{gi} = \frac{\nu_1}{2\mu_3}\mathcal{A}_2 - \mathcal{A}. \tag{3.17}$$

Continuing to follow Ref. [13], we are now in a position to write equations of motion obtained from Eq. (3.4), in a form that is manifestly gauge-invariant for the group of nonlinear gauge transformations. We use the hydrodynamical variables variables ρ^{gi} and $\mathbf{J}^{gi}/\rho^{\mathrm{gi}}$, where the latter has the interpretation of a gauge-invariant velocity field. We then have the interpretation of the (gauge-invariant) magnetic induction as a vorticity field,

$$\nabla \times \left(\frac{\mathbf{J}^{\mathrm{gi}}}{\rho^{\mathrm{gi}}} \right) = -2\,\mathcal{B} = \frac{q}{m}\,\mathbf{B}\,. \tag{3.18}$$

The dynamical equations are the equation of continuity that we wrote above,

$$\frac{\partial \rho^{\mathrm{gi}}}{\partial t} = -\nabla \cdot \mathbf{J}^{\mathrm{gi}}\,, \tag{3.19}$$

together with the equation for the changing velocity field,

$$\begin{aligned}
\frac{\partial}{\partial t}\left(\frac{\mathbf{J}^{\mathrm{gi}}}{\rho^{\mathrm{gi}}} \right) &= \nabla \left[2\tau_1 \nabla \cdot \left(\frac{\mathbf{J}^{\mathrm{gi}}}{\rho^{\mathrm{gi}}} \right) + 2\tau_2 \frac{\nabla^2 \rho^{\mathrm{gi}}}{\rho^{\mathrm{gi}}} + \frac{1}{2}\tau_3 \left(\frac{\mathbf{J}^{\mathrm{gi}}}{\rho^{\mathrm{gi}}} \right)^2 \right] \\
&+ \nabla \left[(2\tau_1 [1+\tau_3] - \tau_4) \left(\frac{\mathbf{J}^{\mathrm{gi}}}{\rho^{\mathrm{gi}}} \right) \cdot \frac{\nabla \rho^{\mathrm{gi}}}{\rho^{\mathrm{gi}}} + 2\tau_5 \left(\frac{\nabla \rho^{\mathrm{gi}}}{\rho^{\mathrm{gi}}} \right)^2 \right] \\
&+ \nabla \left[2\,\frac{\nabla \cdot (\mathcal{A}_1^{\mathrm{gi}} \rho^{\mathrm{gi}})}{\rho^{\mathrm{gi}}} - 2\tau_3\, \mathcal{A}_2^{\mathrm{gi}} \cdot \left(\frac{\mathbf{J}^{\mathrm{gi}}}{\rho^{\mathrm{gi}}} \right) + 2\beta_1 \ln \rho^{\mathrm{gi}} \right] \\
&- \beta_2 \left(\frac{\mathbf{J}^{\mathrm{gi}}}{\rho^{\mathrm{gi}}} \right) + \frac{q}{m}\,\mathbf{E}\,.
\end{aligned} \tag{3.20}$$

In Eq. (3.20) we see that β_2, taken to be positive, has a natural interpretation as a gauge-invariant coefficient of friction—it governs the magnitude of the term in $\partial_t\, (\mathbf{J}^{\mathrm{gi}}/\rho^{\mathrm{gi}})$ that is proportional to $\mathbf{J}^{\mathrm{gi}}/\rho^{\mathrm{gi}}$.

In the framework of the nonlinear quantum mechanics discussed here, the (gauge-invariant) expected values for the position, velocity, and acceleration of the quantum particle (all of which are functions of t) are given respectively by the following expressions:

$$< \mathbf{x} > = \int \mathbf{x}\, \rho^{\mathrm{gi}}(\mathbf{x})\, d\mathbf{x}\,,$$

$$< \mathbf{v} > = \frac{\partial < \mathbf{x} >}{\partial t} = \int \rho^{\mathrm{gi}} \left(\frac{\mathbf{J}^{\mathrm{gi}}}{\rho^{\mathrm{gi}}} \right) d\mathbf{x} = \int \mathbf{J}^{\mathrm{gi}}(\mathbf{x})\, d\mathbf{x}\,, \tag{3.21}$$

$$< \mathbf{a} > = \frac{\partial < \mathbf{v} >}{\partial t} =$$

$$\int \rho^{\,\mathrm{gi}} \left[\frac{1}{2} \nabla \left(\frac{\mathbf{J}^{\mathrm{gi}}}{\rho^{\,\mathrm{gi}}} \right)^2 + \left(\frac{\mathbf{J}^{\mathrm{gi}}}{\rho^{\,\mathrm{gi}}} \right) \times \frac{q}{m} \mathbf{B} + \frac{\partial}{\partial t} \left(\frac{\mathbf{J}^{\mathrm{gi}}}{\rho^{\,\mathrm{gi}}} \right) \right] d\mathbf{x} .$$

In Eqs. (3.20)-(3.21), we see that the laws of force describing the interaction of the charged particle with the $\mathbf{E}$ and $\mathbf{B}$ fields are unchanged from those in linear quantum mechanics.

Now Eq. (3.16) gives us the variation on Maxwell's equations that is the focus of this section. The usual equations for $\mathbf{E}$ and $\mathbf{B}$ are replaced by

$$\nabla \times \mathbf{E} = -\frac{\partial \mathbf{B}}{\partial t} - \beta_2 \mathbf{B} , \quad \nabla \cdot \mathbf{B} = 0 . \tag{3.22}$$

Note that the second of these equations is still consistent with the first. Let us take β_2 to be a constant, independent of t. If $\mathbf{E_0}(\mathbf{x}, t)$, $\mathbf{B_0}(\mathbf{x}, t)$ satisfy the original Maxwell equations (with $\beta_2 = 0$), then fields satisfying Eqs. (3.22) are given by

$$\mathbf{E} = \mathbf{E_0}\, e^{-\beta_2 t}, \quad \mathbf{B} = \mathbf{B_0}\, e^{-\beta_2 t}. \tag{3.23}$$

But nonlinear quantum mechanics alone does not specify the remaining two Maxwell equations. One possibility is to maintain the constitutive equations (1.5), with fixed coefficients ε_0 and μ_0. Then taking $\mathbf{E_0}$, $\mathbf{B_0}$, $\mathbf{D}_0 = \varepsilon_0 \mathbf{E}_0$, $\mathbf{H}_0 = (1/\mu_0)\mathbf{B}_0$, ρ_0 and $\mathbf{j}_0$ to satisfy the usual, linear Maxwell equations (1.1), we must have

$$\mathbf{D} = \mathbf{D_0}\, e^{-\beta_2 t}, \quad \mathbf{H} = \mathbf{H_0}\, e^{-\beta_2 t} , \tag{3.24}$$

and

$$\nabla \cdot \mathbf{D} = \rho , \text{ with } \quad \rho = \rho_0 e^{-\beta_2 t} ,$$

$$\nabla \times \mathbf{H} = \frac{\partial \mathbf{D}}{\partial t} + \beta_2 \mathbf{D} + \mathbf{j} , \text{ with } \quad \mathbf{j} = \mathbf{j}_0 e^{-\beta_2 t} . \tag{3.25}$$

That is, with $\beta_2 > 0$, the magnitudes of all the electric charges and currents are decaying exponentially with time. Of course, the equation of continuity for ρ and $\mathbf{j}$ no longer holds, and net charge is no longer conserved. Instead, we have the equation

$$\frac{\partial \rho}{\partial t} = -\nabla \cdot \mathbf{j} - \beta_2 \rho . \tag{3.26}$$

The nonzero value for β_2, interpreted as a coefficient of friction experienced by a charged particle, has introduced a preferred universal reference frame. The corresponding Maxwell theory is no longer covariant.

Another alternative is to join the *standard* Maxwell equations for **D** and **H** with Eqs. (3.22) by supposing that the coefficients in the constitutive equations—the permittivity and permeability of free space—are time-dependent. With

$$\varepsilon(t) = \varepsilon_0 \, e^{+\beta_2 t}, \quad \mu(t) = \mu_0 \, e^{-\beta_2 t}, \tag{3.27}$$

we may combine Eqs. (3.22) and (3.23) with the constitutive equations,

$$\mathbf{D} = \varepsilon(t)\mathbf{E}, \quad \mathbf{H} = \frac{1}{\mu(t)}\mathbf{B}. \tag{3.28}$$

Then $\mathbf{D} = \mathbf{D}_0$, $\mathbf{H} = \mathbf{H}_0$, $\rho = \rho_0$, and $\mathbf{j} = \mathbf{j}_0$. We have $\varepsilon(t)\mu(t) = 1/c^2$ for all t. The equation of continuity holds for ρ and $\mathbf{j}$, and the net charge is conserved.

Since current systems of units *define* the vacuum permittivity to have a fixed numerical value, the modification embodied in Eqs. (3.28) requires some reexamination of the way in which we define our units of measurement for electromagnetism. We shall return briefly to the idea of time-dependent permittivity and permeability toward the end of the next section.

4. Force Differences Between Like and Unlike Charges

For our final variation on Maxwell's equations, we pursue the suggestion by Lorentz that the electrostatic force of repulsion between like charges is slightly different in magnitude from the electrostatic force of attraction between unlike charges. While Lorentz proposed this idea in 1900 as a way to arrive at a universally attractive (Newtonian) gravitational force, we want to regard it here as just a modification of classical electrodynamics. Let us write a description of the situation in terms of field strengths.

The electrostatic forces among pairs of positively-charged point particles having charges $Q_+ > 0$ and $q_+ > 0$ and/or negatively-charged point particles $Q_- < 0$ and $q_- < 0$, is described by

$$\mathbf{F}_\ell = \frac{1}{4\pi\,\varepsilon_\ell}\,\frac{Q_+ q_+}{r^2}\,\hat{\mathbf{r}} = \frac{1}{4\pi\,\varepsilon_\ell}\,\frac{Q_- q_-}{r^2}\,\hat{\mathbf{r}},$$

$$\mathbf{F}_u = \frac{1}{4\pi\,\varepsilon_u}\,\frac{Q_- q_+}{r^2}\,\hat{\mathbf{r}} \quad \text{or} \quad \mathbf{F}_u = \frac{1}{4\pi\,\varepsilon_u}\,\frac{Q_+ q_-}{r^2}\,\hat{\mathbf{r}}, \tag{4.1}$$

where $\hat{\mathbf{r}}$ is the unit vector at the location of each particle, pointing away from the other, and r is the distance between the pair. The subscripts ℓ and u stand for "like" and "unlike" respectively, and $\varepsilon_\ell \neq \varepsilon_u$. If $|Q_+| = |Q_-|$ and $|q_+| = |q_-|$, then the forces $\mathbf{F}_u$ in the second equation are equal.

Introduce the electric field $\mathbf{E}_+$ exerting force on positive charges, and the electric field $\mathbf{E}_-$ exerting force on negative charges; also the displacement fields $\mathbf{D}_+$ and $\mathbf{D}_-$ produced (respectively) by positive and negative charges. In ordinary electromagnetism, $\mathbf{E} = \mathbf{E}_+ = \mathbf{E}_-$, while $\mathbf{D} = \mathbf{D}_+ + \mathbf{D}_-$. Evidently we must keep track separately of the density $\rho_+ \geqslant 0$ of positive charge and the density $\rho_- \leqslant 0$ of negative charge. From Eqs. (4.1), the force $\mathbf{F}_+$ experienced by the positively charged particle with charge q_+ in the presence of a composite having charges Q_+ and Q_- at a distance r is given by

$$\mathbf{F}_+ = q_+\mathbf{E}_+ = q_+\left(\frac{1}{4\pi\,\varepsilon_\ell}\frac{Q_+}{r^2} + \frac{1}{4\pi\,\varepsilon_u}\frac{Q_-}{r^2}\right)\hat{\mathbf{r}}, \tag{4.2}$$

and similarly for a negatively charge particle,

$$\mathbf{F}_- = q_-\mathbf{E}_- = q_-\left(\frac{1}{4\pi\,\varepsilon_u}\frac{Q_+}{r^2} + \frac{1}{4\pi\,\varepsilon_\ell}\frac{Q_-}{r^2}\right)\hat{\mathbf{r}}. \tag{4.3}$$

There is a net electrostatic force between neutral composites that is attractive if $\varepsilon_u < \varepsilon_\ell$, and repulsive if $\varepsilon_u > \varepsilon_\ell$. The situation thus far is described by the Maxwell equation

$$\nabla\cdot\begin{bmatrix}\mathbf{D}_+\\ \mathbf{D}_-\end{bmatrix} = \begin{bmatrix}\rho_+\\ \rho_-\end{bmatrix}, \tag{4.4}$$

and the constitutive equation

$$\begin{bmatrix}\mathbf{E}_+\\ \mathbf{E}_-\end{bmatrix} = \begin{bmatrix}1/\varepsilon_\ell & 1/\varepsilon_u\\ 1/\varepsilon_u & 1/\varepsilon_\ell\end{bmatrix}\begin{bmatrix}\mathbf{D}_+\\ \mathbf{D}_-\end{bmatrix}. \tag{4.5}$$

Similarly, introduce $\mathbf{B}_+$ and $\mathbf{B}_-$ as the magnetic inductions that exert velocity-dependent forces on positive and negative moving charges q_+ and q_- (respectively); so that the total forces are given (respectively) by

$$F_+ = q_+\mathbf{E}_+ + q_+\mathbf{v}\times\mathbf{B}_+, \quad F_- = q_-\mathbf{E}_- + q_-\mathbf{v}\times\mathbf{B}_-. \tag{4.6}$$

Evidently, we must also keep track separately of the current of positive charge $\mathbf{j}_+$ and the current of negative charge $\mathbf{j}_-$. We introduce $\mathbf{H}_+$ and $\mathbf{H}_-$ as the magnetic fields *produced* (respectively) by electric currents $\mathbf{j}_+$ and $\mathbf{j}_-$, and (respectively) by changing displacement fields $\mathbf{D}_+$ and $\mathbf{D}_-$. Then we obtain, consistent with

Lorentz covariance and the absence of magnetic monopoles, the additional Maxwell equations,

$$\nabla\times\begin{bmatrix}\mathbf{H}_+\\ \mathbf{H}_-\end{bmatrix}=\frac{\partial}{\partial t}\begin{bmatrix}\mathbf{D}_+\\ \mathbf{D}_-\end{bmatrix}+\begin{bmatrix}\mathbf{j}_+\\ \mathbf{j}_-\end{bmatrix},$$

$$\nabla\times\begin{bmatrix}\mathbf{E}_+\\ \mathbf{E}_-\end{bmatrix}=-\frac{\partial}{\partial t}\begin{bmatrix}\mathbf{B}_+\\ \mathbf{B}_-\end{bmatrix},\quad \nabla\cdot\begin{bmatrix}\mathbf{B}_+\\ \mathbf{B}_-\end{bmatrix}=0\,, \tag{4.7}$$

with the constitutive equation

$$\begin{bmatrix}\mathbf{B}_+\\ \mathbf{B}_-\end{bmatrix}=\begin{bmatrix}\mu_\ell & \mu_u\\ \mu_u & \mu_\ell\end{bmatrix}\begin{bmatrix}\mathbf{H}_+\\ \mathbf{H}_-\end{bmatrix}, \tag{4.8}$$

where

$$\varepsilon_\ell\mu_\ell = \varepsilon_u\mu_u = \frac{1}{c^2}\,. \tag{4.9}$$

From Eq. (4.4) and the first of Eqs. (4.7), we have separate continuity equations for ρ_+, $\mathbf{j}_+$ and ρ_-, $\mathbf{j}_-$,

$$\frac{\partial}{\partial t}\begin{bmatrix}\rho_+\\ \rho_-\end{bmatrix}+\nabla\cdot\begin{bmatrix}\mathbf{j}_+\\ \mathbf{j}_-\end{bmatrix}=0\,. \tag{4.10}$$

Thus far we have written in Eqs. (4.4) and (4.7) a *doubled* set of Maxwell equations, indexed by $(+,-)$, that are coupled by the matrix constitutive equations (4.5) and (4.8). To recover the theory as a perturbation of ordinary classical electrodynamics, *define* $\rho=\rho_++\rho_-$ and $\mathbf{D}=\mathbf{D}_++\mathbf{D}_-$; so that ρ is the *net* charge density, and $\nabla\cdot\mathbf{D}=\rho$. Defining $\mathbf{E}=\frac{1}{2}(\mathbf{E}_++\mathbf{E}_-)$, we recover the constitutive equation $\mathbf{E}=(1/\varepsilon_0)\mathbf{D}$ by setting

$$\frac{1}{\varepsilon_0}=\frac{1}{2}\left(\frac{1}{\varepsilon_\ell}+\frac{1}{\varepsilon_u}\right). \tag{4.11}$$

Similarly, defining $\mathbf{j}=\mathbf{j}_++\mathbf{j}_-$ and $\mathbf{H}=\mathbf{H}_++\mathbf{H}_-$, we have the usual Maxwell equation $\nabla\times\mathbf{H}=\partial\mathbf{D}/\partial t+\mathbf{j}$. Letting $\mathbf{B}=\frac{1}{2}(\mathbf{B}_++\mathbf{B}_-)$, we also have the Maxwell equations $\nabla\times\mathbf{E}=-\partial\mathbf{B}/\partial t$ and $\nabla\cdot\mathbf{B}=0$, and we recover the constitutive equation $\mathbf{B}=\mu_0\mathbf{H}$ by setting

$$\mu_0=\frac{1}{2}(\mu_\ell+\mu_u)\,. \tag{4.12}$$

Then it follows from Eqs. (4.9), (4.11), and (4.12) that $\varepsilon_0\mu_0=1/c^2$. In short, we still have the four fields $\mathbf{E}$, $\mathbf{B}$, $\mathbf{D}$, and $\mathbf{H}$, and they still satisfy the usual Maxwell equations incorporating the net charge density and net electric current density, with coefficients ε_0 and $\mu_0 = 1/\varepsilon_0 c^2$.

But we also have an *additional* set of fields, density, and current,

$$\widetilde{\mathbf{D}} = \mathbf{D}_+ - \mathbf{D}_- , \quad \widetilde{\mathbf{E}} = \frac{1}{2}(\mathbf{E}_+ - \mathbf{E}_-),$$
$$\widetilde{\mathbf{H}} = \mathbf{H}_+ - \mathbf{H}_- , \quad \widetilde{\mathbf{B}} = \frac{1}{2}(\mathbf{B}_+ - \mathbf{B}_-), \qquad .(4.13)$$
$$\widetilde{\rho} = \rho_+ - \rho_- , \quad \widetilde{\mathbf{j}} = \mathbf{j}_+ - \mathbf{j}_- .$$

These *also* obey Maxwell's equations; but with new constants in their constitutive equations, $\widetilde{\varepsilon}$ and $\widetilde{\mu}$, that are given by

$$\frac{1}{\widetilde{\varepsilon}} = \frac{1}{2}\left(\frac{1}{\varepsilon_\ell} - \frac{1}{\varepsilon_u}\right), \quad \widetilde{\mu} = \frac{1}{2}(\mu_\ell - \mu_u). \qquad (4.14)$$

In ordinary electromagnetism, $\widetilde{\varepsilon}$ is infinite and $\widetilde{\mu}$ is zero—so that, although $\widetilde{\rho}$, $\widetilde{\mathbf{j}}$, $\widetilde{\mathbf{D}}$ and $\widetilde{\mathbf{H}}$ are defined and nontrivial, $\widetilde{\mathbf{E}} \equiv 0$ and $\widetilde{\mathbf{B}} \equiv 0$.

If it is not actually infinite, the magnitude of $\widetilde{\varepsilon}$ is presumably very large compared with that of ε_0; while if it is not zero, the magnitude of $\widetilde{\mu}$ is small compared with that of μ_0. One verifies straightforwardly that $\widetilde{\epsilon}\,\widetilde{\mu} = 1/c^2$; but the new constants may be of either sign—both positive (for a net attractive force between neutral composites), or both negative (for a net repulsive force). We thus have obtained a pair of fully decoupled Maxwell systems, consistent with the equations set down in Lorentz' original article.

An idealized composite point particle, having positive charge $q_+ \geqslant 0$ and negative charge $q_- \leqslant 0$, may be equivalently described as having *net* charge $q = q_+ + q_-$ and *absolute* charge $\widetilde{q} = q_+ - q_- \geqslant 0$. Such a particle, moving with velocity $\mathbf{v}$, experiences according to Eqs. (4.6) the total force

$$\mathbf{F} = \mathbf{F}_+ + \mathbf{F}_- = q(\mathbf{E} + \mathbf{v} \times \mathbf{B}) + \widetilde{q}(\widetilde{\mathbf{E}} + \mathbf{v} \times \widetilde{\mathbf{B}}). \qquad (4.15)$$

Thus the new fields $\widetilde{\mathbf{E}}$, $\widetilde{\mathbf{B}}$, $\widetilde{\mathbf{D}}$, and $\widetilde{\mathbf{H}}$ couple to the *absolute charge* (which is always positive), and the *absolute current;* while the usual fields $\mathbf{E}$, $\mathbf{B}$, $\mathbf{D}$, and $\mathbf{H}$ still exist in this framework and couple to the *net charge* and the *net current.* One should no longer automatically take positive charge flowing to the right to be indistinguishable mathematically or physically from the same amount of negative charge flowing to the left. The net currents $\mathbf{j}$ are the same in these two cases, but the absolute currents $\widetilde{\mathbf{j}}$ are equal and opposite.

Let us close this section with some comments and speculations, many of them rather obvious, about such a "doubled electromagnetism" theory.

In his original article [18], Lorentz took the new force to be attractive. He sought to identify the absolute charge with mass (and consequently the absolute current with momentum), and to calculate whether the precession of the perihelion of Mercury's orbit could then be understood as due to the (very small) magnetic force that would originate from the absolute current. He concluded that the resulting force would be too weak to explain the astronomical observations, and of course this line of thinking was superseded by the success of Einstein's general relativity.

However, we want to entertain the idea of a modified electrodynamics that does not identify the extra fields with gravity, and that treats absolute charge not as mass but as an additional property of matter. Of course, this does not preclude the possibility that existing measurements of gravitational forces have erroneously incorporated a small extra electrostatic force (attractive or repulsive). It would seem to be an especially interesting conjecture that the new force is repulsive. In any case, we appear to have an additional parameter with which to fit cosmological models, and an additional "test theory" for study through observations in astrophysics.

Since we have a new set of fields obeying Maxwell's equations, we would also need to have a new type of electromagnetic wave (coupling weakly with the absolute charge), a new type of photon, and a new quantum electrodynamics (see below).

Now the absolute electric charge of a system must be at least equal to the net electric charge, and at least equal to the sum of the absolute charges of the system's components. However, it could in principle be greater. The absolute charge of a nucleon, for example, might be the sum of the absolute values of the charges of its constituent quarks; but one could also conjecture additional, unobserved positive and negative charges in equal measure, contributing to a larger value of the overall absolute charge. While net charge is quantized in fixed units, it is plausible but not necessary that absolute charge be similarly quantized. Thus, it does not appear to be inconsistent to take the absolute charge to be proportional to the mass, as Lorentz implicitly did.

However, modern particle physics offers no fundamental theoretical reason to make such an assumption. If we make reference only to constituent quarks and leptons, the absolute charge of a proton (comprised of two up quarks and one down quark) is $5/3$; that of a neutron (comprised of one up quark and two down quarks) is $4/3$; and that of an electron (taken to be fundamental) is 1. Then the absolute charge of a proton together with an electron is $8/3$, double that of the neutron, while the respective masses are very close to equal. Under these assumptions, the absolute charge per gram of electrically neutral matter comprised of heavier elements is macroscopically different from that of matter comprised of lighter elements, and their accelerations under the Earth's absolute electric field would be different in magnitude. Since such differences are not observed, we should take the empirically-determined magnitude of any new inverse-square-law force of "absolute electromagnetism" to be small compared with Newtonian gravity—making it *extremely* small in comparison with ordinary electromagnetism.

Still, one may conjecture that physical "constants" are not actually constant, but change as the universe ages. We discussed in Sec. 3 the possibility of the vacuum permittivity and permeability changing exponentially with t, producing a modification in Maxwell's equations and a universal frictional force that breaks covariance. One may instead take these to be fixed at ε_0 and μ_0, but entertain the possibility that ε_u and ε_ℓ are changing, and that they have not always been as close as they are today. For example, we could have

$$\tilde{\varepsilon}(t) = \tilde{\varepsilon}_0\, e^{+\beta t}\,, \quad \tilde{\mu}(t) = \tilde{\mu}_0\, e^{-\beta t}\,, \tag{4.16}$$

and modify the Maxwell equations for the perturbing fields $\widetilde{\mathbf{E}}$, $\widetilde{\mathbf{B}}$, $\widetilde{\mathbf{D}}$, and $\widetilde{\mathbf{H}}$ as in Sec. 3.

Furthermore, should there be regions of space-time containing plasmas of electrons and positrons, these might contribute proportionally more to absolute charge than to gravitational mass. Such speculations leave open some possibilities for observable effects in astrophysics, even for a small force.

While net charge and absolute charge are both conserved when there is no particle creation or annihilation, it seems clear that absolute charge is *not* conserved by fundamental particle processes. [Of course, at the time of Lorentz' paper, mass, positive charge, and negative charge would all have been taken as separately conserved.] As long as we stay with Lorentz' idea that absolute charge is proportional to mass, then (as mass is transformed into energy during fundamental particle processes) we have a conservation law. But if we take absolute charge to be an independent quantity with which the new electromagnetic fields couple, it becomes an unlikely, speculative possibility that absolute charge transforms into something previously unknown during annihilation processes, so as to maintain a conservation law. More likely, one should take the new $U(1)$ gauge symmetry to be broken outside the classical domain that is governed by Maxwell's equations, requiring a different quantum electrodynamics for absolute electromagnetism.

We have not discussed the question of absolute charge from virtual particle-antiparticle pairs, or vacuum polarization. It appears that, unlike the situation for net charge, it should be possible to have a coherent superposition of quantum states having different absolute charges. For example, the neutral pion, written as a linear combination of up and down quark-antiquark pairs, $(\bar{u}u - \bar{d}d)/\sqrt{2}$, would combine states of absolute charges $4/3$ and $2/3$ (in units of the electron's charge).

If Lorentz' conjecture breaks no known physical principle, then the question of a discrepancy in magnitude between the electrostatic forces between like and unlike charges is purely an empirical one, and the best we can do *in principle* is to establish an experimental upper bound to this discrepancy (or, equivalently, to the ratio $\varepsilon_0/\tilde{\varepsilon}$). For example, it is clear that his conjecture continues to respect the charge conjugation invariance of electromagnetism. However, it should be noted that current physics tends to *assume* the equality in magnitude between these forces. Thus the present, official definition of the coulomb is effectively as a unit of *net* charge, while the value of ε_0 is not measured but fixed by definition. Just as we have come to distinguish (theoretically) gravitational mass from inertial mass, and to regard their proportionality as a question to be determined by experiment, so may it be necessary to distinguish (theoretically) ε_ℓ from ε_u, and to regard their closeness as having a value to be bounded by experiment.

Possible further development of a "doubled electromagnetism" theory could entail its involvement in electroweak unification or in the standard model, its nonlinear modifications (as discussed for Maxwell's equations in Sec 2 of this article), its coupling with linear and nonlinear Schrödinger quantum mechanics (as discussed for Maxwell fields in Sec. 3 of this article), its non-Abelian generalizations, and its quantum electrodynamics.

Acknowledgments

One of the authors (G. Goldin) thanks H.-D. Doebner, J. Lucido, V. Shtelen, and G. Svetlichny for interesting discussions on topics included in this article.

References

[1] J. D. Jackson, *Classical Electrodynamics* (3rd ed.), New York: John Wiley & Sons (1999).

[2] M. Born & L. Infeld, *Proc. R. Soc. London* **144** (1934), 425.

[3] C. N. Yang and R. L. Mills, *Phys. Rev.* **95** (1954), 631.

[4] R. Roskies, Phys. Rev. D, **14**, N6, (1977), 1722.

[5] A. A. Tseytlin, *Nuclear Phys. B* **501** (1997), 41.

[6] J.-H. Park, A study of a non-Abelian generalization of the Born-Infeld action. *Phys. Lett. B* **458** (1999), 471.

[7] D. Gal'tsov and R. Kerner, Classical glueballs in non-Abelian Born-Infeld theory. *Phys. Rev. Lett.* **84** (2000), 5955.

[8] E. Serié, T. Masson, and R. Kerner, Non-Abelian generalization of Born-Infeld theory inspired by non-commutative geometry. *arXiv:hep-th/030710* **v2** (1 Sep 2003)

[9] G. Amelino-Camelia, C. Lämmerzahl, A. Macias, and H. Müller, The search for quantum gravity signals. arXiv: gr-qc/0501053 (2005).

[10] C. Laemmerzahl, A. Macias, and H. Mueller, Lorentz invariance violation and charge (non-)conservation: A general theoretical frame for extensions of the Maxwell equations. arXiv: gr-qc/0501048 (2005). To appear in *Phys. Rev. D* **71** (2005) 025007.

[11] G. A. Goldin & V. M. Shtelen, On Galilean invariance and nonlinearity in electrodynamics and quantum mechanics. *Phys. Lett. A* **279** (2001), 321.

[12] G. A. Goldin & V. M. Shtelen, Generalizations of Yang-Mills theory with nonlinear constitutive equations. *J. Phys. A.: Math. Gen.* **37** (2004), 10711.

[13] G. A. Goldin, "Perspectives on Nonlinearity in Quantum Mechanics," in H.-D. Doebner, J.-D. Hennig, W. Lücke & V. K. Dobrev, *Quantum Theory and Symmetries: Procs. of the Int'l Symposium, Goslar, Germany, 18–22 July 1999,* Singapore: World Scientific (2000), pp. 111–123.

[14] H.-D. Doebner & G. A. Goldin, *Phys. Lett. A* **162** (1992), 397.

[15] H.-D. Doebner & G. A. Goldin, Introducing nonlinear gauge transformations in a family of nonlinear Schrödinger equations. *Phys. Rev. A* **54** (1996), 3764.

[16] H.-D. Doebner, G. A. Goldin & P. Nattermann, *J. Math. Phys.* **40** (1999), 49.

[17] M. D. Kostin, *J. Chem. Phys.* **57** (1972), 3589.

[18] H. A. Lorentz, "Considerations on Gravitation", in *Koninklijke Akademie van Wetenschappen te Amsterdam,* Proceedings of the Section of Sciences, vol. II (Amsterdam, Johannes Müller, July 1900), pp. 559–574.

[19] W. I. Fushchich, V. M. Shtelen, and N. I. Serov, *Symmetry Analysis and Exact Solutions of Equations of Nonlinear Mathematical Physics,* Kluwer Acad. Publ., Dordrecht (1993).

[20] M. Le Bellac and J.-M. Levy-Leblond, Galilean electromagnetism, Nuovo Cim. **14B** (1973) 217.

[21] R. P. Feynman and A. R. Hibbs, *Quantum Mechanics and Path Integrals*, New York: McGraw-Hill (1965).

[22] B. Mielnik, Commun. Math. Phys. **37**, 221 (1974).

[23] G. A. Goldin and G. Svetlichny, J. Math. Phys. **35**, 3322 (1994).

[24] I. Bialynicki-Birula and J. Mycielski, Ann. Phys. **100**, 62 (1976).

[25] R. Haag and U. Bannier, Commun. Math. Phys. **60**, 1 (1978).

[26] G. A. Goldin, Nonlinear Math. Phys. **4**, 6 (1997).

Krasnow Institute for Advanced Study, George Mason University, Fairfax, VA 22030, USA *E-mail:* ascoli@gmu.edu

Departments of Mathematics and Physics, Rutgers University, New Brunswick, NJ 08903, USA *E-mail:* geraldgoldin@dimacs.rutgers.edu

Physical Applications of Algebras of Unbounded Operators

F. Bagarello

After a brief historical introduction on the standard algebraic approach to quantum mechanics of large systems, QM_∞, we review some basic mathematical aspects and few physical applications of algebras of unbounded operators.

This paper is dedicated to
Gerard Emch for his 70th birthday

1. Introduction

During the past 20 years a long series of papers concerning algebras of unbounded operators appeared in the literature, papers which, though being originally motivated by physical arguments, contain almost no physics at all. On the contrary the mathematical aspects of these algebras have been analyzed in many details and this analysis produced, up to now, the monographes [32] and [2]. Some physics appeared first in [28] and [31], in the attempt to describe systems with a *very large* (10^{24}) number of degrees of freedom, following some general ideas originally proposed in the famous Haag and Kastler's paper, [27], on QM_∞.

More recently other physical applications of algebras of unbounded operators have been proposed, see [4, 5, 7, 8, 10, 11, 12, 13] for instance.

Here we review some basic mathematical results as well as some physical applications. In particular, in Section II we give an introduction to non relativistic quantum mechanics for systems with infinite degrees of freedom, focusing on some physically relevant results and on open problems. In Section III we introduce some mathematical definitions and results concerning algebras of unbounded operators, while their physical applications are given in Section IV.

2. A Short Review of Non Relativistic QM_∞

It is a well known fact that for systems with infinite degrees of freedom the uniqueness von Neumann theorem does not hold, in the sense that the same physical system may have several inequivalent representations. For this reason a purely algebraic or an Hilbert space descriptions of such a system need not to be equivalent, as it happens for finite systems.

This kind of physical systems can be properly discussed within the framework of C*-algebras, as first proposed by Haag and Kastler, [27], whose construction goes as follows.

The algebra. Let Σ be a physical system with infinite degrees of freedom, $V \subset \mathbb{R}^d$ a finite d-dimensional region, $\mathcal{H}_V$ the related Hilbert space (whose construction depends in general on Σ, [33]), $\mathfrak{A}_V = B(\mathcal{H}_V)$ the associated C*-algebra of bounded operators acting on $\mathcal{H}_V$ and let finally H_V be the self-adjoint energy operator for Σ_V, the restriction of Σ in V.

The family of algebras $\{\mathfrak{A}_V\}$ satisfies the following properties:

- *isotony*: if $V_1 \subset V_2$ then $\mathfrak{A}_{V_1} \subset \mathfrak{A}_{V_2}$. Moreover $\|.\|_2 \lfloor_{V_1} = \|.\|_1$ $(\Rightarrow \mathfrak{A}_{V_1}, \mathfrak{A}_{V_2} \subset \mathfrak{A}_{V_1 \cup V_2})$;
- if $V_1 \cap V_2 = \emptyset$ then $[\mathfrak{A}_{V_1}, \mathfrak{A}_{V_2}] = 0$.

In particular this last property clearly shows the non relativistic regime we are working with here, since it simply means that two operators localized in disjoint spatial regions are necessarily independent, i.e. they must commute. Then we define $\mathfrak{A} = \overline{\mathfrak{A}_0}^{\|.\|}$, where $\mathfrak{A}_0 = \cup_V \mathfrak{A}_V$ and $\mathfrak{A}$ is the *quasi-local C*-algebra of the* **bounded** *observables.*

On this algebra we can introduce the *spatial translations* $\{\gamma_x\}$, which is a group of *-automorphisms of $\mathfrak{A}$ satisfying the following: $\gamma_x \mathfrak{A}_V = \mathfrak{A}_{V+x}$, $\gamma_{x_1}\gamma_{x_2} = \gamma_{x_1+x_2}$.

The states. The states of Σ are positive, normalized linear functionals on $\mathfrak{A}$ which, when restricted to V, reduces to the states over the *finite* system Σ_V and, therefore, over the finite volume algebra $\mathfrak{A}_V$. In other words, they corresponds to a family of density matrices ρ_V: $\hat{\rho}(A) = tr_V(\rho_V A)$ for each $A \in \mathfrak{A}_V$. Here tr_V is the trace in $\mathcal{H}_V$. These states satisfy the following *consistency condition*: $tr_V(\rho_V A) = tr_{V'}(\rho_{V'} A)$ $\forall A \in \mathfrak{A}_V$, $V \subset V'$.

They have a physical interpretation which is given by the Ruelle, Dell'Antonio and Doplicher's theorem: these states have zero probability to describe an infinite number of particles in a finite region. They are usually called in the literature *locally finite states.*

Among all the states a particular role is played by the so called *pure states*: ρ is *pure* if it is not a convex combination of other states, i.e. if there are no ρ_1, ρ_2 and $\lambda \in]0, 1[$ such that $\rho = \lambda\rho_1 + (1-\lambda)\rho_2$. Their relevance is due to the fact that, how we will discuss in the following, they are related to the pure thermodynamical phases of a certain physical system.

The dynamics. The next step in our analysis is related to the description of the *time evolution* of the physical system Σ. This is obtained from the dynamics of Σ_V in the Heisemberg representation as follows:

first we define the time evolution of the element $A \in \mathfrak{A}_V$ in the volume V as follows: $\mathfrak{A}_V \ni A \to \alpha_V^t(A) := e^{iH_V t/\hbar} A e^{-iH_V t/\hbar}$.

secondly we use $\alpha_V^t(A)$ to define $\alpha^t(A)$ as follow $\alpha^t(A) = \tau - \lim_V \alpha_V^t(A)$, where τ is a *reasonable* topology of $\mathfrak{A}$, i.e. a topology usually related to Σ itself. Possible topologies are the following:

for short range interactions and discrete systems τ is usually the *uniform* topology, [26];

for long range interactions it is known that α_V^t is not $\|.\|-$converging: a possible alternative for τ is the *strong* topology (restricted to a *relevant family of states*). This different topology has been used by many authors, e.g. [6, 29, 37] and references therein. In this case a state ρ must be chosen in such a way that

$$\rho(\alpha_V^t(A)) \to \rho(\alpha^t(A)) =: \rho_t(A),$$

and this limit defines the time evolution of the state ρ, ρ_t, by means of $\rho_t(A) := \rho(\alpha^t(A))$. It is clear that the existence of (sufficiently many) such ρ's has to be checked in each model.

The symmetry. It is possible to introduce the concept of symmetry related to Σ: an automorphism of $\mathfrak{A}$, γ, is a *symmetry* of the system Σ if $\alpha^t(\gamma(A)) = \gamma(\alpha^t(A))$ and is a *local symmetry* if $\gamma : \mathfrak{A}_V \to \mathfrak{A}_V$ and if $\gamma(H_V) = H_V$.

Moreover, the automorphism γ is a *symmetry of the state* ρ if $\rho_\gamma(A) := \rho(\gamma(A)) = \rho(A), \forall A \in \mathfrak{A}$.

Representations and GNS-construction. A crucial notion, also in view of its physical applications, is that of a **-representation* of a *-algebra. This is essentially a map $\pi : \mathfrak{A} \to B(\mathcal{H})$, for a certain $\mathcal{H}$, which preserves the algebraic structure of $\mathfrak{A}$:

$$\pi(A+B) = \pi(A) + \pi(B), \quad \pi(\lambda A) = \lambda\, \pi(A),$$

$$\pi(AB) = \pi(A)\pi(B), \qquad \pi(A^*) = \pi(A)^*.$$

It is clear that $\pi(\mathfrak{A})$ is a *-algebra as well.

It is a well known fact that any state ρ over an abstract C*-algebra $\mathfrak{A}$ produces a unique (but for equivalence) triplet $(\mathcal{H}_\rho, \pi_\rho, \Omega_\rho)$, where $\mathcal{H}_\rho$ is an Hilbert space, π_ρ is a *-representation, in the sense just discussed, and Ω_ρ is a cyclic vector of $\mathcal{H}_\rho$, i.e. $\pi_\rho(\mathfrak{A})\Omega_\rho$ is dense in $\mathcal{H}_\rho$. Moreover we have, $\forall A \in \mathfrak{A}$,

$$\rho(A) = <\Omega_\rho, \pi_\rho(A)\Omega_\rho> .$$

Also, π_ρ is irreducible if and only if ρ is pure.

Remark. (1) The first obvious remark is that GNS representations generated by different states need not be unitarily equivalent!

(2) Each (GNS) representation corresponds to a *phase* of the physical system. In particular, GNS representations generated by pure states correspond to pure phases [30].

(3) States which are only *locally different* are *macroscopically indistinguishable*: all the macroscopic observables have the same expectation values. This has a clear physical interpretation: equal values of the macroscopic observables (the so-called *order parameters*) label unitarely equivalent representations, which are interpreted as the same phase of the matter. In other words: two different phases of the matter correspond to two representations in which some macroscopic observable assumes different values.

(4) An interesting result is the following: even if the algebraic dynamics for Σ cannot be given in an hamiltonian form, nevertheless, under certain assumption on Σ, the dynamics in each representation π_ρ is *hamiltonian*: there exists a s.a. operator $\hat{H}_\rho$ such that, $\forall A \in \mathfrak{A}$,

$$\frac{d}{dt}\alpha_\rho^t(\pi_\rho(A)) = i[\hat{H}_\rho, \alpha_\rho^t(\pi_\rho(A))],$$

see [33] and references therein. $\hat{H}_\rho$ is what is often called in literature *the effective hamiltonian.*

Also, this result has a clear physical interpretation: different phases of Σ may have, and they usually have, different dynamical behaviors, and this is reflected in the different possible expressions for H_ρ.

Continuing our brief review of standard results in QM_∞, we discuss now some results on *equilibrium states* and on the associated *phase structure* for quantum systems with infinite degrees of freedom.

It is well known that, for finite systems, there exists a single thermodynamical phase and an unique equilibrium state corresponding to some fixed thermodynamical conditions. The situation is completely different for infinite systems. For these systems the role of the thermodynamical limit is crucial, and not only for the existence of the *time evolution*, as we will discuss in a moment.

In order to keep the analysis simple, it is convenient to make the following assumptions on the finite volume hamiltonian H_V:

(1) first we require that H_V is such that $H_{V_1 \cup V_2} - H_{V_1} - H_{V_2}$ is a surface effect. This condition holds, for instance, for short range forces;

(2) there exists $c > 0$ such that $\|H_V\| \leqslant c|V|$, where $|V|$ is the volume of V or, for discrete systems, the number of lattice sites inside V.

These assumptions imply, first of all, that $\alpha^t(A) = \| \, \| - \lim_{V \nearrow} \alpha_V^t(A)$.

Now, let us define the following functionals:

$$\begin{cases} E_V(\rho_V) = tr_V(\rho_V H_V), \\ S_V(\rho_V) = -k \, tr_V(\rho_V \log(\rho_V)), \\ F_V(\rho_V) = E_V(\rho_V) - T S_V(\rho_V). \end{cases}$$

These are called the local *energy*, the *entropy* and the *free energy* functionals.

In [33] it is proven, because of the assumptions on H_V and of the subadditivity of the entropy, that the following *global density functionals*

$$e(\rho) = \lim_{V \nearrow} \frac{E_V(\rho_V)}{|V|}, \quad s(\rho) = \lim_{V \nearrow} \frac{S_V(\rho_V)}{|V|}, \quad f(\rho) = \lim_{V \nearrow} \frac{F_V(\rho_V)}{|V|}$$

exist, as well as the following *incremental functionals*

$$\begin{cases} \Delta E(\rho|\rho') = \lim_{V \nearrow} \left(E_V(\rho'_V) - E_V(\rho_V)\right), \\ \Delta S(\rho|\rho') = \lim_{V \nearrow} \left(S_V(\rho'_V) - S_V(\rho_V)\right), \\ \Delta F(\rho|\rho') = \lim_{V \nearrow} \left(F_V(\rho'_V) - F_V(\rho_V)\right), \end{cases}$$

where ρ' is a *local* modification of ρ, i.e. a state which differs from ρ only on a volume of finite size and ρ_V is the restriction of ρ to V.

The role of these functionals is crucial in the analysis of the equilibrium states. A state $\tilde{\rho}$ is *globally thermodynamically stable* (GTS) if it is invariant under translations and if it minimizes $f(\rho)$. It is *locally thermodynamically stable* (LTS) if $\Delta F(\tilde{\rho}|\rho') \geqslant 0$ for all ρ', local modification of $\tilde{\rho}$. Then, [33], it is proved that a GTS state is a LTS state, while an LTS state which is invariant under translations is also a GTS state for systems with short range interactions.

Again, this result has a physical interpretation, which can be deduced also from the explicit solution of some easy physical models: a GTS state is an equilibrium state. The LTS states are, for systems with long range interactions, only *metastable* states (i.e. states with a long mean life and *good* thermodynamical properties). In [33] it is also proven that the LTS and the KMS states are exactly the same objects. This implies, of course, that a KMS state for an infinite system is not an equilibrium, in general, but only a metastable state. We recall here that ρ is a β-KMS state if for each $A, B \in \mathfrak{A}$ there exists a complex function $F_{AB}(z)$ which is analytical in the strip $\Im(z) \in [0, \hbar\beta]$, continuous on the boundaries, and is such that

$$F_{AB}(t) = \rho(BA_t), \qquad F_{AB}(t + i\hbar\beta) = \rho(A_t B)$$

What makes the difference between finite and infinite systems is that, while the potential $\hat{F}_V(\rho)$ is convex, and therefore admits an unique minimum, the free energy density functional $f(\rho)$ is affine, so that more than a single minimum may be achieved.

If this is the case, Σ admits different thermodynamical phases under the same thermodynamical conditions, each corresponding to a different GTS state. We say that the system possesses *macroscopic degeneracy*: these different equilibrium states (and the related physical phases) are labeled by the (different) values of some *macroscopic observables* (like the magnetization in the case of ferromagnetic materials), [33].

This fact has several related consequences, the first being that we can use this approach to discuss in a natural settings the so called *spontaneous breaking of a symmetry*:

> suppose that Σ has a local symmetry γ and let $\Delta = \{\rho \in \mathfrak{A}' : \rho \text{ is GTS}\}$, be the set of GTS states. Then, since necessarily we have $f(\rho) = f(\rho_\gamma)$, it follows that for any $\rho \in \Delta$ also $\rho_\gamma \in \Delta$: the symmetry γ maps Δ into itself.

Therefore we see that, if Δ consists of a single element ρ_1, γ is necessarily a symmetry of ρ_1: *the symmetry is unbroken*. In other words, in this case it is clear that $\rho_1 = (\rho_1)_\gamma$.

If, on the contrary, $\Delta = \{\rho_1, \ldots, \rho_n\}$, then, in general, we can only say that $(\rho_i)_\gamma = \rho_j$, for some i and j not necessarily coincident: if this is the case, then *the symmetry is spontaneously broken*.

Of course, whenever a system exhibits a spontaneously symmetry breaking, a related result on the spectrum of the theory can be deduced by making use of the *non relativistic Goldstone's theorem* which has proved to be a very important tool both in condensed matter and in quantum field theory, [35].

We want to end this list of results related to our algebraic approach by mentioning a very interesting relation between KMS states and the Tomita-Takesaki theory:

> let ρ be a -1-KMS state (i.e. a KMS state corresponding to $\beta = -1$. This is what it is usually called simply a KMS state). This state generates a GNS representation $(\mathcal{H}_\rho, \pi_\rho, \Omega_\rho)$. Let $U_\rho(t)$ the unitary operator which implements α^t in this representation. Then Ω_ρ is cyclic and, since ρ is KMS, is also separating for $\pi_\rho(\mathfrak{A})''$, i.e. $\pi_\rho(X)\Omega_\rho = 0$ implies that $\pi_\rho(X) = 0$, [17].

Then we are in the assumptions of Tomita-Takesaki's construction, so that we can introduce a *modular conjugation* J_ρ and a *modular operator* Δ_ρ associated to $(\pi_\rho(\mathfrak{A})'', \Omega_\rho)$. Calling H_ρ the generator of $U_\rho(t)$, we find that

$$\Delta_\rho = e^{H_\rho}.$$

In other words: given a system Σ an *effective hamiltonian* surely exists in any representation GNS-constructed by a given KMS-state.

It may be worth stressing that these are only few results which can be obtained within an algebraic frameworks. More results on phase transitions, applications to quantum field theory, statistical mechanics etc. can be found in many specialized textbooks, among which we only cite [36, 17, 18, 33, 34, 22].

We devote the second part of this section to a brief discussion of some limits which are, in our opinion, intrinsic with the approach discussed so far, and which suggest the construction of a slightly generalized algebraic framework. These conclusions are based on a simple remark: the main results which have been given in this section are obtained under some requirements, which may not be necessarily satisfied in many relevant conditions. For instance, we have assumed that *the norm of the local hamiltonian H_V does not grow faster than $|V|$: $\|H_V\| \leqslant c|V|$.*

However, this is not always true: actually, it is false quite often! For instance, this inequality is violated already by a gas of free bosons, for which $H_V = \sum_{j\in V} a_j^\dagger a_j$, since each creation and annihilation operator is such that $\|a_j\| = \|a_j^\dagger\| = \infty$.

The second assumption considered above is that *the interactions are short ranged.*

However this is not the case in many situations. For instance, the Coulomb interaction is long ranged, while in the mean field models the *real forces* are replaced by interactions with an infinite range: this means that, given two particles localized for instance in the two lattice sites i and j, they feel the same strength independently of the difference $|i-j|$. However, many results can be obtained even under these conditions. In particular we find that:

- α_V^t is not norm convergent to an algebraic dynamics α^t, but, as we have already sketched before, we can (often) find a different topology which makes of $\alpha_V^t(A)$ a Cauchy sequence for each (or for many) $A \in \mathfrak{A}$, see [6, 37, 21] just to cite few authors.

- We have already mentioned that, in general, KMS states are not equilibrium states (i.e. GTS states). Moreover, they are not even limits of Gibbs state, [25], in general.

- surface effects become volume effects, so that *variables at infinity* (i.e. completely delocalized operators) appear in the dynamics of strictly localized operators. These are related to the *order parameters* used to describe different phases, see [29] and references therein;

- the density functionals $e(\rho)$ and $f(\rho)$ do not necessarily exist, since in the proof of their existence, the assumption that the forces are short ranged is crucial, see [33];
- the Goldstone's theorem holds only in a modified form [29].

We see that a new world appears whenever the range of the interactions appearing in the physical system modify their range. We also observe that many things can be said but many other aspects are still to be clarified.

A third assumption which is usually somehow hidden in the C*-algebraic approach to QM_∞ is related again to the presence on the (almost) unavoidable unbounded operators. Consider, for instance, the position and momentum operators $\hat{q}$ and $\hat{p}$. As we have already mentioned, they satisfy the following commutation relations: $[\hat{q}, \hat{p}] = i\,\mathbb{I}$ (in convenient units) and, as a consequence, it is an easy exercise to check that at least one of them must be unbounded. Actually, it is well known that they are both unbounded on $\mathcal{L}^2(\mathbb{R})$. In the literature three main possible ways to deal with unbounded operators have been proposed: the first one consists in restricting the action of the operators on some (possibly dense) subset of a given Hilbert space, a sort of *common domain* of all the operators. A second possibility is to exponentiate these unbounded and self-adjoint operators in order to define unitary (and therefore bounded) operators. The original operators can be recovered by taking suitable derivatives of the unitary maps on certain relevant sets of vectors. A third possibility is the following: we could replace, say, the self-adjoint operator $\hat{p}$ with a bounded operator $\hat{p}_N$ whose spectrum coincides with the one of $\hat{p}$ inside a compact interval $[-N, N]$, and is zero outside this set. It is clear that $\hat{p}_N$ is bounded and, as $N \to \infty$, approaches $\hat{p}$ *in some sense*. Then one considers only those states ω_N on $\mathfrak{A}$, N-depending as well, such that $\omega_N(\hat{p}_N)$ converges in the limit $N \to \infty$ to some finite quantity, which is interpreted as the mean value of the original operator $\hat{p}$ on a state which can be seen as the *limit* of the family of states ω_N. An example of this procedure can be found in [1].

However, quite often none of the above procedures works as one would like. As an example, we cite the Lindblad expression for the generator L of a completely positive semigroup (describing the time evolution of a quantum open system). These structures play a very important role in the analysis of order-disorder transitions out of equilibrium, [33, 34].

More in details, let $\mathfrak{A}$ and $\mathcal{B}$ be C*-algebras. We recall that a map $f : \mathfrak{A} \to \mathcal{B}$ is *positive* if $f(A) > 0$ for each $A > 0$. It is *completely positive*, CP, if, for any finite matrix algebra $\mathcal{M}$, the mapping $f \otimes \mathbb{I} : \mathfrak{A} \otimes \mathcal{M} \to \mathcal{B} \otimes \mathcal{M}$ is positive.

A *quantum dynamical semi-group* is a set $\{T_t : t \geqslant 0\}$ of completely positive, identity preserving maps of $\mathfrak{A}$ such that $T_s T_t = T_{s+t}$ for all $s, t \geqslant 0$ and $T_0 = 1\!\!1$. If T_t is normwise continuous in t for all $A \in \mathfrak{A}$, then there exists an infinitesimal generator L defined by the formula

$$\frac{d}{dt} T_t A = L T_t A = T_t L A, \qquad \forall A \in \mathfrak{A}$$

Lindblad proved that, if $\mathfrak{A} = B(\mathcal{H})$ (for some $\mathcal{H}$), L has necessarily the following expression:

$$LA = i[H, A] + \sum_j \left(V_j^* A V_j - \frac{1}{2}\{V_j^* V_j, A\} \right),$$

where H is self-adjoint and $V_j, \sum_j V_j^* V_j \in \mathfrak{A}$.

Very few results exist for unbounded operators, [24, 20, 14], mainly because of the following technical difficulty: if T_t is a semigroup (and not a group) it follows that $T_t(AB) \neq T_t(A)T_t(B)$.

To avoid this kind of difficulties, in many physical applications involving open systems the boson reservoir is replaced by a fermionic one, as for instance, in [19] for the open BCS-model, changing a realistic into a, somehow, toy model. This suggests that an alternative procedure should be considered, and this will be the output of the next section.

3. Algebras of Unbounded Operators

In this section we will briefly introduce different examples of what we generically call *algebras of unbounded operators*, giving only those mathematical results and definitions which are relevant for our purposes. A much deeper analysis of these aspects can be found, for instance, in [2] or in [32].

Let $\mathfrak{A}$ be a linear space, $\mathfrak{A}_0 \subset \mathfrak{A}$ a $*$-algebra with unit $\mathbb{1}$ (otherwise we can always add it): $\mathfrak{A}$ is a *quasi $*$-algebra over* $\mathfrak{A}_0$ if

[i] the right and left multiplications of an element of $\mathfrak{A}$ and an element of $\mathfrak{A}_0$ are always defined and linear;

[ii] $x_1(x_2a) = (x_1x_2)a$, $(ax_1)x_2 = a(x_1x_2)$ and $x_1(ax_2) = (x_1a)x_2$, for each $x_1, x_2 \in \mathfrak{A}_0$ and $a \in \mathfrak{A}$;

[iii] an involution $*$ (which extends the involution of $\mathfrak{A}_0$) is defined in $\mathfrak{A}$ with the property $(ab)^* = b^*a^*$ whenever the multiplication is defined.

A quasi $*$-algebra $(\mathfrak{A}, \mathfrak{A}_0)$ is *locally convex* (or *topological*) if in $\mathfrak{A}$ a locally convex topology τ is defined such that (a) the involution is continuous and the left and right multiplications are separately continuous; and (b) $\mathfrak{A}_0$ is dense in $\mathfrak{A}[\tau]$.

Let $\{p_\alpha\}$ be a directed set of seminorms which defines τ. The existence of such a directed set can always be assumed. We can further assume that $\mathfrak{A}[\tau]$ is *complete*. Indeed, if this is not so, then the τ-completion $\tilde{\mathfrak{A}}[\tau]$ is again a topological quasi $*$-algebra over the same $*$-algebra $\mathfrak{A}_0$.

Example. Let $\mathcal{H}$ be a separable Hilbert space and N an unbounded, densely defined, self-adjoint operator. Let $D(N^k)$ be the domain of the operator N^k, $k \in \mathbb{N}_0$, and

$\mathcal{D}$ the domain of all the powers of N: $\mathcal{D} \equiv D^\infty(N) = \cap_{k\geqslant 0} D(N^k)$. This set is dense in $\mathcal{H}$. Let us now introduce $\mathcal{L}^\dagger(\mathcal{D})$, the *-algebra of all the closable operators defined on $\mathcal{D}$ which, together with their adjoints, map $\mathcal{D}$ into itself. Here the adjoint of $X \in \mathcal{L}^\dagger(\mathcal{D})$ is $X^\dagger = X^*_{\restriction \mathcal{D}}$.

In $\mathcal{D}$ the topology is defined by the following N-depending seminorms: $\phi \in \mathcal{D} \to \|\phi\|_n \equiv \|N^n\phi\|$, $n \in \mathbb{N}_0$, while the topology τ_0 in $\mathcal{L}^\dagger(\mathcal{D})$ is introduced by the seminorms

$$X \in \mathcal{L}^\dagger(\mathcal{D}) \to \|X\|^{f,k} \equiv \max\left\{\|f(N)XN^k\|, \|N^k X f(N)\|\right\},$$

where $k \in \mathbb{N}_0$ and $f \in \mathcal{C}$, the set of all the positive, bounded and continuous functions on $\mathbb{R}_+$, which are decreasing faster than any inverse power of x: $\mathcal{L}^\dagger(\mathcal{D})[\tau_0]$ is a complete *-algebra.

It is clear that $\mathcal{L}^\dagger(\mathcal{D})$ contains unbounded operators. Indeed, just to consider the easiest example, it contains all the positive powers of N, but it does not contain elements like e^N. Moreover, if N is the closure of $N_o = a^\dagger a$, with $[a, a^\dagger] = \mathbb{I}$, $\mathcal{L}^\dagger(\mathcal{D})$ also contains all positive powers of a and $a^\dagger$.

Let further $\mathcal{L}(\mathcal{D}, \mathcal{D}')$ be the set of all continuous maps from $\mathcal{D}$ into $\mathcal{D}'$, with their topologies (in $\mathcal{D}'$ this is the strong dual topology), and let τ denotes the topology defined by the seminorms

$$X \in \mathcal{L}(\mathcal{D}, \mathcal{D}') \to \|X\|^f = \|f(N)Xf(N)\|,$$

$f \in \mathcal{C}$. Then $\mathcal{L}(\mathcal{D}, \mathcal{D}')[\tau]$ is again a complete vector space.

In this case $\mathcal{L}^\dagger(\mathcal{D}) \subset \mathcal{L}(\mathcal{D}, \mathcal{D}')$ and the pair

$$(\mathcal{L}(\mathcal{D},\mathcal{D}')[\tau], \mathcal{L}^\dagger(\mathcal{D})[\tau_0])$$

is a *concrete realization* of a locally convex quasi *-algebra.

Remark. let us now suppose that $\mathcal{D} \equiv \mathcal{S}(\mathbb{R})$, the set of the test functions, and $\mathcal{D}' = \mathcal{S}'(\mathbb{R})$. Since $\mathcal{S}(\mathbb{R}) \subset \mathcal{S}'(\mathbb{R})$, it is easy to check that $\mathcal{L}^\dagger(\mathcal{S}) \subset \mathcal{L}(\mathcal{S}, \mathcal{S}')$. Let $\Psi(x) \in \mathcal{S}'(\mathbb{R})$. We define the map Z_Ψ as follows: $(Z_\Psi f)(x) = \Psi(x)f(x)$, $\forall f(x) \in \mathcal{S}(\mathbb{R})$. Since $\Psi(x)f(x) \in \mathcal{S}'(\mathbb{R})$, and since Z_Ψ is continuous, we conclude that $Z_\Psi \in \mathcal{L}(\mathcal{S}, \mathcal{S}')$. It is clear that, for instance, Z_Ψ^2, does not exist for generic Ψ, and this reflects the fact that $\mathcal{L}(\mathcal{S}, \mathcal{S}')$ is not an algebra.

Using a quasi *-algebra is not the only possibility to include unbounded operators in a reasonable algebraic framework. For completeness we briefly mention now two other possibilities which, however, will play no major role in the physical applications considered in the next section.

We begin recalling that a *partial *-algebra* [3] is a complex vector space $\mathfrak{A}$ with involution * (with the usual properties) and a subset $\Gamma \subset (\mathfrak{A}, \mathfrak{A})$ such that

$(x, y) \in \Gamma$ iff $(y^*, x^*) \in \Gamma$;

if $(x, y), (x, z) \in \Gamma$ then $(x, \lambda y + \mu z) \in \Gamma$ for all $\lambda, \mu \in \mathbb{C}$;

if $(x, y) \in \Gamma$ then there exists an element $x \cdot y \in \mathfrak{A}$. This multiplication satisfies the following properties: $x \cdot (y + \lambda z) = x \cdot y + \lambda x \cdot z$ and $(x \cdot y)^* = y^* \cdot x^*$, $\forall (x, y), (x, z) \in \Gamma$.

It is easy to check that each quasi *-algebra is also a partial *-algebra, but not vice-versa.

Other examples of algebras of unbounded operators are the so-called *CQ*-algebras* [9], which can be seen as particular cases of topological quasi *-algebras. We do not give here the general definition but only its simplest version, referring to [9, 15, 16] for more details.

A (proper) CQ*-algebra is a quasi *-algebra such that: $\mathfrak{A}_0[\|.\|_0]$ is a C*-algebra; $\mathfrak{A}[\|.\|]$ is a Banach space in which $\mathfrak{A}_0$ is dense; the two norms are related as follows: $\|x\|_0 = \max\left\{\sup_{\|a\|\leqslant 1}\|ax\|, \sup_{\|a\|\leqslant 1}\|xa\|\right\}, \forall x \in \mathfrak{A}_0$.

This is a natural generalization of C*-algebras: indeed the completion of any C*-algebra $(\mathfrak{A}_0, \|\;\|_0)$ with respect to a weaker norm $\|\;\|$ satisfying: (i) $\|A^*\| = \|A\|$, $\forall A \in \mathfrak{A}_0$ and (ii) $\|AB\| \leqslant \|A\|\|B\|_0, \forall A, B \in \mathfrak{A}_0$, is a CQ*-algebra.

Remark. These structures have been proved to be useful in relation with the Tomita-Takesaki's theory, [16].

Let us now go back to quasi *-algebras. It is not surprising that, analogously to what happens for C*-algebras, a crucial role for physical application is played by the **-representations of a quasi *-algebra*:

let $(\mathfrak{A}, \mathfrak{A}_0)$ be a quasi *-algebra, $\mathcal{D}_\pi$ a dense domain in a certain Hilbert space $\mathcal{H}_\pi$, and π a linear map from $\mathfrak{A}$ into $\mathcal{L}^\dagger(\mathcal{D}_\pi, \mathcal{H}_\pi)$, where

$$\mathcal{L}^\dagger(\mathcal{D}_\pi, \mathcal{H}_\pi) = \{X \text{ closable in } \mathcal{H}_\pi : D(X) = \mathcal{D}_\pi \text{ and } D(X^*) \supseteq \mathcal{D}_\pi\}.$$

This is a *partial *-algebra* with the usual operations $X+Y, \lambda X$, the involution $X^\dagger = X^*_{\restriction \mathcal{D}_\pi}$ and the weak product $X \,\Box\, Y \equiv X^{\dagger *}Y$(defined whenever $Y\mathcal{D}_\pi \subset D(X^{\dagger *})$ and $X^\dagger \mathcal{D}_\pi \subset D(Y^*)$. Notice that these conditions produce the definition of the set $\Gamma \subset (\mathcal{L}^\dagger(\mathcal{D}_\pi, \mathcal{H}_\pi), \mathcal{L}^\dagger(\mathcal{D}_\pi, \mathcal{H}_\pi)))$. Let furthermore

$$\mathcal{L}^\dagger(\mathcal{D}_\pi) = \{A \in \mathcal{L}^\dagger(\mathcal{D}_\pi, \mathcal{H}_\pi) :\; A,\; A^\dagger : \mathcal{D}_\pi \to \mathcal{D}_\pi\}.$$

$\mathcal{L}^\dagger(\mathcal{D}_\pi)$ is a *-algebra and for its elements the weak multiplication $\Box$ reduces to the ordinary multiplication of operators.

A **-representation of* $\mathfrak{A}$ is a linear map from $\mathfrak{A}$ into $\mathcal{L}^\dagger(\mathcal{D}_\pi, \mathcal{H}_\pi)$ such that:

(i) $\pi(a^*) = \pi(a)^\dagger, \quad \forall a \in \mathfrak{A}$;

(ii) if $a \in \mathfrak{A}$, $x \in \mathfrak{A}_0$, then $\pi(a) \Box \pi(x)$ is well defined and $\pi(ax) = \pi(a) \Box \pi(x)$.

Moreover, if

(iii) $\pi(\mathfrak{A}_0) \subset \mathcal{L}^\dagger(\mathcal{D}_\pi)$,
then π is said to be a **-representation of the quasi *-algebra* $(\mathfrak{A}, \mathfrak{A}_0)$.

As for C*-algebras, we see here that a *-representation preserves the algebraic structure of the abstract quasi *-algebra $(\mathfrak{A}, \mathfrak{A}_0)$.

Remark. It may be worth noticing that it might appear more natural to represent $(\mathfrak{A}, \mathfrak{A}_0)$ in another quasi *-algebra $(\mathcal{L}(\mathcal{D}_\pi, \mathcal{D}'_\pi), \mathcal{L}^\dagger(\mathcal{D}_\pi))$, analogous to the one constructed above with $\mathcal{D}_\pi$ instead of $\mathcal{D} = D^\infty(N)$. We will return on this quasi *-algebra in the following. Nevertheless, it is usually more convenient to use $\mathcal{L}^\dagger(\mathcal{D}_\pi, \mathcal{H}_\pi)$ for the following reasons:

(1) if $a \in \mathfrak{A}$ then $\pi(a) \in \mathcal{L}^\dagger(\mathcal{D}_\pi, \mathcal{H}_\pi)$. Therefore, $\forall \varphi \in \mathcal{D}_\pi$, $\pi(a)\varphi \in \mathcal{H}_\pi$. Of course, if we decide to represent a as an element of $\mathcal{L}(\mathcal{D}_\pi, \mathcal{D}'_\pi)$, we go out of the Hilbert space ($\pi(a)\varphi \notin \mathcal{H}_\pi$, in general)! This is not exactly what one expects from a *representation* of a *-algebra, since the abstract elements of the algebra are usually represented acting and living on some Hilbert space;

(2) we also have a technical reason to use $\mathcal{L}^\dagger(\mathcal{D}_\pi, \mathcal{H}_\pi)$, which will appear clear in a moment: in the theorem on the derivations given in the next section the topology τ_s plays a role, and this can be defined on $\mathcal{L}^\dagger(\mathcal{D}_\pi, \mathcal{H}_\pi)$ but not on $\mathcal{L}(\mathcal{D}_\pi, \mathcal{D}'_\pi)$;

(3) finally we can observe that any partial *-algebra is a quasi *-algebra: therefore the choice of representing a quasi *-algebra into a partial *-algebra of operators is, in any case, absolutely consistent.

The *-representation π is called *ultra-cyclic* if there exists $\xi_0 \in \mathcal{D}_\pi$ such that $\pi(\mathfrak{A}_0)\xi_0 = \mathcal{D}_\pi$. π is *faithful* if $\pi(x) = 0$ implies $x = 0$.

As we have anticipated, we use π to introduce a certain topology on $\pi(\mathfrak{A})$: let π be a *-representation of $\mathfrak{A}$. *The strong topology* τ_s on $\pi(\mathfrak{A})$ is defined by the seminorms: $\{p_\xi(.);\ \xi \in \mathcal{D}_\pi\}$, where $p_\xi(\pi(a)) \equiv \|\pi(a)\xi\|$, $a \in \mathfrak{A}$, $\xi \in \mathcal{D}_\pi$. This will be used in the next section. It may be worth noticing again that $\|\pi(a)\xi\|$ would make no sense in general if $\pi(a)$ was an element of $\mathcal{L}^\dagger(\mathcal{D}, \mathcal{D}')$.

As for ordinary C*-algebras, even now it is possible to give a *GNS-like construction*. As a matter of fact, several possible extensions of this construction exist, but we will here mention only one, [38], where the topology τ is taken as a norm $\|.\|$. Therefore $\mathfrak{A}$ is a Banach space and we suppose, for simplicity, that $(\mathfrak{A}, \mathfrak{A}_0)$ has a unit $\mathbb{I}$. Let φ be a sesquilinear form on $\mathfrak{A} \times \mathfrak{A}$ such that

(i) $\varphi(x, x) \geqslant 0, \quad \forall x \in \mathfrak{A}$;

(ii) $\varphi(ax, y) = \varphi(x, a^* y), \forall a \in \mathfrak{A}, x, y \in \mathfrak{A}_0$;

(iii) there exists $\gamma > 0$ such that $|\varphi(x, y)| \leqslant \gamma \|x\| \|y\|, \forall x, y \in \mathfrak{A}_0$.

These conditions imply that

$$N_\varphi = \{a \in \mathfrak{A} : \varphi(a, a) = 0\} = \{a \in \mathfrak{A} : \varphi(a, b) = 0, \ \forall b \in \mathfrak{A}\}.$$

Let $\lambda_\varphi(\mathfrak{A}) = \mathfrak{A}/N_\varphi$ and let us introduce a scalar product on this vector space as follows: $< \lambda_\varphi(a), \lambda_\varphi(b) >= \varphi(a, b)$.

Let further $\mathcal{H}_\varphi$ be the completion of $\lambda_\varphi(\mathfrak{A})$ wrt the norm inherited by this scalar product. One can check that $\lambda_\varphi(\mathfrak{A}_0)$ is dense in $\mathcal{H}_\varphi$.

Let us finally define a map π_φ^o:

$$\pi_\varphi^o(a)\lambda_\varphi(x) = \lambda_\varphi(ax), \quad a \in \mathfrak{A}, x \in \mathfrak{A}_0 :$$

Then π_φ^o is a *-representation of $\mathfrak{A}$ in $\mathcal{L}^\dagger(\lambda_\varphi(\mathfrak{A}_0), \mathcal{H}_\varphi)$. Moreover,

(i) $\lambda_\varphi(\mathfrak{A}_0) = \pi_\varphi^o(\mathfrak{A}_0)\lambda_\varphi(\mathbb{I})$ (i.e. $\lambda_\varphi(\mathbb{I})$ is ultra-cyclic);

(ii) $\varphi(a, b) =< \pi_\varphi^o(a)\lambda_\varphi(\mathbb{I}), \pi_\varphi^o(b)\lambda_\varphi(\mathbb{I}) >, \forall a, b \in \mathfrak{A}$.

Remark. (1) this approach uses *sesquilinear forms* instead of *linear* functionals: indeed, while $\omega(a^* b)$ is not well defined for generic $a, b \in \mathfrak{A}$, independently of the choice of the linear functional ω, $\varphi(a, b)$ surely makes sense if φ is a generic sesquilinear form on $\mathfrak{A} \times \mathfrak{A}$. This allows to define a scalar product on $\mathfrak{A} \times \mathfrak{A}$, as shown above;

(2) the *physical interpretation* is analogous to that discussed in the previous section: different sesquilinear forms produce different representations which can still be interpreted as different phases of the matter.

4. Physical Applications

In this section we will show how the algebraic framework discussed so far can be applied in the rigorous treatment of some physical systems.

4.1. Existence of An Effective Hamiltonian

We begin with reviewing some recent results obtained in collaboration with A. Inoue and C. Trapani, [12, 13], and concerning the possibility of introducing, under certain conditions on Σ, an effective hamiltonian.

Definition. Let $(\mathfrak{A}[\tau], \mathfrak{A}_0)$ be a quasi *-algebra. A **-derivation of* $\mathfrak{A}_0$ is a linear map $\delta : \mathfrak{A}_0 \to \mathfrak{A}$ with the following properties:

(i) $\delta(x^*) = \delta(x)^*, \ \forall x \in \mathfrak{A}_0;$

(ii) $\delta(xy) = x\delta(y) + \delta(x)y, \ \forall x, y \in \mathfrak{A}_0.$

As we see, a *-derivation of $\mathfrak{A}_0$ does exactly what one expects from a similar object: it is a linear map, it preserves the adjoint, and it satisfies the Leibnitz rule, of course only for those elements for which this can be defined. From a physical point of view we know that it is quite hard for a derivation to be *implemented* by an hamiltonian at an algebraic level: even if a local hamiltonian H_V does exist (i.e. the energy for finite V), usually the sequence $\{H_V\}$ do not converge to a self-adjoint operator H in most topologies, even when the sequence $e^{iH_Vt}\, X\, e^{-iH_Vt}$ converges for each observable X. This means that the dynamics is, in general, *hamiltonian* only at a local level. However, as we have already discussed in Section II, under some conditions of Σ an *effective hamiltonian* exists in $B(\mathcal{H}_\rho)$, i.e. in the C*-algebra obtained, via GNS-representation, from some state ρ. The role of the representation is evident, and we will recover the relevance of certain representations even in our settings. In particular, we will restrict to those *-representations π of $(\mathfrak{A}, \mathfrak{A}_0)$ such that, whenever $x \in \mathfrak{A}_0$ satisfies $\pi(x) = 0$, then $\pi(\delta(x)) = 0$. Under this rather natural assumption, the linear map

$$\delta_\pi(\pi(x)) = \pi(\delta(x)), \qquad x \in \mathfrak{A}_0,$$

is well-defined on $\pi(\mathfrak{A}_0)$ with values in $\pi(\mathfrak{A})$ and it is a *-derivation of $\pi(\mathfrak{A}_0)$. We call δ_π the **-derivation induced by* π.

Given such a representation π and its dense domain $\mathcal{D}_\pi \subset \mathcal{H}_\pi$, we consider the graph topology $t_\dagger$ generated by the seminorms

$$\xi \in \mathcal{D}_\pi \to \|A\xi\|, \qquad A \in \mathcal{L}^\dagger(\mathcal{D}_\pi).$$

Let $\mathcal{D}'_\pi$ be the conjugate dual space of $\mathcal{D}_\pi$ and $t'_\dagger$ the strong dual topology of $\mathcal{D}'_\pi$, i.e. the topology generated by the seminorms

$$\mathcal{D}'_\pi \ni z \to \rho_{\mathcal{E}}(z) := \sup_{x\in\mathcal{E}} |<x, z>|,$$

where $<,>$ is the form which puts in duality $\mathcal{D}_\pi$ and $\mathcal{D}'_\pi$ and $\mathcal{E}$ is a bounded set in $\mathcal{D}_\pi$. Then we get the rigged Hilbert space

$$\mathcal{D}_\pi[t_\dagger] \subset \mathcal{H}_\pi \subset \mathcal{D}'_\pi[t'_\dagger].$$

Let $\mathfrak{L}(\mathcal{D}, \mathcal{D}^\times)$ denote the space of all continuous linear maps from $\mathcal{D}_\pi[t_\dagger]$ into $\mathcal{D}'_\pi[t'_\dagger]$ and let $\mathcal{L}^\dagger(\mathcal{D}_\pi)$ be as usual. Then one has

$$\mathcal{L}^\dagger(\mathcal{D}_\pi) \subset \mathfrak{L}(\mathcal{D}, \mathcal{D}^\times).$$

Each operator $A \in \mathcal{L}^\dagger(\mathcal{D}_\pi)$ can be extended to an operator $\hat{A}$ on the whole $\mathcal{D}'_\pi$ in the following way:

$$< \hat{A}\xi', \eta > = < \xi', A^\dagger \eta >, \quad \forall \xi' \in \mathcal{D}'_\pi, \ \eta \in \mathcal{D}_\pi.$$

Therefore the left and right multiplication of $X \in \mathcal{L}(\mathcal{D}_\pi, \mathcal{D}'_\pi)$ and $A \in \mathcal{L}^\dagger(\mathcal{D}_\pi)$ can always be defined:

$$(X \circ A)\xi = X(A\xi), \text{ and } (A \circ X)\xi = \hat{A}(X\xi), \ \forall \xi \in \mathcal{D}_\pi,$$

and for that we can conclude that $(\mathcal{L}(\mathcal{D}_\pi, \mathcal{D}'_\pi), \mathcal{L}^\dagger(\mathcal{D}_\pi))$ is a quasi *-algebra.

Let δ be a *-derivation of $\mathfrak{A}_0$ and π an ultra-cyclic *-representation of $(\mathfrak{A}, \mathfrak{A}_0)$ with ultra-cyclic vector ξ_0. Then $\pi(\mathfrak{A}_0) \subset \mathcal{L}^\dagger(\mathcal{D}_\pi)$. We say that the *-derivation δ_π induced by π is *spatial* if there exists $H_\pi = H_\pi^\dagger \in \mathcal{L}(\mathcal{D}_\pi, \mathcal{D}'_\pi)$ such that $H_\pi\xi_0 \in \mathcal{H}_\pi$ and

$$\delta_\pi(\pi(x)) = i\{H_\pi \circ \pi(x) - \pi(x) \circ H_\pi\}, \quad \forall x \in \mathfrak{A}_0.$$

The meaning of this definition is clear: a derivation produces in a representation π a spatial induced derivation if this can be *implemented by* a symmetric operator H_π: the way in which δ_π is, in a certain obvious sense, described by H_π is via a generalized commutator, i.e. a commutator where we may need to consider some opportune extensions of the the operators involved. To be more explicit, if $x \in \mathfrak{A}_0$, then $\pi(x) \in \mathcal{L}^\dagger(\mathcal{D}_\pi)$, so that, since by definition $H_\pi \in \mathcal{L}(\mathcal{D}_\pi, \mathcal{D}'_\pi)$, it is clear that $H_\pi \circ \pi(x)\psi = H_\pi\, \pi(x)\psi \in \mathcal{D}'_\pi$ for each $\psi \in \mathcal{D}_\pi$. Viceversa, in general we have $\pi(x) \circ H_\pi \psi = \hat{\pi}(x) H_\pi \psi \in \mathcal{D}'_\pi$, since $H_\pi \psi \in \mathcal{D}'_\pi$ for each $\psi \in \mathcal{D}_\pi$, so that $\pi(x)(H_\pi \Psi)$ is not well defined. This kind of difficulties, however, is not a big surprise since they are very common whenever one deals with unbounded operators, and it is easily overcome here by means of the $\circ$ multiplication.

The main result concerning spatial derivations is contained in the following theorem, [12], which extends and analogous result for C*-algebras which can be found, for instance, in [17].

Theorem 4.1. *Let $(\mathfrak{A}[\tau], \mathfrak{A}_0)$ be a locally convex quasi *-algebra with identity and δ be a *-derivation of $\mathfrak{A}_0$.*

Then the following statements are equivalent:

(i) *There exists a $(\tau - \tau_s)$-continuous, ultra-cyclic *-representation π of $\mathfrak{A}$, with ultra-cyclic vector ξ_0, such that the *-derivation δ_π induced by π is spatial.*

(ii) *There exists a positive linear functional f on $\mathfrak{A}_0$ such that $f(x^*x) \leqslant p(x)^2$, $\forall x \in \mathfrak{A}_0$, for some continuous seminorm p of τ and, denoting with $\tilde{f}$ the continuous extension of f to $\mathfrak{A}$, the following inequality holds:*

$$|\tilde{f}(\delta(x))| \leqslant C\left(\sqrt{f(x^*x)} + \sqrt{f(xx^*)}\right), \quad \forall x \in \mathfrak{A}_0,$$

for some positive constant C.

(iii) *There exists a positive sesquilinear form φ on $\mathfrak{A} \times \mathfrak{A}$ such that:*

φ is invariant, i.e. $\varphi(ax, y) = \varphi(x, a^*y)$, for all $a \in \mathfrak{A}$ and $x, y \in \mathfrak{A}_0$;

φ is τ-continuous, i.e. $|\varphi(a, b)| \leqslant p(a)p(b)$, for all $a, b \in \mathfrak{A}$, for some continuous seminorm p of τ;

φ satisfies the following inequality:

$$|\varphi(\delta(x), \mathbb{1})| \leqslant C\left(\sqrt{\varphi(x, x)} + \sqrt{\varphi(x^*, x^*)}\right), \quad \forall x \in \mathfrak{A}_0,$$

for some positive constant C.

Remark. (1) even if δ cannot be written as $\delta(x) = i[H, x]$, for any $H \in \mathfrak{A}$, if the above theorem can be applied, then δ_π is, essentially, the commutator with a certain symmetric operator, H_π. Again, as we expect for physical reasons, *the dynamics depends on the representation π and, as a consequence, on the phase of the matter.*

(2) The above theorem can be used to answer to the following question: suppose we add to a spatial *-derivation δ_0 a *perturbation* δ_p such that $\delta = \delta_0 + \delta_p$ is again a *-derivation. Under which conditions δ is still spatial? A sufficient condition for this to be true, [12], is that $|\tilde{f}(\delta_p(x))| \leqslant |\tilde{f}(\delta_0(x))|$, for all $x \in \mathfrak{A}_0$, which is exactly what we expect since this means simply that δ_p is *smaller than* δ_0. If we call H_π, $H_{\pi,0}$ and $H_{\pi,p}$ the operators which implement δ, δ_0 and δ_p, we can also prove that $i[H_\pi, A]\psi = i[H_{\pi,0} + H_{\pi,p}, A]\psi$, for all $A \in \mathcal{L}^\dagger(\mathcal{D}_\pi)$ and $\psi \in \mathcal{D}_\pi$.

Theorem 1 is the starting point to consider the problem of the removal of the cutoff for infinite systems. This means that we are assuming, as usual, that the dynamical behavior of the infinite system is obtained as a suitable limit of its restriction to a finite volume V_L. At the infinitesimal level, this means that we have a family of inner derivations δ_L, with h_L their associated energies, (i.e. $\delta_L(x) = i[h_L, x]$ for all $x \in \mathfrak{A}_0$ and for all L) but we don't now if the limit of these derivations is still inner or, at least, if the induced limiting derivation is spatial in some particular representations. Let now be more precise.

Let $\mathcal{S}$ be a physical system, $(\mathfrak{A}, \mathfrak{A}_0)$ a quasi *-algebra and $\{\mathcal{S}_L = \{\mathfrak{A}_L \subset \mathfrak{A}_0, \Sigma, \alpha_L^t\}, L \in \Lambda\}$ a family of *regularized* systems, i.e. $\mathcal{S}_L$ is the restriction of $\mathcal{S}$ to some finite volume V_L. Here Λ is a set of indexes, used to label the finite volume systems S_L. We suppose that, for each fixed L, the dynamics α_L^t is generated by a *-derivation δ_L:

$$\alpha_L^t(x) = \tau_0 - \sum_{k=0}^{\infty} \frac{t^k}{k!} \delta_L^k(x) = e^{ih_L t} x e^{-ih_L t}, \qquad \forall x \in \mathfrak{A}_L.$$

Here τ_0 is a topology on $\mathfrak{A}_0$. Actually, this assumption is not necessary even if this is really what we have in mind, and what actually happens for ordinary C*-algebras at least for discrete systems, where each h_L is a bounded operator and τ_0 is the uniform topology.

Definition 4.2. The family $\{\mathcal{S}_L, L \in \Lambda\}$ is said to be c-representable if there exists a *-representation π of $(\mathfrak{A}, \mathfrak{A}_0)$ into some $(\mathcal{L}(\mathcal{D}_\pi, \mathcal{D}'_\pi), \mathcal{L}^\dagger(\mathcal{D}_\pi))$ such that:

(i) π is $(\tau - \tau_s)$-continuous;

(ii) π is ultra-cyclic with ultra-cyclic vector ξ_0;

(iii) when $\pi(x) = 0$, then $\pi(\delta_L(x)) = 0, \forall L \in \Lambda$.

Any such representation π is a c-representation.

Making use of this definition we can prove the following Proposition, [13]:

Proposition 4.3. *Let $\{S_L, L \in \Lambda\}$ be a c-representable family and π a c-representation. Let $h_L = h_L^* \in \mathfrak{A}_L$ be the element which implements δ_L: $\delta_L(x) = i[h_L, x]$, $\forall x \in \mathfrak{A}_0$, $\forall L \in \Lambda$. Suppose that $\delta_L(x)$ is τ-Cauchy $\forall x \in \mathfrak{A}_0$ and that* $\sup_L \|\pi(h_L)\xi_0\| < \infty$.

Then, one has

*(a) $\delta(x) = \tau - \lim_L \delta_L(x)$ exists in $\mathfrak{A}$ and is a *-derivation of $\mathfrak{A}_0$;*

*(b) δ_π, the *-derivation induced by π, is well defined and spatial.*

Remark. (1) It is clear that if the sequence $\{h_L\}$ is τ-convergent, then $\delta_L(x)$ is automatically τ-Cauchy $\forall x \in \mathfrak{A}_0$.

(2) This Proposition implies that any physical system $\mathcal{S}$ with a c-representable regularized family $\{\mathcal{S}_L, L \in \Lambda\}$ *admits an effective hamiltonian in the sense of* [10, 37, 33, 28].

(3) This is our version of Sewell's result on the existence of different *effective hamiltonians* in different (GNS-like) representations: a physical system Σ exhibits different dynamics in its different thermodynamical phases).

(4) We want to cite here an open problem which, in our opinion, deserves a deeper analysis: what does it happen if we consider a representation π' *globally equivalent* but locally different from a given c-representation π? Is π' still a c-representation? Do the related effective hamiltonians coincide? This is indeed what one expects in connection with what has been discussed in the C*-algebraic approach, even if no explicit proof of this claim exists at this stage.

4.2. The Time Evolution α^t

We discuss here the following problem: suppose that we have been able to prove that the derivation δ exists. Nevertheless, in general we have no information about $\delta^2, \delta^3, \ldots$. Moreover, even if all these maps do exist, this does not mean that the series $\sum_{k=0}^{\infty} \frac{t^k}{k!}\delta^k(x)$, which can be taken as the definition of $\alpha^t(x)$ when x and H are both bounded, exists as well, for a generic $x \in \mathfrak{A}_0$. In other words, the existence of $\delta(x)$ does not imply, by no means, the existence of $\alpha^t(x)$ for $x \in \mathfrak{A}$ or even for $x \in \mathfrak{A}_0$.

Furthermore, the effective hamiltonian H_π, whose existence has been proved in the previous subsection, is symmetric but not self-adjoint, and therefore the spectral theorem cannot be used to define $e^{iH_\pi t}$ and, as a consequence, to define $\pi(\alpha^t(x))$ as $e^{iH_\pi t}\pi(x)e^{-iH_\pi t}$! So the following crucial problem arises: how to define a *time evolution* in this case?

This is a rather hard problem already in a standard setting, when there is no problem in multiplying elements of the algebra. Here this is quite a dangerous operation and the difficulties are even more than before. We will devote the rest of this subsection to some partial results which can be used to analyze certain classes of physical systems. It may be stressed that no general result really exists up to now.

4.2.1. *From δ to α^t: the first way*

We begin considering a class of models which are suggested by the *mean field spin models*, reviewing some results first obtained in [13].

Let us assume that the finite volume hamiltonians h_L can be written in terms of some s.a. elements s_L^α, $\alpha = 1, 2, .., N$, which are τ-converging to some elements $s^\alpha \in \mathfrak{A}$, commuting with all the elements of $\mathfrak{A}_0$:

$$s^\alpha = \tau - \lim_L s_L^\alpha, \qquad [s^\alpha, x] = 0, \ \forall x \in \mathfrak{A}_0.$$

For mean field spin models s^α is the *magnetization* and τ is the strong topology restricted to a relevant family of states, [6], or, alternatively, the so-called physical topology, [8, 10, 11].

We say that $\{s_L^\alpha\}$ is *uniformly τ-continuous* if, for each continuous seminorm p of τ and for all $\alpha = 1, 2, \ldots, N$, there exists another continuous seminorm q of τ and a positive constant $c_{p,q,\alpha}$ such that

$$p(s_L^\alpha b) \leqslant c_{p,q,\alpha} q(b), \ \forall b \in \mathfrak{A}, \ \forall L \in \Lambda.$$

From this definition it also follows that $p(as_L^\alpha) \leqslant c_{p,q,\alpha} q(b)$, $\forall b \in \mathfrak{A}$, and that the same inequalities can be extended to s^α. Then we have

Lemma 4.4. *If $\{s_L^\alpha\}$ is a uniformly τ-continuous sequence and if $\tau - \lim_L s_L^\alpha = s^\alpha$, $\forall \alpha$, then $\tau - \lim_L \left(s_L^\alpha\right)^k = (s^\alpha)^k$, $\forall \alpha$ and for $k = 1, 2, \ldots$.*

Proposition 4.5. *Suppose that (1) $\forall x \in \mathfrak{A}_0$ $[h_L, x]$ depends on L only through s_L^α and (2) $s_L^\alpha \xrightarrow{\tau} s^\alpha$ and $\{s_L^\alpha\}$ is a uniformly τ-continuous sequence.*

Then, for each $k \in \mathbb{N}$, the following limit exists

$$\tau - \lim_L i^k [h_L, x]_k \stackrel{\circ}{=} \tau - \lim_L \delta_L^k(x), \ \forall x \in \mathfrak{A}_0,$$

and defines an element of $\mathfrak{A}$ which we call $\delta^{(k)}(x)$.

Remark. The reason why we prefer to use $\delta^{(k)}(x)$ instead of $\delta^k(x)$ is just to stress in this way that it is not possible to write $i^k[h, x]_k$ since, first of all, no global h does exist and, secondly, even if it does, $[h, x]_k$ is not well defined in general because of domain difficulties.

Once we have obtained conditions for all the multiple commutators to exist in some reasonable sense, we still need to find conditions for which the infinite series which defines $\alpha^t(x)$ do converge. For that it is convenient to introduce here the following definition:

Definition 4.6. We say that $x \in \mathfrak{A}_0$ *is a generalized analytic element of δ* if, for all t, the series $\sum_{k=0}^{\infty} \frac{t^k}{k!} \pi(\delta^{(k)}(x))$ is τ_s-convergent. The set of all generalized analytic elements is denoted with $\mathcal{G}$.

Therefore we have, [13],

Proposition 4.7. *Let x_γ be a net of elements of $\mathfrak{A}_0$ and suppose that, whenever $\pi(x_\gamma) \xrightarrow{\tau_s} \pi(x)$ then $x_\gamma \xrightarrow{\tau} x$. Then, $\forall x \in \mathcal{G}$ and $\forall t \in \mathbb{R}$, the series $\sum_{k=0}^{\infty} \frac{t^k}{k!} \delta^{(k)}(x)$ converges in the τ-topology to an element of $\mathfrak{A}$ which we call $\alpha^t(x)$.*

Moreover, α^t can be extended to the τ-closure $\overline{\mathcal{G}}$ of $\mathcal{G}$.

It may be worth noticing that, even if the assumptions are rather strong, they are satisfied, for instance, by mean field spin models!

4.2.2. From δ to α^t: the second way

Let π be a faithful *-representation of the quasi *-algebra $(\mathfrak{A}, \mathfrak{A}_0)$ and δ a *-derivation of $\mathfrak{A}_0$ such that δ_π, is well-defined on $\pi(\mathfrak{A}_0)$ with values in $\pi(\mathfrak{A})$.

We define the following subset of $\mathfrak{A}_0$ (a *domain of regularity* of δ)

$$\mathfrak{A}_0(\delta) := \{x \in \mathfrak{A}_0 : \quad \delta^k(x) \in \mathfrak{A}_0, \quad \forall k \in \mathbb{N}_0\}.$$

Whenever δ is *regular* the set $\mathfrak{A}_0(\delta)$ is large. For instance, if δ is inner in $\mathfrak{A}_0$ with an implementing element $h \in \mathfrak{A}_0$, then $\mathfrak{A}_0(\delta) = \mathfrak{A}_0$. For general δ, $\mathfrak{A}_0(\delta)$ is not empty since it contains, at least, all the multiples of the identity $\mathbb{1}$ of $\mathfrak{A}_0$.

$\mathfrak{A}_0(\delta)$ is a *-algebra which is mapped into itself by δ. Moreover it is easy to check that $\pi(\delta^k(x)) = \delta_\pi^k(\pi(x))$, $\forall x \in \mathfrak{A}_0(\delta)$ and $\forall k \in \mathbb{N}_0$. Therefore it follows that $\delta_\pi^k(\pi(x)) \in \pi(\mathfrak{A}_0)$.

Let σ_s be the topology on $\mathfrak{A}$ defined via τ_s in the following way:

$$\mathfrak{A} \ni a \to q_\xi(a) = p_\xi(\pi(a)) = \|\pi(a)\xi\|, \quad \xi \in \mathcal{D}_\pi.$$

Then we have the following theorem, [13]:

Theorem 4.8. *Let $(\mathfrak{A}, \mathfrak{A}_0)$ be a quasi *-algebra with identity, δ a *-derivation on $\mathfrak{A}_0$ and π a faithful *-representation of $(\mathfrak{A}, \mathfrak{A}_0)$ such that the induced derivation δ_π is well defined. Then, we have:*

(1) if the following inequality holds

$$\forall \eta \in \mathcal{D}_\pi\ \exists c_\eta > 0 : p_\eta(\delta_\pi(\pi(x))) \leqslant c_\eta p_\eta(\pi(x)), \quad \forall x \in \mathfrak{A}_0(\delta),$$

then $\sum_{k=0}^{\infty} \frac{t^k}{k!}\delta^k(x)$ converges for all t in the topology σ_s to an element of $\overline{\mathfrak{A}_0(\delta)}^{\sigma_s}$ which we call $\alpha^t(x)$; α^t can be extended to $\overline{\mathfrak{A}_0(\delta)}^{\sigma_s}$. Moreover $\alpha^t : \overline{\mathfrak{A}_0(\delta)}^{\sigma_s} \to \overline{\mathfrak{A}_0(\delta)}^{\sigma_s}$ and

$$\alpha^{t+\tau}(x) = \alpha^t(\alpha^\tau(x)), \quad \forall t, \tau,\ \forall x \in \mathfrak{A}_0(\delta);$$

(2) Suppose that the following inequality holds

$$\exists c > 0 : \forall \eta_1 \in \mathcal{D}_\pi\ \exists A_{\eta_1} > 0,\ n \in \mathbb{N} \text{ and } \eta_2 \in \mathcal{D}_\pi :$$
$$p_{\eta_1}(\delta_\pi^k(\pi(x))) \leqslant A_{\eta_1} c^k k! k^n p_{\eta_2}(\pi(x)), \quad \forall x \in \mathfrak{A}_0(\delta),\ \forall k \in \mathbb{N}_0,$$

then $\sum_{k=0}^{\infty} \frac{t^k}{k!}\delta^k(x)$ converges, for $t < \frac{1}{c}$ in the topology σ_s to an element of $\overline{\mathfrak{A}_0(\delta)}^{\sigma_s}$ which we call $\alpha^t(x)$; α^t can be extended to $\overline{\mathfrak{A}_0(\delta)}^{\sigma_s}$. Moreover α^t maps $\overline{\mathfrak{A}_0(\delta)}^{\sigma_s}$ into itself for $t < \frac{1}{c}$ and, $\forall x \in \mathfrak{A}_0(\delta)$,

$$\alpha^{t+\tau}(x) = \alpha^t(\alpha^\tau(x)), \quad \forall t, \tau, \text{ with } t + \tau < \frac{1}{c}.$$

Remark. (1) As we see, this theorem gives sufficient conditions for α^t to be defined (as a converging series) at least on a certain subset of $\mathfrak{A}_0$.

(2) Here and in the previous approach the spatiality of the derivation is not used. It is obvious that, when H_π exists as a **self-adjoint** operator mapping $\mathcal{D}_\pi$ into $\mathcal{H}_\pi$, we could use the spectral theorem to define $\pi(\alpha^t(x)) := e^{iH_\pi t}\pi(x)e^{-iH_\pi t}$;

4.2.3. A different point of view

In a recent paper, [11], we have considered the problem of the existence of α^t from a slightly different point of view, which is maybe more suitable for systems with a finite number of degrees of freedom. This is because we have assumed that the energy operator of our quantum system system does exist as a self-adjoint, unbounded and densely defined operator $H_0 \geqslant \mathbb{1}$. Then, it is known that the operator e^{iH_0t}, and therefore the time evolution of an observable X, can be defined via the spectral theorem. However, but for finite dimensional Hilbert spaces, our claim is that the *natural algebraic framework* to discuss the dynamical behavior of the system is $\mathcal{L}_{\rm w}^{\dagger}(\mathcal{D},\mathcal{H})[\tau_0]$, where $\mathcal{D} = D^\infty(H_0)$, rather than $B(\mathcal{H})$. Indeed, if $dim(\mathcal{H}) = \infty$, it is clear that in general $H_0 \notin B(\mathcal{H})$ and that δ does not map $B(\mathcal{H})$ into itself. On the other hand, it is evident that $H_0 \in \mathcal{L}_{\rm w}^{\dagger}(\mathcal{D},\mathcal{H})[\tau_0]$ and that $\delta : \mathcal{L}_{\rm w}^{\dagger}(\mathcal{D},\mathcal{H})[\tau_0] \to \mathcal{L}_{\rm w}^{\dagger}(\mathcal{D},\mathcal{H})[\tau_0]$.

These claims are based on the following natural procedure:

let $H_0 = \int_1^\infty \lambda dE(\lambda)$ be the spectral decomposition of H_0. We define, for $L \geqslant 1$, the projectors $Q_L^0 = \int_1^L dE(\lambda)$ and we introduce the *regularized hamiltonian* $H_L = Q_L^0 H_0 Q_L^0$.

For each L, we see that $Q_L^0, H_L \in B(\mathcal{H}) \bigcap \mathcal{L}_{\rm w}^{\dagger}(\mathcal{D},\mathcal{H})$. Furthermore, we have $[Q_L^0, H_L] = [Q_L^0, H_0] = [H_0, H_L] = 0$.

If τ_0 is the topology on $\mathcal{L}_{\rm w}^{\dagger}(\mathcal{D},\mathcal{H})$ generated by the seminorms

$$\mathcal{L}_{\rm w}^{\dagger}(\mathcal{D},\mathcal{H}) \ni A \mapsto \|A\|^{f,k} = \max\{\|H_0^k A f(H_0)\|, \|f(H_0) A H_0^k\|\},$$

then we have:

(i) $H_L \to H_0$ with respect to the topology τ_0;

(ii) $\{e^{itH_L}\}$ is τ_0-Cauchy in $\mathcal{L}_{\rm w}^{\dagger}(\mathcal{D},\mathcal{H})$ and converges to e^{iH_0t}

(iii) $\forall A \in \mathcal{L}_{\rm w}^{\dagger}(\mathcal{D},\mathcal{H})$, the sequence $\{e^{itH_L}Ae^{-itH_L}\}$ is τ_0-Cauchy in $\mathcal{L}_{\rm w}^{\dagger}(\mathcal{D},\mathcal{H})$ and converges to $e^{itH_0}Ae^{-itH_0}$.

We can therefore conclude that H_0, e^{iH_0t}, and $\alpha^t(A) := \mathrm{e}^{itH_0}A\mathrm{e}^{-itH_0}$ all belong to $\mathcal{L}_{\mathrm{w}}^{\dagger}(\mathcal{D},\mathcal{H})$, $\forall A \in \mathcal{L}_{\mathrm{w}}^{\dagger}(\mathcal{D},\mathcal{H})$. Moreover we can also show that

$$\alpha^t(A) = \tau_0 - \lim_L \mathrm{e}^{itH_L}A\mathrm{e}^{-itH_L} = \left(\tau_0 - \lim_L \mathrm{e}^{itH_L}\right) A \left(\tau_0 - \lim_L \mathrm{e}^{-itH_L}\right).$$

This suggests the use of $\mathcal{L}_{\mathrm{w}}^{\dagger}(\mathcal{D},\mathcal{H})[\tau_0]$ as a natural algebraic and topological framework for the analysis of the time evolution of, at least, finite quantum systems. Of course, a similar construction can be repeated also for QM_∞ systems, at least for those systems for which an unbounded, self-adjoint and densely defined operator M exists such that $[M, H_L] = 0$ (on a dense domain), [28].

In the same paper we have considered the role of a perturbation in this approach: let $H = H_0 + B$, and suppose that the spectral decomposition of the *free hamiltonian* H_0 is explicitly known while the spectral decomposition of the *perturbed hamiltonian* H cannot be exactly found, which is exactly what usually happens in concrete situations. We have shown that the *convenient algebraic structure* is again $\mathcal{L}_{\mathrm{w}}^{\dagger}(\mathcal{D},\mathcal{H})$, with $\mathcal{D} = D^\infty(H_0)$, (since, if H_0 has discrete spectrum, we know an o.n. set in $\mathcal{D}$ and, as a consequence, we know $\mathcal{D}$) but the *technically convenient* topology, τ, is that given by the seminorms

$$\mathcal{L}_{\mathrm{w}}^{\dagger}(\mathcal{D},\mathcal{H}) \ni A \mapsto \|A\|_+^{f,k} = \max\{\|H^k A f(H)\|, \|f(H) A H^k\|\},$$

because with this choice some of the above convergence results (i)-(iii) can still be established. Moreover this apparent difference between the algebraic and the topological frameworks, can be easily controlled. Indeed we have proven in [11] that, if (1) $D(H_0) \subseteq D(B)$ and if $H = H_0 + B$ is self-adjoint on $D(H_0)$, and (2) if $\mathcal{D}^\infty(H_0) = \mathcal{D}^\infty(H)$ (hypothesis for which we gave necessary and sufficient conditions), then $\tau_0 \equiv \tau$.

Under these assumptions we can therefore undertake a deeper analysis of the existence of the algebraic dynamics for a perturbed hamiltonian. We refer to [11] for a detailed analysis, which eventually produces a rigorous definition of the Schrödinger dynamics.

4.3. Fixed Point Results

This is an alternative procedure which again produces a rigorous definition of the dynamics of a (closed) physical system, [5], and which is based on a generalization of well known fixed point theorems.

Let $\mathcal{B}$ be a τ-complete subspace of $\mathcal{L}^\dagger(\mathcal{D})$ and T a map from $\mathcal{B}$ into $\mathcal{B}$. We say that T is a *weak τ strict contraction over* $\mathcal{B}$, briefly a wτsc($\mathcal{B}$), if there exists a constant $c \in]0,1[$ such that, for all $(h,k) \in \mathcal{C}_N := (\mathcal{C}, \mathfrak{N})$, there exists a pair $(h', k') \in \mathcal{C}_N$

satisfying

$$\|Tx - Ty\|^{h,k} \leqslant c\|x - y\|^{h',k'} \qquad \forall\, x, y \in \mathcal{B}. \tag{4.1}$$

In what follows we will consider equations of the form $Tx = x$, T being a $\mathrm{w}\tau\,\mathrm{sc}(\mathcal{B})$. The first step consists in introducing the following subset of $\mathcal{B}$:

$$\mathcal{B}_L \equiv \left\{ x \in \mathcal{B} : \sup_{(h,k)\in\mathcal{C}_N} \|Tx - x\|^{h,k} \leqslant L \right\}, \tag{4.2}$$

L being a fixed positive real number.

Lemma 4.9. *Let T be a $\mathrm{w}\tau\,sc(\mathcal{B})$. Then*

(a) *if $T0 = 0$ then any $x \in \mathcal{B}$ such that $\sup_{(h,k)\in\mathcal{C}_N} \|x\|^{h,k} \leqslant L_1$ belongs to $\mathcal{B}_L$ for $L \geqslant L_1(1 + c)$;*

(b) *if $\|T0\|^{h,k} \leqslant L_2$ for all $(h, k) \in \mathcal{C}_N$, then any $x \in \mathcal{B}$ such that $\sup_{(h,k)\in\mathcal{C}_N} \|x\|^{h,k} \leqslant L_1$ belongs to $\mathcal{B}_L$ for $L \geqslant L_1(1 + c) + L_2$;*

(c) *if $x \in \mathcal{B}_L$ then $T^n x \in \mathcal{B}_L$, for all $n \in \mathfrak{N}$;*

(d) *$\mathcal{B}_L$ is τ-complete;*

(e) *if $\mathcal{B}_L$ is not empty, then T is a $\mathrm{w}\tau\,sc(\mathcal{B}_L)$.*

$\mathcal{B}_L$ is non empty, see [5]. The existence of a fixed point is ensured by the following Proposition:

Proposition 4.10. *Let T be a $\mathrm{w}\tau\,sc(\mathcal{B})$. Then*

(a) *$\forall x_0 \in \mathcal{B}_L$ the sequence $\{x_n \equiv T^n x_0\}_{n\geqslant 0}$ is τ-Cauchy in $\mathcal{B}_L$. Its τ-limit, $x \in \mathcal{B}_L$, is a fixed point of T;*

(b) *if $x_0, y_0 \in \mathcal{B}_L$ satisfy the condition $\sup_{(h,k)\in\mathcal{C}_N} \|x_0 - y_0\|^{h,k} < \infty$, then $\tau - \lim_n T^n x_0 = \tau - \lim_n T^n y_0$.*

For physical applications we need to consider the case in which these maps depend on an external parameter:

let $I \subset \mathbb{R}$ be a set such that 0 is one of its accumulation points. A family of weak τ strict contractions $\{T_\alpha\}_{\alpha\in I}$ is said *uniform* if

1) $T_\alpha : \mathcal{B} \to \mathcal{B}$ $\forall \alpha \in I$, $\mathcal{B}$ being a τ-complete subspace of $\mathcal{L}^+(\mathcal{D})$;

2) $\forall (h,k) \in \mathcal{C}_N$ and $\forall \alpha \in I$ there exist $(h',k') \in \mathcal{C}_N$, independent of α, and $c_\alpha \in]0,1[$, independent of (h,k), such that

$$\|T_\alpha x - T_\alpha y\|^{h,k} \leqslant c_\alpha \|x-y\|^{h',k'}, \quad \forall x, y \in \mathcal{B}; \tag{4.3}$$

3) $c_- \equiv \lim_{\alpha,0} c_\alpha \in]0,1[$.

We further say that the family $\{T_\alpha\}_{\alpha\in I}$ is τ-strong Cauchy if, for all $(h,k) \in \mathcal{C}_N$ and $\forall y \in \mathcal{B}$,

$$\|T_\alpha y - T_\beta y\|^{h,k} \overset{\alpha,\beta\to 0}{\longrightarrow} 0. \tag{4.4}$$

We call $\mathcal{B}_L^{(\alpha)}$ the following set $\mathcal{B}_L^{(\alpha)} \equiv \left\{x \in \mathcal{B} : \sup_{(h,k)\in\mathcal{C}_N} \|T_\alpha x - x\|^{h,k} \leqslant L\right\}$.

Proposition 4.11. *Let $\{T_\alpha\}_{\alpha\in I}$ be a τ-strong Cauchy uniform family of $w\tau sc(\mathcal{B})$. Then*

(1) There exists a $w\tau sc(\mathcal{B})$, T, which satisfies the following relations:

$$\|Ty - T_\alpha y\|^{h,k} \to 0 \quad \forall y \in \mathcal{B},\ \forall (h,k) \in \mathcal{C}_N$$

and

$$\|Ty - Tz\|^{h,k} \leqslant c_- \|y-z\|^{h',k'} \quad \forall y, z \in \mathcal{B},$$

where (h',k') are those of inequality (4.3).

(2) let $\{x_\alpha\}_{\alpha\in I}$ be a family of fixed points of the net $\{T_\alpha\}_{\alpha\in I}$: $T_\alpha x_\alpha = x_\alpha$, $\forall \alpha \in I$. If $\{x_\alpha\}_{\alpha\in I}$ is a τ-Cauchy net then, calling x its τ-limit in $\mathcal{B}$, x is a fixed point of T.

(3) If the set $\cap_{\alpha\in I}\mathcal{B}_L^{(\alpha)}$ is not empty and if the following commutation rule holds

$$T_\alpha(T_\beta y) = T_\beta(T_\alpha y), \quad \forall \alpha, \beta \in I \text{ and } \forall y \in \mathcal{B},$$

then, calling

$$x_\alpha = \tau - \lim_{n\to\infty} T_\alpha^n x^0, \text{ where } x^0 \in \cap_{\alpha\in I}\mathcal{B}_L^{(\alpha)},$$

each x_α is a fixed point of T_α, $T_\alpha x_\alpha = x_\alpha$ and $\{x_\alpha\}_{\alpha\in I}$ is a τ-Cauchy net. Moreover $\tau - \lim_{\alpha\to 0} x_\alpha$ is a fixed point of T.

As an application we have proven in [5] that, under certain technical assumptions, the α-time evolution of a given operator x,

$$x_\alpha(t) = x + i\int_0^t ds[H_\alpha, x_\alpha(s)],$$

is associated with an uniform family of $\text{w}\tau\,\text{sc}(\mathcal{L}^\dagger)$, $\{U_\alpha\}$, which is also τ-strong Cauchy. This implies that, because of the Proposition above, the dynamics for the physical system can be obtained as a τ-limit of the regularized dynamics $x_\alpha(t)$, which is a fixed point of $U := \lim_\alpha U_\alpha$.

4.4. Explicit Estimates

We end this excursus of (class of) models for which the time evolution is under control, by considering the so-called almost mean field Ising model, defined by the following finite volume hamiltonian

$$H_V = \frac{J}{|V|^\gamma}\sum_{i,j\in V}\sigma_i^3\sigma_j^3, \tag{4.5}$$

with $0 < \gamma \leqslant 1$, [7]. Particularly relevant in the mathematical description of this model is the *almost magnetization* operator $S_V^3 := \frac{1}{|V|^\gamma}\sum_{p\in V}\sigma_p^3$. In fact, if A is a local observable, its regularized time evolution $\alpha_V^t(A) := e^{iH_Vt}Ae^{-iH_Vt}$ in general depends on t, A and S_V^3.

We refer to the original paper, [7], for the mathematical details. Here we introduce a different topology τ on the C*-spin algebra $\mathfrak{A}_s$, which has proved to be of some usefulness, as follows:

$$\tau: \qquad \|A\|_{\{n\}}^f := \|f(M_{\{n\}})\pi_{\{n\}}(A)f(M_{\{n\}})\|,$$

where f belongs to $\mathcal{C}$. With these definitions, calling $\mathfrak{A}_0$ the τ_0-completion of $\mathfrak{A}_s$ and $\mathfrak{A}$ the τ-completion of $\mathfrak{A}_s$, we proved in [7] that:

- $(\mathfrak{A}[\tau], \mathfrak{A}_0[\tau_0])$ is a topological quasi *-algebra;
- all the powers of the almost magnetization S_3^V are τ_0-converging in $\mathfrak{A}$;
- the finite volume dynamics α_V^t τ_0-converges to a one-parameter group of automorphisms α^t of $\mathfrak{A}_0$;
- α^t solves the τ_0-limit of the finite volume Heisenberg equation of motion.

Another spin model which can be analyzed within the same algebraic framework is the *almost mean field Heisenberg model*,

$$H_V = \frac{J}{|V|^\gamma} \sum_{i,j\in V} \sum_{\alpha=1}^{3} \sigma_i^\alpha \sigma_j^\alpha,$$

with $\frac{1}{2} < \gamma \leqslant 1$, see [8], which differs from the Ising model because it is intrinsically three-dimensional.

A different class of models that we have considered using the same approach involves free and interacting bosons, [4]. The formal hamiltonian H for the one mode free bosons is simply the number operator $N = a^\dagger a$, a and $a^\dagger$ being the annihilation and creation operators for the bosons. They satisfy the canonical commutation relation $[a, a^\dagger] = \mathcal{I}$. (More properly, N is the unique self-adjoint extension of the symmetric operator $a^\dagger a$.)

The construction of the topological quasi *-algebra is the usual one. Let $\mathcal{D} := D^\infty(N) = \cap_{k\geqslant 0} D(N^k)$. This set is dense in the Fock-Hilbert space $\mathcal{H}$ constructed in the standard way. Starting from $\mathcal{D}$ we can define the *-algebra $\mathcal{L}^\dagger(\mathcal{D})$. It is clear that all powers of a and $a^\dagger$ belong to this set. The topology in $\mathcal{L}^\dagger(\mathcal{D})$ is the usual quasi-uniform topology:

$$X \in \mathcal{L}^\dagger(\mathcal{D}) \to \|X\|^{f,k} := \max\left\{\|f(N)XN^k\|, \|N^kXf(N)\|\right\}, \qquad (4.6)$$

where $f \in \mathcal{C}$ and $k \geqslant 0$.

Let $\mathcal{E}_l$ be the subspace of $\mathcal{H}$ generated by all the vectors which are proportional to $(a^\dagger)^l\Phi_0$. Let also $\mathcal{F}_L$ be the direct sum $\mathcal{F}_L := \mathcal{E}_0 \oplus \mathcal{E}_1 \oplus \ldots\ldots \oplus \mathcal{E}_L$. Finally, let $N = \sum_{l=0}^{\infty} l\Pi_l$ be the spectral decomposition of the number operator N. The operators Π_l are projection operators, as well as the operators $Q_L = \sum_{l=0}^{L} \Pi_l$. The following properties are obvious:

$$\Pi_k\Pi_l = \delta_{kl}\Pi_l, \quad \Pi_k^\dagger = \Pi_k; \qquad Q_LQ_M = Q_L, \quad \text{if } L \leqslant M, \qquad Q_L^\dagger = Q_L.$$

It is clear that $\Pi_k : \mathcal{H} \to \mathcal{E}_k$, and $Q_L : \mathcal{H} \to \mathcal{F}_L$. The operator Q_L is used to cut-off the hamiltonian, by replacing a with $a_L := Q_L a Q_L$. The regularized hamiltonian is simply $H_L = Q_L N Q_L = N Q_L$ and the related time evolution is $\alpha_L^t(X) = e^{iH_Lt}Xe^{-iH_Lt}$. This *occupation number cut-off* produces a self adjoint bounded operator H_L and we have shown in [4] that *the limits of* $\alpha_L^t(a^n)$ *and* $\alpha_L^t((a^\dagger)^n)$ *exist in* $\mathcal{L}^\dagger(\mathcal{D})[\tau_0]$ *for all* $n \in \mathbb{N}$. Incidentally, let us remind that we have just seen that this result can be further generalized, see [11].

The same algebraic framework turns out to be useful also in the analysis of the thermodynamical limit of the interacting model described by the following formal hamiltonian:

$$H_V = \frac{J}{|V|} \sum_{i,j \in V} \sigma_i^3 \sigma_j^3 + a^\dagger a + \gamma (a + a^\dagger) \sigma_V^3,$$

where $\sigma_V^3 = \frac{1}{|V|} \sum_{i \in V} \sigma_i^3$. Here the algebra $\mathcal{L}^\dagger(\mathcal{D})$ must be replaced by $\mathfrak{A} = B(\mathcal{H}_{spin}) \otimes \mathcal{L}^\dagger(\mathcal{D})$. The topology on $\mathfrak{A}$, τ_{comp}, is generated by the following seminorms: $\|XA\|^{f,k,\Psi} \equiv \|X\|^{f,k} \|A\Psi\|$, $X \in \mathcal{L}^\dagger(\mathcal{D})$ and $A \in B(\mathcal{H}_{spin})$. It is worthwhile to remind also that Ψ cannot be a generic vector in $\mathcal{H}_{spin}$, but must belong to the set

$$\mathcal{F} = \left\{ \Psi \in \mathcal{H}_{spin} \; : \; \lim_{|V|,\infty} \frac{1}{|V|} \sum_{p \in V} \sigma_p^3 \Psi = \sigma_\infty^3 \Psi, \;\; \|\sigma_\infty^3\| \leqslant 1 \right\}.$$

As before, the regularized hamiltonian is obtained by replacing a with $a_L := Q_L a Q_L$, so that the new hamiltonian $H_{V,L}$ depends on two, in principle, unrelated cutoffs. The existence of the limit of $\alpha_{V,L}^t(X) = e^{iH_{V,L}t} X e^{-iH_{V,L}t}$ is ensured by the following result, [4]: *the limit of* $\alpha_{V,L}^t(a)$ *for* $|V|$ *and* L *both diverging exists in* $\mathfrak{A}[\tau_0]$. *Moreover, if the two cutoffs satisfy the relation* $|V| = L^r$, *for a certain integer* $r > 1$, *the same holds true also for* $\alpha_{V,L}^t(\sigma_\alpha^i)$.

Acknowledgement

This work has been financially supported in part by M.U.R.S.T., within the project *Problemi Matematici Non Lineari di Propagazione e Stabilità nei Modelli del Continuo*, coordinated by Prof. T. Ruggeri.

References

[1] G. Alli and G. L. Sewell, *New methods and structures in the theory of the multi-mode Dicke laser model*, J. Math. Phys. **36**, 5598 (1995)

[2] J.-P. Antoine, A. Inoue and C. Trapani *Partial *-algebras and Their Operator Realizations*, Kluwer, Dordrecht, 2002

[3] J.-P. Antoine and W. Karwowski, *Partial *-Algebras of Closed Linear Operators in Hilbert Space*, Publ.RIMS, Kyoto Univ. 21, 205-236 (1985).

[4] F. Bagarello, *Applications of Topological *-Algebras of Unbounded Operators*, J. Math Phys., **39**, 2730 (1998)

[5] F. Bagarello, *Fixed Points in Topological *-Algebras of Unbounded Operators*, Publ. RIMS, Kyoto Univ. 37, (2001), 397-418.

[6] F. Bagarello, G. Morchio, *Dynamics of mean field spin models from basic results in abstract differential equations*, J. Stat. Phys. **66**, 849 (1992)

[7] F. Bagarello and C. Trapani, *'Almost' Mean Field Ising Model: an Algebraic Approach*, J.Statistical Phys. 65, (1991), 469-482.

[8] F. Bagarello and C. Trapani, *A note on the algebraic approach to the "almost" mean field Heisenberg model*, Il Nuovo Cimento B 108, (1993), 779-784.

[9] F. Bagarello and C. Trapani, *States and representations of CQ^*-algebras*, Ann. Inst. H. Poincaré 61, (1994), 103-133.

[10] F. Bagarello, C. Trapani, *The Heisenberg Dynamics of Spin Sistems: a Quasi*-Algebras Approach*, J. Math. Phys. 37, (1996), 4219-4234.

[11] F. Bagarello, C. Trapani, *Algebraic dynamics in O*-algebras: a perturbative approach* , J. Math. Phys. 43, (2002), 3280-3292.

[12] F. Bagarello, A. Inoue, C. Trapani, *Derivations of quasi $*$-algebras*, Int. Jour. Math. and Math. Sci., **21**, 1077-1096 (2004)

[13] F. Bagarello, A. Inoue, C. Trapani, *Exponentiating derivations of quasi *-algebras: possible approaches and applications*, Int. Jour. Math. and Math. Sci., **17**, 2805-2820 (2005)

[14] F. Bagarello, G. L. Sewell, *New Structures in the Theory of the Laser Model II: Microscopic Dynamics and a Non-Equilibrim Entropy Principle*, J. Math. Phys., **39** (1998), 2730-2747.

[15] F. Bagarello, C. Trapani *Morphisms of Certain Banach C*-Algebras*, Publ. RIMS, Kyoto Univ., **36**, No. 6, 681-705, (2000)

[16] F. Bagarello, A. Inoue, C. Trapani, *Some classes of topological quasi $*$-algebras*, Proc. Amer. Math. Soc., **129**, 2973-2980 (2001).

[17] O. Bratteli and D.W. Robinson, *Operator algebras and Quantum statistical mechanics 1*, Springer-Verlag, New York, 1987.

[18] O. Bratteli and D.W. Robinson, *Operator algebras and Quantum statistical mechanics 2*, Springer-Verlag, New York, 1987.

[19] E. Buffet, P.A. Martin, *Dynamics of the Open BCS Model*, J. Stat. Phys., **18**, No. 6, 585-632, (1978)

[20] A.M. Chebotarev, *Lectures on quantum probability*, Sociedad Matemática Mexicana (2000)

[21] D.A. Dubin, G.L. Sewell, *Time-translations in the algebraic formulation of statistical mechanics*, J. Math. Phys. **11**, (1970), 2990-2998.

[22] G.G. Emch, *Algebraic Methods in Statistical Mechanics and Quantum Field Theory*, Wiley, New York, (1972)

[23] G.G. Emch, H.J.F. Knops, *Pure thermodynamical phases as extremal KMS states*, J. Math. Phys. **11**, (1970), 3008-3018.

[24] F. Fagnola, *Quantum Markov Semigroups and Quantum Flows*, Proyecciones, **18**, no. 3, 1-144, (1999)

[25] R. Haag, *Local Quantum Physics*, Springer, Berlin (1992)

[26] R. Haag, N.M. Hugenholtz, M. Winnink, *On the equilibrium states in quantum statistical mechanics*, Comm. Math. Phys., **5**, 215-236, (1967)

[27] R. Haag and D. Kastler, *An Algebraic Approach to Quantum Field Theory*, J.Math.Phys. 5, (1964), 848-861.

[28] G. Lassner, *Topological algebras and their applications in Quantum Statistics*, Wiss. Z. KMU-Leipzig, Math.-Naturwiss. R. 30, (1981), 572–595.

[29] G. Morchio and F. Strocchi, *Mathematical structures for long range dynamics and symmetry breaking* J. Math. Phys. 28, (1987), 622-635.

[30] D. Ruelle, *Statistical Mechanics*, W.A. Benjamin, New York (1969)

[31] M. Schröder and W. Timmermann, *Invariance of domains and automorphisms in algebras of unbounded operators*, in Proc. Int. Conf. on Operator Algebras and Group Representations, Romania (1980), 134-139.

[32] K. Schmüdgen, *Unbounded operator algebras and Representation theory*, Birkhäuser, Basel, 1990

[33] G.L. Sewell, *Quantum Theory of Collective Phenomena*, Oxford University Press, Oxford (1989)

[34] G.L. Sewell, *Quantum Mechanics and its Emergent Macrophysics*, Princeton University Press, (2002)

[35] F. Strocchi, *Elements of Quantum Mechanics of Infinite Systems*, World Scientific, (1985)

[36] W. Thirring, *A Course in Mathematical Physics 4: Quantum Mechanics of Large Systems*, Springer-Verlag, Wien (1980)

[37] W. Thirring and A. Wehrl, *On the Mathematical Structure of the B.C.S.-Model*, Commun.Math.Phys. 4, (1967), 303-314.

[38] C. Trapani, *Bounded elements and spectrum in Banach quasi *-algebras*, Studia Matematica, in press

Dipartimento di Metodi e Modelli Matematici, Facoltà di Ingegneria, Università di Palermo, Viale delle Scienze, I-90128 Palermo, Italy *E-mail:* bagarell@unipa.it home page: www.unipa.it/~bagarell

Where's That Quantum?

Stephan De Bièvre

The nature and properties of the vacuum as well as the meaning and localization properties of one or many particle states have attracted a fair amount of attention and stirred up sometimes heated debate in relativistic quantum field theory over the years. I will review some of the literature on the subject and will then show that these issues arise just as well in non-relativistic theories of extended systems, such as free bose fields. I will argue they should as such not have given rise either to surprise or to controversy. They are in fact the result of the misinterpretation of the vacuum as "empty space" and of a too stringent interpretation of field quanta as point particles. I will in particular present a generalization of an apparently little known theorem of Knight on the non-localizability of field quanta, Licht's characterization of localized excitations of the vacuum, and explain how the physical consequences of the Reeh-Schlieder theorem on the cyclicity and separability of the vacuum for local observables are already perfectly familiar from non-relativistic systems of coupled oscillators.

1. Introduction

Quantum field theory is the study of quantum systems with an infinite number of degrees of freedom. The simplest quantum field theories are the free bose fields, which are essentially assemblies of an infinite number of coupled oscillators. Examples include the quantized electromagnetic field, lattice vibrations in solid state physics, and the Klein-Gordon field, some of which are relativistic while others are not. Particles show up in quantum field theory as "field quanta." As we will see, these do not quite have all the properties of the usual particles of Newtonian physics or of Schrödinger quantum mechanics. I will in particular show (Section 3) that the quanta of free bose fields, relativistic or not, can not be perfectly localized in a bounded subset of space. This, in my opinion, shows conclusively that the difficulties encountered when attempting to define a position observable for the field quanta of relativistic fields, that continue to be the source of regular debate in the literature [20] [21] [22] [3] [10] [15] [18] [19] [48] [47], do not find their origin in any form of causality violation, as seems to be generally thought. Instead, they result from an understandable but ill-fated attempt to force too stringent a particle interpretation on the states of the quantum field containing a finite number of quanta.

States of free bose fields that are perfectly localized in bounded sets exist, but necessarily contain an infinite number of quanta. They can be classified quite easily, following the ideas of Licht [28]. This is done in Sections 5 and 6. It turns out that those states do not form a vector subspace of the quantum Hilbert space. This has on occasion been presented as surprising or paradoxical within the context of relativistic quantum field theory, but it isn't: these properties of localized states are familiar already from finite systems of coupled harmonic oscillators.

The previous issues are intimately related to certain properties of the vacuum of free bose fields, that also have stirred up animated debate in the literature. I will in particular explain the physical implications of the Reeh-Schlieder theorem for free bose fields and their link with the localization issue. The Reeh-Schlieder theorem has been proven in the context of relativistic field theory, but holds equally well for (finite or infinite) non-relativistic systems of coupled oscillators. By studying it in that context, one easily convinces oneself that its implications for the quantum theory of measurement, for example, have nothing particularly surprising or paradoxical, and do not lead to causality violation, but are the result of the usual "weirdness" of quantum mechanics, since they are intimately linked to the observation that the vacuum is an entangled state. This thesis is developed in Section 7.

The paper is organized as follows. After defining the class of free bose fields under consideration (Section 2), I will briefly discuss the question of the "localizability" of states with a finite number of quanta in Section 3. I will in particular recall from [7] a generalization to free bose fields of a little known result of Knight about the Klein-Gordon field [27], which shows they can not be perfectly localized excitations of the vacuum. I will then discuss the various, sometimes contradictory viewpoints on the question of particle localization prevalent in quantum field theory textbooks in Section 4 and compare them to the one presented here. I will argue in detail that the latter could help to clarify the situation considerably.

Section 5 contains the precise definition of "localized excitation of the vacuum" that I am using and some technical material needed for the presentation of Licht's theorem on the characterization of local states in Section 6. Section 7 explains the physical implications of the Reeh-Schlieder theorem for free bose fields and their link with the localization issue.

Some of the proofs missing here, as well as a detailed critique of the Newton-Wigner position operator from the present viewpoint on localization can be found in [7] which, together with the present work, is itself a short version of Chapter 6 of [5]. I refer to these references for more details.

Intrigued by a talk by G. Hegerfeldt at the University of Rochester, where I was a graduate student, I have touched upon the particle localization issue in the last chapter of my PhD thesis [6], written under Gérard Emch's supervision, and it has continued to intrigue me. In recent times, I have come back to this and to the related issues discussed here, from quite a different angle. Knowing Gérard's lifelong fondness for the algebraic approach to quantum field theory and statistical mechanics as well as for foundational issues in physics, I feel I can safely assume that the mixture of both I am presenting in these pages will provide him with some reading pleasure, particularly because there is plenty of material here for him to disagree with.

2. Free Bose Fields

The simple physical systems under study here obey an equation of the form

$$\ddot{q} + \Omega^2 q = 0, \tag{2.1}$$

where Ω is a self-adjoint, positive operator on a dense domain $\mathcal{D}(\Omega)$ in a real Hilbert space $\mathcal{K}$ and having a trivial kernel. I will always suppose $\mathcal{K}$ is of the form $\mathcal{K} = L^2_{\mathrm{r}}(K, \mathrm{d}\mu)$, where K is a topological space and μ a Borel measure on K. Here the subscript "r" indicates that we are dealing with the real Hilbert space of real-valued functions. In fact, all examples of interest I know of are of this type. Those include:

(i) Finite dimensional systems of coupled oscillators, where $\mathcal{K} = \mathbb{R}^n$ and Ω is a positive definite matrix;

(ii) Lattices or chains of coupled oscillators, where $\mathcal{K} = \ell^2(\mathbb{Z}^d, \mathbb{R})$ and Ω^2 is usually a bounded finite difference operator with a possibly unbounded inverse;

(iii) The wave and Klein-Gordon equations, where $\mathcal{K} = L^2(K, \mathbb{R})$, $K \subset \mathbb{R}^d$ and $\Omega^2 = -\Delta + m^2$ with suitable boundary conditions. More precisely, the Klein-Gordon equation is the equation

$$\partial_t^2 q(t,x) = -(-\Delta + m^2) q(t,x).$$

Equation (2.1) can be seen as a Hamiltonian system with phase space

$$\mathcal{H} = \mathcal{K}_{1/2} \oplus \mathcal{K}_{-1/2}, \quad \text{where} \quad \mathcal{K}_{\pm 1/2} = [\mathcal{D}(\Omega^{\pm 1/2})].$$

Here the notation [ ] means that we completed $\mathcal{D}$ in the topology induced by $\| \Omega^{\pm 1/2} q \|$ where $\| \cdot \|$ is the Hilbert space norm of $\mathcal{K}$. On $\mathcal{H}$, the Hamiltonian $(X = (q, p) \in \mathcal{H})$

$$H(X) = \frac{1}{2} p \cdot p + \frac{1}{2} q \cdot \Omega^2 q, \tag{2.2}$$

defines a Hamiltonian flow with respect to the symplectic structure

$$s(X, X') = q \cdot p' - q' \cdot p.$$

The Hamiltonian equations of motion $\dot{q} = p$, $\dot{p} = -\Omega^2 q$ are equivalent to (2.1). Note that I use $\cdot$ for the inner product on $\mathcal{K}$.

The quantum mechanical description of these systems can be summarized as follows. Given a harmonic system determined by $\mathcal{K}$ and Ω, one chooses as the quantum Hilbert space of such a system the symmetric Fock space $\mathcal{F}^+(\mathcal{K}^{\mathbb{C}})$, and as quantum Hamiltonian the second quantization of Ω: $H = \mathrm{d}\Gamma(\Omega)$. Note that this is a positive operator and that the Fock vacuum is its ground state, with eigenvalue 0. In terms of the standard creation and annihilation operators on this Fock space, the quantized fields and their conjugates are then *defined* by ($\eta \in \mathcal{K}^{\mathbb{C}}_{-1/2}$):

$$\eta \cdot Q := \frac{1}{\sqrt{2}}(a(\Omega^{-1/2}\overline{\eta}) + a^{\dagger}(\Omega^{-1/2}\eta)), \tag{2.3}$$

and, similarly ($\eta \in \mathcal{K}^{\mathbb{C}}_{1/2}$),

$$\eta \cdot P := \frac{\mathrm{i}}{\sqrt{2}}(a^{\dagger}(\Omega^{1/2}\eta) - a(\Omega^{1/2}\overline{\eta})). \tag{2.4}$$

For later purposes, I define, for each $\xi \in \mathcal{K}^{\mathbb{C}}$, the Weyl operator

$$W_{\mathrm{F}}(\xi) = \exp(a^{\dagger}(\xi) - a(\xi)). \tag{2.5}$$

I will also need, for each $X = (q, p) \in \mathcal{H}$

$$z_{\Omega}(X) = \frac{1}{\sqrt{2}}(\Omega^{1/2}q + \mathrm{i}\Omega^{-1/2}p) \in \mathcal{K}^{\mathbb{C}}. \tag{2.6}$$

It follows easily that

$$W_{\mathrm{F}}(z_{\Omega}(X)) = \exp -\mathrm{i}(q \cdot P - p \cdot Q).$$

Loosely speaking, the observables of the theory are "all functions of Q and P." For mathematical precision, one often uses various algebras (called CCR-algebras) generated by the Weyl operators $W_{\mathrm{F}}(z_{\Omega}(X))$, $X \in \mathcal{H}$, as we will see in some more detail below.

Things are particularly simple when the set K is discrete, as in examples (i) and (ii) above. One can then define the displacements Q_j and momenta P_j of the individual oscillators, with $j = 1, \ldots, n$ in the first case and $j \in \mathbb{Z}^d$ in the second case. These examples are a helpful guide to the intuition, as we will see below.

Apart from the vacuum $|0\rangle$, which is the ground state of the system, excited states of the form

$$a^{\dagger}(\eta_1)\ldots a^{\dagger}(\eta_k)|0\rangle,$$

play a crucial role and are referred to as states with k "quanta." The quanta of the Klein-Gordon field, for example are thought of as spinless particles of mass m. In the case of an oscillator chain or lattice, they are referred to as "phonons." To the extent that these quanta are thought of as "particles," the question of their whereabouts is a perfectly natural one. It is to its discussion I turn next.

3. So, Where's That Quantum?

Let me start with an informal discussion of the issue under consideration. Among the interesting observables of the oscillator systems we are studying are certainly the "local" ones. I will give a precise definition in Section 5, but thinking for example of a finite or infinite oscillator chain, "the displacement q_7 or the momentum p_7 of the seventh oscillator" is certainly a "local" observable. In the same way, if dealing with a wave equation, "the value $q(x)$ of the field at x" is a local observable. Generally, "local observables" are functions of the fields and conjugate fields in a bounded region of space. In the case of the oscillator chain or lattice, space is $K = \mathbb{Z}^d$ $(d \geqslant 1)$, and for a finite chain, space is simply the index set $K = \{1, \ldots n\}$.

I find this last example personally most instructive. It forces one into an unusual point of view on a system of n coupled oscillators that is well suited for making the transition to the infinite dimensional case. Think therefore of a system of n oscillators characterized by a positive n by n matrix Ω^2. A local observable of such a system is a function of the positions and momenta of a fixed finite set B of oscillators. In this case, $\mathcal{K} = \mathbb{R}^n$, which I view as $L^2_{\mathrm{r}}(K)$, where K is simply the set of n elements. Indeed, $q \in \mathbb{R}^n$ can be seen as a function $q : j \in \{1, \ldots n\} \mapsto q(j) \in \mathbb{R}$, obviously square integrable for the counting measure.

Consider now a subset B of K, say $B = \{3, 6, 9\}$. A local observable over B is then a finite linear combination of operators on $L^2(\mathbb{R}^n)$ of the form $(a_j, b_j \in \mathbb{R}, j \in B)$:

$$\exp -\mathrm{i}\left(\sum_{j\in B}(a_j P_j - b_j Q_j)\right).$$

Note that those form an algebra. More generally it is an operator of the form

$$\int \left(\Pi_{j\in B}\mathrm{d}a_j\mathrm{d}b_j\right)\ f(a_j, b_j)\,\mathrm{e}^{-\mathrm{i}\left(\sum_{j\in B}(a_j P_j - b_j Q_j)\right)},$$

for some function f in a reasonable class. In other words, it is a function of the position and momentum operators of the oscillators inside the set B.

Better yet, if you write, in the Schrödinger representation,

$$L^2(\mathbb{R}^n) \cong L^2(\mathbb{R}^{\sharp B}, \prod_{j\in B}\mathrm{d}x_j) \otimes L^2(\mathbb{R}^{n-\sharp B}\prod_{j\notin B}\mathrm{d}x_j),$$

then it is clear that the weak closure of the above algebra of local observables is

$$\mathcal{B}(L^2(\mathbb{R}^{\sharp B}, \prod_{j\in B}\mathrm{d}x_j)) \otimes \mathbb{1}.$$

So, indeed, a local observable is clearly one that acts only on the degrees of freedom indexed by elements of B. The definition of a local observable over a finite subset B of any oscillator lattice is perfectly analogous, where this time the Q_j, P_j are defined on Fock space, as explained in the previous section. This definition is natural and poses no problems.

Now what is a local state of such a system? More precisely, I want to define what a "strictly local excitation of the vacuum" is. The equivalent classical notion is readily described. The vacuum, being the ground state of the system, is the quantum mechanical equivalent of the global equilibrium $X = 0$, which belongs of course to the phase space $\mathcal{H}$, and a local perturbation of this equilibrium is an initial condition $X = (q, p)$ with the support of q and of p contained in a subset B of K. An example of a local perturbation of an oscillator lattice is a state $X \in \mathcal{H}$ where only q_0 and p_0 differ from 0. In the classical theory, local perturbations of the equilibrium are therefore states that differ from the equilibrium state only inside a bounded subset B of K. It is this last formulation that is readily adapted to the quantum context, through the use of the notion of "local observable" introduced previously. Turning again to the above example, let me write $|0, \Omega\rangle$ for the ground state of the oscillator system in the Schrödinger representation; then a strictly local excitation of the vacuum over B is a state $\psi \in L^2(\mathbb{R}^n, \mathrm{d}x)$ so that, for all $a_j, b_j \in \mathbb{R}$,

$$\langle\psi|\exp -\mathrm{i}\sum_{j\notin B}(a_jP_j - b_jQ_j)|\psi\rangle = \langle 0, \Omega|\exp -\mathrm{i}\sum_{j\notin B}(a_jP_j - b_jQ_j)|0, \Omega\rangle.$$

In other words, outside B, the states ψ and $|0, \Omega\rangle$ coincide. Any measurement performed on a degree of freedom outside B gives the same result, whether the system is in the vacuum state or in the state ψ. The mean kinetic or potential energy of any degree of freedom outside B is identical in both cases as well. All this certainly expresses the intuitive notion of "localized excitation of the vacuum". Note that it is based on the idea of viewing the full system as composed of two subsystems: the degrees of freedom inside B and the degrees of freedom outside B. The analogous definition of strictly local excitation of the vacuum for the general class of bose fields considered in the previous section is easily guessed and given in Section 5.

The following question arises naturally. Let's consider a free bose field and suppose we consider some state containing one quantum, meaning a state of the form $a^\dagger(\xi)|0\rangle$. Can such a state be perfectly localized in a bounded set B? Since we like to think of these quanta as particles, one could a priori expect the answer to be positive, but the answer is simply: "NO, not in any model of interest." This is the content of the generalization of Knight's theorem proven in [7] (see Theorem 7.1 (iv)). States containing only one quantum (or even a finite number of them), cannot be strictly localized excitations of the vacuum. This is a little surprising at first, but perfectly natural. In fact, it is true even in finite chains of oscillators, as I will now show.

A special case of the result is indeed easily proven by hand, and clearly brings out the essential ingredient of the general phenomenon. Consider a finite oscillator chain, and suppose simply Ω^2 does not have any of the canonical basis vectors e_i of $\mathbb{R}^n$ as an eigenmode. This means that each degree of freedom is coupled to at least one other one and is certainly true for the translationally invariant finite chain, to give a concrete example. I will show by a direct computation that in this situation there does not exist a one quantum state $a^\dagger(\xi)|0\rangle$ ($\xi \in \mathbb{C}^n, \overline{\xi} \cdot \xi = 1$), that is a perturbation of the vacuum strictly localized on one of the degrees of freedom, say the first one $i = 1$. In other words, there is no such state having the property that, for all $Y = (a, b) \in \mathbb{R}^{2n}$, $a_1 = 0 = b_1$, one has

$$\langle 0|a(\xi) \exp -\mathrm{i} \sum_{j=2}^{n} (a_j P_j - b_j Q_j) a^\dagger(\xi)|0\rangle = \langle 0| \exp -\mathrm{i} \sum_{j=2}^{n} (a_j P_j - b_j Q_j)|0\rangle. \tag{3.1}$$

Now, a simple computation with the Weyl operators shows that this last condition is equivalent to

$$\overline{\xi} \cdot z_\Omega(Y) = 0,$$

for all such Y. Here $z_\Omega(Y) = \frac{1}{\sqrt{2}}(\Omega^{1/2}a + \mathrm{i}\Omega^{-1/2}b)$, and simple linear algebra then implies that this last condition can be satisfied for some choice of ξ if and only if e_1 is an eigenvector of Ω. But this implies it is an eigenvector of Ω^2, which is a situation I excluded. Hence $a^\dagger(\xi)|0\rangle$ is not a strictly localized excitation of the vacuum at site $i = 1$ for any choice of ξ.

Of course, if the matrix Ω^2 is diagonal, this means that the degrees of freedom at the different sites are not coupled, and then the result breaks down. But in all models of interest, the degrees of freedom at different points in space are of course coupled.

The main ingredient for the proof of the generalization of Knight's theorem to free bose fields given in [7] is the non-locality of Ω. A precise definition will follow below, but the idea is that the operator Ω, which is the square root of a finite difference or of a second order differential operator in all models of interest, does not preserve supports. The upshot is that states of free bose fields with a finite number of particles, and a fortiori, one-particle states, are never strictly localized in a bounded set B. This gives a precise sense in which the elementary excitations of the vacuum in a bosonic field theory (relativistic or not) differ from the ordinary point particles of non-relativistic mechanics: their Hilbert space of states contains no states in which they are perfectly localized.

Having decided that one-quantum states cannot be strictly localized excitations of the vacuum on bounded sets, the question arises if such strictly localized states exist. Sticking to the simple example of the chain, any excitation of the vacuum strictly localized on the single site $i = 1$ can be proven to be of the type $\exp \mathrm{i}F(Q_1, P_1)|0\rangle$, where $F(Q_1, P_1)$ is a self-adjoint operator, function of Q_1 and P_1 alone. For a precise statement, see Theorem 7.1. Coherent states $\exp -\mathrm{i}(a_1 P_1 - b_1 Q_1)|0\rangle$ are of this type. So there are plenty such states. Note however that the linear superposition of two such states is not usually again such a state: the strictly localized excitations of the vacuum on the site $i = 1$ do not constitute a vector subspace of the space of all states. This is in sharp contrast to what happens when, in the non-relativistic quantum mechanics of a system of n particles moving in $\mathbb{R}$, we ask the question: "What are the states $\psi(x_1, \ldots, x_n)$ for which all particles are in some interval $I \subset \mathbb{R}$?" These are all wave functions supported on $I \times \cdots \times I$, and they clearly form a vector space. In that case, to the question "Are all particles inside I?" corresponds therefore a projection operator P_I with the property that the answer is "yes" with probability $\langle \psi | P_I | \psi \rangle$. But in oscillator lattices there is no projection operator corresponding to the question "Is the state a strictly local excitation of the vacuum inside B?". This situation reproduces itself in relativistic quantum field theory, and does not any more constitute a conceptual problem there as in the finite oscillator chain. I will discuss this point in more detail in Section 7.

In view of the above, it is clear that no position operator for the quanta of, for example, the Klein-Gordon field can exist. Those quanta simply do not have all attributes of the point particles of our classical mechanics or non-relativistic quantum mechanics courses. But, since the same conclusion holds for the quanta of a lattice vibration field, this has nothing to do with causality or relativity, as seems to be generally believed. It nevertheless seems that this simple lesson of quantum field theory has met and still continues to meet with a lot of resistance, as we will see in Section 4.

So, to sum it all up, one could put it this way. To the question

> Why is there no sharp position observable for particles?

the answer is

> It is the non-locality of Ω, stupid!

More details on this aspect of the story have been published in [7], where it is in particular argued more forcefully that the Newton-Wigner position operator is not a good tool for describing sharp localization properties of the quanta of quantum fields.

4. Various Viewpoints on Localization

Now, what is currently the standard view in the physics community on the question of localization of particles or quanta in field theory? Let me first point out that this is not necessarily easy to find out from reading the textbooks on quantum field theory or relativistic quantum physics. Indeed, the least one can say is that the whole localization issue does not feature prominently in these books. One seems to be able to detect three general attitudes. First of all, some books make no mention of it at all: [31] [9] [11] [26] [12]. In [9], for example, when discussing the attributes of particles in an introductory chapter, the authors identify those as rest mass, spin, charge and lifetime, but do not mention any notion of position or localizability.

A second group of authors give an intuitive discussion, based on the uncertainty principle, together with the non-existence of superluminal speeds, to explain single massive particles should not be localizable within a region smaller than their Compton wavelength: [4] [36] [8] are examples. The idea is that, since $\Delta X \Delta P \geqslant \hbar$, whenever ΔX is of the order of the Compton wavelength $\hbar/mc$, $\Delta P \geqslant mc$. Since the gap between positive and negative eigenstates is $2mc^2$, this is indicating that to obtain such sharp localization, negative energy eigenstates are "needed." Let me point out that this reasoning does not by any means exclude the possibility of having one-particle states strictly localized in regions bigger than the Compton wavelength. The argument is then further used to support the idea that in a fully consistent relativistic quantum theory, one is unavoidably led to a many body theory with particle/anti-particle creation, and to field theory.

The essential idea underlying this type of discussion is that, even though the solutions to the relevant wave equation (Klein-Gordon or Dirac, mostly) do not, as such, have a satisfactory probabilistic interpretation, the coupling to other (classical or quantum) fields is done via the solution and so the particle is present essentially where this wave function is not zero: the particle is where its energy density is. In this spirit, Björken and Drell, for example write, at the end of the section where they address some of the problems associated with the single particle interpretation of the Klein-Gordon and Dirac equations (including a brief discussion of the Klein paradox at a potential step): "We shall tackle and resolve these questions in Chapter 5. Before doing this let us look in the vast, if limited, domain of physical problems where the application forces are weak and smoothly varying on a scale whose energy unit is mc^2 and whose distance unit is $\frac{\hbar}{mc}$. Hence we may expect to find fertile fields for application of the Dirac equation for positive energy solutions."

This attitude means that whenever a conceptual problem arises, it is blamed on the single-particle approach taken. The trouble is that, once these authors treat the full field theory, they do not come back to the localization issue at all, be it for particle states or for general states of the field.

A third group of authors give a sligthly more detailed discussion, including of the Newton-Wigner position operator, but fail to indicate the problems associated with the latter, leaving the impression that strict particle localization is quite possible after all: [37] [16] [41] [42]. Sterman, for example, when discussing one-quantum states of the field, writes: "To merit the term 'particle', however, such excitations [of the quantum field] should be localisable." He then discusses the Newton-Wigner operator as the solution to this last problem, without pointing out the difficulties associated with it. Something similar happens in Schweber who writes: "By a single particle state we mean an entity of mass m and spin 0 which has the property that the events caused by it are localized in space." Interestingly, neither of them makes any mention of the contradicting intuitive argument which completely rules out perfect single particle localization. This is in contrast to Greiner, who does give this argument, but does not point the apparent contradiction with the notion of a position operator, the (generalized) eigenstates of which are supposed to correspond to a perfectly localized particle. Of course, this last point is obscured, as in many cases, by the observation that the Klein-Gordon wave function corresponding to such a state is exponentially decreasing with a localization length which is the Compton wavelength.

Let me note in passing that it is generally admitted that a photon is not localisable at all, not even approximately. This is sometimes argued by pointing out it admits no Newton-Wigner position operator, or intuitively, based on the uncertainty principle argument which blames this on its masslessness: its Compton wavelength is infinite. Note however that massless spinless particles *are* localisable in the Newton-Wigner sense, so that this argument is not convincing. In fact, it was proven in [3] that one-photon states can be localized with sub-exponential tails in the sense that the field energy density of the state decreases sub-exponentially away from a localization center.

To summarize, it seems there is no simple, generally agreed upon and clearly argued textbook viewpoint on the question of localization in quantum field theory, even for free fields. The general idea seems to be that the above intuitive limitations on particle localization give a sufficient understanding since, at any rate, the whole issue is not very important. As Bacry puts it, not without irony, in [2]: "The position operator is only for students and . . . for people interested in the sex of the angles, this kind of people you find among mathematical physicists, even among the brightest ones such as Schrödinger and Wigner."

In my view, it is the absence of a clear definition of "localized state" – such as the one provided by Knight – that has left the field open for competing speculations on how to circumvent the various problems with the idea of sharp localization of particles and in particular with the Newton-Wigner operator. Some authors seem to believe strongly in the need for a position observable for particles, claiming the superluminal speeds it entails do not constitute a problem after all, essentially because the resulting causality violation is too small to be presently observable [35] [15]. Others have provided alternative constructions with non-commuting components [2] [15].

To me, this is similar to clinging at all cost to ether theory in the face of the "strange" properties of time implied by Einstein's special relativity. Between giving up causality or giving up position operators for field quanta, I have made my choice. This, together with the definition of Knight and the accompanying theorem, which could easily be explained in simple terms in physics books, as Section 3 shows, would go a long way in clarifying the situation. On top of that, it would restore the "democracy between particles" [2]. In fact, contrary to what is usually claimed and although I have not worked out this in detail here, photons are no worse or better than electrons when it comes to localization, a point of view that I am not the first one to defend. Peierls, for example, in [32], compares photon and electron properties with respect to localization and although he starts of with the statement "On the other hand, one of the essentially particle-like properties of the electron is that its position is an observable, there is no such thing as the position of the photon," he concludes the discussion as follows, after a more careful analysis of the relativistic regime: "If we work at relativistic energies, the electron shows the same disease. So in this region, the electron is as bad a particle as the photon."

At any rate, if you find my point of view difficult to accept and are reluctant to do so, you are in good company. Here is what Wigner himself says about it in [49], almost forty years after his paper with Newton: "One either has to accept this [referring to non-causality] or deny the possibility of measuring position precisely or even giving significance to this concept: a very difficult choice." In spite of this, in the conclusion of this same article, he writes, apparently joining my camp: "Finally, we had to recognize, every attempt to provide a precise definition of a position coordinate stands in direct contradiction to relativity."

Having advocated Knight's definition of "strictly local excitations of the vacuum", I turn in Section 6 to their further study. First, we need some slightly more technical material.

5. All Things Local

Let's recall we consider harmonic systems over a real Hilbert space $\mathcal{K}$ of the form $\mathcal{K} = L^2_{\mathrm{r}}(K, \mathrm{d}\mu)$, where K is a topological space and μ a Borel measure on K. I need to give a precise meaning to "local observables in $B \subset K$", for every Borel subset of K. To do that, I introduce the notion of "local structure", which is a little abstract, but the examples given below should give you a good feel for it.

Definition 5.1. A local structure for the oscillator system determined by Ω and $\mathcal{K} = L^2_{\mathrm{r}}(K, \mathrm{d}\mu)$ is a subspace $\mathcal{S}$ of $\mathcal{K}$ with the following properties:

(1) $\mathcal{S} \subset \mathcal{K}_{1/2} \cap \mathcal{K}_{-1/2}$;

(2) Let B be a Borel subset of K, then $\mathcal{S}_B := \mathcal{S} \cap L^2_{\mathrm{r}}(B, \mathrm{d}\mu)$ is dense in $L^2_{\mathrm{r}}(B, \mathrm{d}\mu)$.

In addition, we need

$$\mathcal{H}(B, \Omega) \stackrel{\text{def}}{=} \mathcal{S}_B \times \mathcal{S}_B.$$

Note that, thanks to the density condition in the definition, this is a symplectic subspace of $\mathcal{H}$. This is a pretty strange definition, and I will turn to the promised examples in a second, but let me first show how to use this definition to define what is meant by "local observables".

Definition 5.2. Let $\mathcal{K} = L^2_{\mathrm{r}}(K, \mathrm{d}\mu)$, Ω, $\mathcal{S}$ be as above and let B be a Borel subset of K. The algebra of local observables over B is the algebra

$$\mathrm{CCR}_0(\mathcal{H}(B,\Omega)) = \mathrm{span}\,\{W_{\mathrm{F}}(z_\Omega(Y)) \mid Y \in \mathcal{S}_B \times \mathcal{S}_B\}.$$

Here "CCR" stands for Canonical Commutation Relations. The algebras $\mathrm{CCR}_0(\mathcal{H}(B,\Omega))$ form a net of local algebras in the usual way [14] [24] [17]. Note that Ω plays a role in the definition of $\mathcal{S}$ through the appearance of the spaces $\mathcal{K}_{\pm 1/2}$. The first condition on $\mathcal{S}$ guarantees that $\mathcal{S} \times \mathcal{S} \subset \mathcal{H}$ so that, in particular, for all $Y \in \mathcal{S} \times \mathcal{S}$, $s(Y, \cdot)$ is well defined as a function on $\mathcal{H}$ which is important for the definition of the local observables to make sense.

For the wave or Klein-Gordon equation, one can choose $\mathcal{S}$ to be either the space of Schwartz functions or $C_0^\infty(\mathbb{R}^d)$. Similarly, on a lattice, one can use the space of sequences of rapid decrease or of finite support. In the simple example of a finite system of oscillators, $\mathcal{S} = \mathbb{R}^n$ will do.

Finally, I need the following definition:

Definition 5.3. Ω is said to be strongly non-local on B if there does not exist a non-vanishing $h \in \mathcal{K}_{1/2}$ with the property that both h and Ωh vanish outside B.

Here I used the further definition:

Definition 5.4. Let $h \in \mathcal{K}_{\pm 1/2}$ and $B \subset K$. Then h is said to vanish in B if for all $\eta \in \mathcal{S}_B$, $\eta \cdot h = 0$. Similarly, it is said to vanish outside B, if for all $\eta \in \mathcal{S}_{B^c}$, $\eta \cdot h = 0$.

Intuitively, a strongly non-local operator is one that does not leave the support of any function h invariant. In the examples cited, this is always the case (see [5] for details).

Finally, we can give the general definition of "strictly local state," which goes back to Knight [27] for relativistic fields, and generalizes (3.1).

Definition 5.5. If B is a Borel subset of K, a strictly local excitation of the vacuum with support in B is a normalized vector $\psi \in \mathcal{F}^+(\mathcal{K}^{\mathbb{C}})$, different from the vacuum itself such that

$$\langle\psi|W_{\mathrm{F}}(z_\Omega(Y))|\psi\rangle = \langle 0|W_{\mathrm{F}}(z_\Omega(Y))|0\rangle \tag{5.1}$$

for all $Y = (q, p) \in \mathcal{H}(B^c, \Omega)$.

So it is a state which is indistinguishable from the vacuum outside B.

6. Characterizing The Stricly Local Excitations

Having established in Section 3 that in no models of interest finite particle states can be strictly localized excitations of the vacuum, it is natural to wonder which states do have this property. I mentioned that coherent states are in this class, as Knight already pointed out. Knight also conjectured that all states that are strictly localized excitations of the vacuum over some open set B, are obtained by applying a unitary element of the local algebra to the vacuum. This was subsequently proven for relativistic fields by Licht in [28]. Here is a version of this result adapted to our situation [5].

Theorem 6.1. *Suppose we are given a harmonic system determined by* Ω *and* $\mathcal{K} = L^2_{\mathrm{r}}(K, \mathrm{d}\mu)$*, and with a local structure* $\mathcal{S}$*. Let* $B \subset K$ *and suppose* Ω *is strongly non-local over* B*. Let* $\psi \in \mathcal{F}^+(\mathcal{K}^{\mathbb{C}})$*. Then the following are equivalent:*

- *(i)* $\psi \in \mathcal{F}^+(\mathcal{K}^{\mathbb{C}})$ *is a strictly local excitation of the vacuum inside* B*;*
- *(ii) There exists a partial isometry* U*, belonging to the commutant of* $\mathrm{CCR}_0(\mathcal{H}(B^{\mathrm{c}}, \Omega))$ *so that*

$$\psi = U|0\rangle.$$

As we have seen, Ω tends to be strongly non-local over bounded sets in all examples of interest, so the result gives a complete characterization of the localized excitations of the vacuum over bounded sets in those cases.

Since the condition in the definition of localized excitation is quadratic in the state, there is no reason to expect the set of localized excitations inside B to be closed under superposition of states. Of course it is closed under the taking of convex combinations (mixtures). Licht gives a simple criterium in the cited 1963 paper allowing to decide whether the linear combination of two stricly local excitations is still a strictly local excitation. Both the statement and the proof are again easily adapted to our situation [5].

Theorem 6.2. *Suppose we are given a harmonic system determined by* Ω *and* $\mathcal{K} = L^2_{\mathrm{r}}(K, \mathrm{d}\mu)$*, and with a local structure* $\mathcal{S}$*. Let* $B \subset K$ *and suppose* Ω *is strongly non-local over* B*. Let* $\psi_1, \psi_2 \in \mathcal{F}^+(\mathcal{K}^{\mathbb{C}})$ *be strictly local excitations of the vacuum inside* B*, so that* $\psi_i = U_i|0\rangle$*, with* $U_1, U_2 \in (\mathrm{CCR}_0(\mathcal{H}(B^{\mathrm{c}}, \Omega)))'$*. Then*

$$\psi = (\alpha\psi_1 + \beta\psi_2)/(\| \alpha\psi_1 + \alpha\psi_2 \|)$$

is a stricly local excitation of the vacuum inside B *for all choices of* $(\alpha, \beta) \neq (0, 0)$ *iff* $U_2^* U_1$ *is a multiple of the identity operator.*

The conclusion is then clear. Whenever Ω is strongly non-local over B, the superposition of strictly localized states does typically *not* yield a strictly localized state over B. In fact, taking $U_1 = W_F(z_\Omega(Y_1))$, and $U_2 = W_F(z_\Omega(Y_2))$, with

$Y_1, Y_2 \in \mathcal{H}(B, \Omega)$, it is clear that $U_2^* U_1$ is a multiple of the identity only if $Y_1 = Y_2$ so that the localized states do certainly not form a vector space in that situation. This is in sharp contrast to what we are used to in the non-relativistic quantum mechanics of systems of a finite number N of particles. In that case, a wave function $\psi(x_1, \ldots x_N)$ describes a state of the system with all the particles in a subset B of $\mathbb{R}^3$ iff it vanishes as soon as one of the variables is outside of B. The corresponding states make up the subspace $L^2(B^N)$ of $L^2(\mathbb{R}^N)$. In that case, to the question "Are all the particles in the set B?" corresponds a projection operator P_B with the property that the answer is "yes" with probability $\langle\psi|P_B|\psi\rangle$. To the question "Is the state a stricly local excitation of the vacuum in B?" cannot correspond such a projection operator! I will belabour this point in Section 7.

7. Surprises?

Here is a list of three mathematical truths that have originally been proven in the context of relativistic quantum field theory, and that seem to have generated a fair amount of surprise and/or debate:

(1) The vacuum is a cyclic vector for the local algebras over open regions.

(2) The vacuum is a separating vector for the local algebras over open regions.

(3) The set of stricly local states (in the sense of Knight) over an open region is not closed under superposition of states.

In that context, the open regions referred to are open regions of Minkowski space time. And by relativistic, I mean of course invariant under the Poincaré group. Recall that a vector ϕ in a Hilbert space $\mathcal{V}$ is cyclic for an algebra $\mathcal{A}$ of bounded operators on $\mathcal{V}$ if $\overline{\text{span}\mathcal{A}\phi} = \mathcal{V}$. And it is separating if $A \in \mathcal{A}$, $A\phi = 0$, implies $A = 0$.

It turns out that, as soon as Ω is a non-local operator, suitably adapted analogous statements hold for the harmonic systems that are the subject of this paper. We already saw this for the third statement in the previous section. The results are summed up in the following theorem (proven in [5]).

Theorem 7.1. *Let* $\mathcal{K} = L^2(K, \mathrm{d}\mu)$, $\Omega^2 \geqslant 0$ *and* $\mathcal{S}$ *be as before. Suppose that, for some* $B \subset K$, Ω *is strongly non-local over* B. *Then*

(i) The vacuum is a cyclic vector for $\mathrm{CCR}_0(\mathcal{H}(B^c, \Omega))$;

(ii) The vacuum is a separating vector for $\mathrm{CCR}_w(\mathcal{H}(B, \Omega))$;

(iii) The set of stricly local states over B *do NOT form a vector space.*

(iv) There do not exist finite particle states that are stricly local states over B.

Since, as we pointed out, the hypotheses of the theorem hold in large classes of examples, and for various sets B, so do the conclusions. Note that they therefore do not have a particular link with relativistic invariance. Note also that for lattices, the vacuum cannot be cyclic for a bounded set, since then $\mathcal{H}(B, \Omega)$ is finite dimensional so that $\overline{\text{span}_{\mathbb{C}} z_\Omega(\mathcal{H}(B, \Omega))}$ is a strict subspace of $\mathcal{K}^{\mathbb{C}}$. However, it is easily shown in translationally invariant models, for example, that the vacuum is cyclic for the the CCR-algebra over the complement of any bounded set B (See [5] for details).

My goal in this section is to explain in each case why the above statements have generated surprise in the context of relativistic field theory, then to argue that none of these properties should have surprised anyone precisely since they hold for simple systems of n coupled oscillators, and for free bose fields in rather great generality as the previous theorem shows. In particular, they therefore hold for the Klein-Gordon equation on Minkowski spacetime, which happens to be relativistic (meaning here Poincaré invariant). Why should it then be contrary to anyone's physical intuition if they continue to hold for interacting relativistic fields?

Statement 1. This was proven for relativistic quantum fields by Reeh and Schlieder in [34] and has been a well-known feature of *relativistic* quantum field theory ever since. As already mentioned, in that context, the open regions referred to are open regions of Minkowski space time. For a textbook formulation in the context of axiomatic, respectively algebraic relativistic quantum field theory you may consult [43], respectively [17] [24].

Rather than giving the full proof of Theorem 7.1 (i), I will once again restrict myself to a system of n oscillators and consider the subspace of the state space $L^2(\mathbb{R}^n)$ containing all vectors of the form

$$\sum_{j=1}^{L} c_j \exp -\mathrm{i}(a_j P_1 - b_j Q_1)|0, \Omega\rangle,$$

for all choices of $c_1 \ldots c_L$, $L \in \mathbb{N}$. Note that, since only the operators Q_1 and P_1 occur in the exponents, it follows from the considerations of the previous sections that such vector is a linear combination of excitations of the vacuum that are strictly localized at site $i = 1$. Nevertheless, one can easily prove that that those vectors form a dense subset of the full Hilbert space, which means that the vacuum is a cyclic vector for the local algebra over that site. A very pedestrian proof goes as follows: thinking for simplicity of the case $n = 2$, it is easy to see, taking limits, that the vectors $Q_1^k|0, \Omega\rangle$ and $P_1^\ell|0, \Omega\rangle$ belong to the closure of the span of the above vectors. Now, using that Ω is not diagonal, one easily concludes that therefore all wave functions of the form

$$p(x_1, x_2) \exp -\frac{1}{2} x \cdot \Omega x,$$

with $p(x_1, x_2)$ any polynomial belong to the space. Taking Hermite polynomials, for example, one obtains a basis for the full state space $L^2(\mathbb{R}^2)$. Note that, again, Ω

has to be non-diagonal for this to work. So indeed, the vacuum is a cyclic vector for the algebra of local observables over B (here $B = \{1\}$). It holds in much more generality, provided Ω is non-local: this is the content of Theorem 7.1. There are various poetic and misleading ways to express this result, for example by saying: "Local operations on the vacuum can produce instantaneous and arbitrary changes to the state vector arbitrarily far away." Lest one enjoys confusing oneself, it is a good idea to stay clear of such loose talk, as I will further argue below.

Let me now corroborate my claim that the cyclicity of the vacuum came as a surprise when it was proven for relativistic fields. Segal writes in [39] it is "particularly striking" and Segal and Goodman say it is "quite surprising" and a "bizarre phenomenon" in [40]. Streater and Wightman call it "a surprise" in [43]. Even rather recently, Haag refers to it as "startling" in [17] (p. 102), although he reduces this qualification to "(superficially) paradoxical" later on in his book (p. 254). Redhead in [33] similarly calls it "surprising, even paradoxical".

The surprise finds its origin in an apparent contradiction between the above mathematical statement and some basic physical intuition on the behaviour of quantum mechanical systems. Segal for example writes in [39] that Reeh-Schlieder is particularly striking because it apparently means that "the entire state vector space of the field could be obtained from measurements in an arbitrarily small region of space-time." This, he argues, is "quite at variance with the spirit of relativistic causality." Similar arguments can be found for example in [33] or in [17] and in [15], the authors write that "it is hard to square with naïve, or even educated, intuitions about localization."

This supposed contradiction is, as we shall see, directly related to the misconception of the vacuum as "empty space," which already was part of the problem with the debate surrounding the Newton-Wigner position operator. To understand this, let me explain what the contradiction translates to in our present context of harmonic systems: it is the too naive and, as we shall see, erroneous expectation that the mathematical operation of applying to the vacuum vector a local observable A belonging to the local algebra over some subset B of K yields a state $A|0\rangle/\langle 0|A^*A|0\rangle^{1/2}$ which is a strictly localized excitation of the vacuum in B in the sense of Knight's Definition 5.5 (for brevity called "local states" in what follows) [25]. This, if it were true, would of course be in blatant contradiction with the cyclicity of the vacuum. Indeed, if the vacuum is cyclic, any state of the system, including one that differs from the vacuum very far away from B can be approximated by one of the above form. Fortunately, we know from Licht's result that applying a local observable to the vacuum yields a stricly local excitation only if the local observable is a partial isometry. Of course, this only shifts the paradox, because one may choose to find Licht's result paradoxical. Indeed, when A is a projector, one can, according to the standard interpretational rules of quantum mechanics, and in particular the "collapse of the wave function" prescription, prepare (in principle!) an ensemble of systems, all in the state $A|0\rangle/\langle 0|A^*A|0\rangle^{1/2}$. Now, if the projector A is a local observable, these measurements can correctly be thought of as being executed within B. Now, *if you think of the vacuum as empty space*, it is inconceivable on physical grounds that such a measurement could instantaneously change something outside B. But this is then in contradiction with the mathematical result of Licht which asserts that if A is a projector, the state $A|0\rangle/\langle 0|A^*A|0\rangle^{1/2}$ is *not* local and therefore *does* differ from the vacuum outside B. The way out is obvious: the vacuum is not empty space but the ground state of an extended system. To see what is happening, the example of the chain of n coupled oscillators of is again instructive. Concentrate for example on the seventh oscillator of the chain and consider the question: "Does the displacement of the seventh oscillator fall within the interval $[a, b]$?" To this corresponds the projector $\chi_{[a,b]}(Q_7)$. The outcome of the corresponding preparation procedure will be an ensemble of systems, all in the state

$$\chi_{[a,b]}(Q_7)|0\rangle/\langle 0|\chi_{[a,b]}(Q_7)|0\rangle$$

with the non-vanishing probability $\langle 0|\chi_{[a,b]}(Q_7)|0\rangle$. But it is obvious that this state differs from the vacuum on the neighbouring site 8 and even on very far away sites! So even though the above projector corresponds to a local physical operation it does nevertheless not lead to a local excitation of the vacuum because even a local physical operation (here a measurement of the displacement of a single oscillator on one site) on the vacuum will instantaneously change the state of the system everywhere else. This is obviously the result here of the fact that the vacuum exhibits correlations between (commuting!) observables at different sites along the ring, a perfectly natural and expected phenomenon. After all, the oscillators on different sites are connected by springs. The ultimate reason for this phenomenon is therefore that the ground state of a typical oscillator system characterized by an n by n matrix Ω^2 is an "entangled" state in $L^2(\mathbb{R}^n) \cong L^2(\mathbb{R}) \otimes L^2(\mathbb{R}) \cdots \otimes L^2(\mathbb{R})$, unless of course Ω^2 is diagonal so that the oscillators are uncoupled to begin with. This entanglement can be seen in the fact that the ground state is a Gaussian with correlation matrix Ω, which is not diagonal. The change far away that the vacuum undergoes in the above measurement process is therefore nothing new, but a version of the usual "weirdness" of quantum mechanics, at the origin also of the EPR paradox.

In conclusion, I would therefore like to claim that in oscillator systems such as the ones under study here, one should expect that physical changes (such as measurements) operated on the vacuum vector inside some set $B \subset K$ (for example as the result of a measurement) will alter the state of the system outside B instantaneously. This is already true for finite dimensional systems as explained above and remains true for infinite dimensional ones such as oscillator lattices or the Klein-Gordon equation. That the latter has the additional feature of being Poincaré invariant does not in any way alter this conclusion nor does it lead to additional paradoxes. It does in particular not cause any causality problems in the sense that the phenomenon cannot be used to send signals, as one can easily see.

In short, the cyclicity of the vacuum does not lead to any contradictions with basic physical intuition, provided the latter is correctly used. It is perhaps interesting to note that Licht's 1963 result which is of great help in understanding the situation, is not cited in any of the other works on the subject I mentioned (although his paper sometimes is to be found in the bibliography . . .).

Statement 2. What about the separability of the vacuum? The separability also gives rise to an apparent contradiction that is equally easily dispelled with. Indeed, the "surprise" can be formulated as follows. If a non-trivial projector P belongs to the local algebra $\mathrm{CCR}_w(\mathcal{H}(B,\Omega)$, then $P|0\rangle \neq 0$ since the vacuum is a separating vector. Now this means that if the system is in its vacuum state and one measures locally some property of the system, such as, for example, whether the displacement of the oscillator on site 69 has a value between 7 and 8, then the answer is "yes" with non-zero probability. This consequence of separability (possibly first mentioned in [23]) can be paraphrased suggestively as follows:

> "When the system is in the vacuum state, anything that can happen will happen,"

or, alternatively, as in [44], "every local detector has a non-zero vacuum rate." This result is certainly paradoxical if you think of the vacuum as being empty space. Indeed, how can any measurement, local or not, give a non-trivial result in empty space?

To see why there is no reason to be surprised, let us look again at my favourite example, a system of n coupled oscillators. Its ground state is a Gaussian with correlation matrix Ω, so the probability that the displacement of the oscillator on site 69 has a value between 7 and 8 is obviously non-zero! That a similar property survives in the quantized Klein-Gordon field does not strike me as particularly odd, since the mathematical structure of both models is exactly identical, as should be clear from the previous sections.

Statement 3. In the context of relativistic quantum field theory, this was proven by Licht in his cited 1963 paper. He uses as a basic ingredient the result of Reeh and Schlieder. In our context here, it is the content of Theorem 6.2, which is a consequence of the strong non-locality of Ω.

Now, Theorem 6.2 has an interesting consequence for quantum measurement theory. Indeed, given a set B so that Ω is strongly non-local over B, there cannot exist a projection operator P_B on $\mathcal{F}^+(\mathcal{K}^{\mathbb{C}})$ with the property that $P_B\psi = \psi$ if and only if $\psi \in \mathcal{F}^+(\mathcal{K}^{\mathbb{C}})$ is a strictly local excitation of the vacuum over B. Indeed, if such an operator existed, the strictly localized excitations would be stable under superposition of states, of course. So there is no projector associated to the "yes-no" question: "Is the system strictly localized in B?" This is different from what we are used to in the non-relativistic quantum mechanics of a finite number of particles. There questions such as "Are all particles in B?" have a projector associated to them. But of course, we are dealing here with extended systems, such as oscillator lattices, and asking questions about excitations of the vacuum, not about the whereabouts of the individual oscillators, for example! It is only when you forget that, and try to interpret all statements about the fields in terms of particles that you run into trouble with your intuition.

As I pointed out before, this third statement above follows from the second, the second from the first and the main ingredient of the proof of the first by Reeh and Schlieder is relativistic invariance and the spectral property. This seems to have lead to the impression that these three properties, and especially the last one, are typical of relativistic fields, and absent in non-relativistic ones. For example, Redhead says in [33] that "to understand why the relativistic vacuum behaves in such a remarkable way, let us begin by contrasting the situation with nonrelativistic quantum field theory". It is furthermore said in [25] that "in a relativistic field theory it is not possible to define a class of states strictly localized in a finite region of space within a given time interval if we want to keep all the properties which one would like to associate with localization." The authors include in those properties the fact that it has to be a linear manifold. Quite recently still, a similar argument is developed in some detail in [10]. They explain that "there are marked differences between non-relativistic and relativistic theories, which manifest themselves in the following alternative structure of the set of vectors" ψ representing states that are strictly localized excitations of the vacuum. They then go on to explain that, in the non-relativistic case, the set of localized states are closed under superposition, whereas in relativistic quantum field theory, they are not.

But these statements are potentially misleading since precisely the same phenomena produce themselves in the eminently non-relativistic systems of coupled oscillators that I have been describing as is clear from the results of the previous sections. These phenomena are a consequence of the strong non-locality of Ω, and have nothing to do with relativistic invariance. The problem is that one has the tendency to compare relativistic field theories with the second quantization of the Schrödinger field, which is of course a non-relativistic field theory. In that context, the above three statements do not hold and in particular, the set of local states is a linear subspace of the Hilbert space. But if you compare, as you should, relativistic field theories to the equally non-relativistic harmonic systems, such as lattices of coupled oscillators, you remark that many of the features of the relativistic fields are perfectly familiar from the non-relativistic regime. They should therefore not come as a surprise, and not generate any paradoxes. It should be noted that already in [40] the source of the Reeh-Schlieder properties for free relativistic fields is identified to be the non-locality of $(-\Delta + m^2)^{1/2}$. This is further exploited in [29] [30] and [46] where the anti-locality of potential perturbations of $-\Delta$, respectively of the

Laplace-Beltrami operator on Riemannian manifolds is proven, which then yields a proof of the Reeh-Schlieder property in these situations.

As a further remark along those lines, I would like to point out that the fact that the stricly local states are not stable under superposition of states is sometimes related to the type of the local algebras of observables [28] [10]. In relativistic quantum field theory, they are known to be of type III [13], a result that generated a fair amount of excitement when it was discovered, since type III factors were thought to be esoteric objects [39]. It should be noted, however, that the local algebras of observables in oscillator lattices are type I factors, and that the stricly local excitations nevertheless are not stable under superposition. In addition, just as in relativistic quantum field theory, pure states look locally like mixtures: this is a consequence of entanglement, not of relativity.

A further paradox related to the third statement is the following. Suppose you have a strictly local excitation of the vacuum ψ over some set B. Now ask yourself the question if a local measurement inside B can prepare this state. In other words, is the projector onto ψ a local observable? In relativistic theories, the answer is in the negative [33]. Indeed, since the algebras are of type III, they contain no finite dimensional projectors. But even in oscillator lattices, the answer is negative, since the local algebras do not contain finite dimensional projectors either, and this despite the fact that they are of type I. Of course, this is not in the least little bit surprising: intuitively also, to fix the state of an extended system such as an oscillator lattice, you expect to need to make measurements on every site of the lattice. In particular, if the state is a stricly local excitation of the vacuum on the fifth site, you should check it coincides with the vacuum on all other sites. So you can neither measure nor prepare such a state by working only on a few lattice sites. Again, the situation is very different from the one of, for example, a one-electron system, where local states can be prepared locally.

Of course, by now, I hope I have brainwashed you into agreeing that it is really a bad idea to think of the vacuum as empty space. But should you not be convinced, again, you are in excellent company. This is how Schwinger talks about the vacuum in [38]: "With [quantum field theory] the vacuum becomes once again a physically reasonable state with no particles in evidence. The picture of an infinite sea of negative energy electrons is now but regarded as an historical curiosity, and forgotten. Unfortunately, this episode, and discussions of vacuuum fluctuations, seem to have left people with the impression that the vacuum, the physical state of nothingness (under controlled physical circumstances), is actually the scene of wild action." And a bit further down in the same article, he insists again: "I recall that for us the vacuum is the state in which no particles exist. It carries no physical properties: it is structureless and uniform. I emphasize this by saying that the vacuum is not only the state of minimal energy, it is the state of *zero* energy, zero momentum, zero charge, zero whatever. Physical properties, structure, come into existence only when we disturb the vacuum, when we excite it."

8. Conclusions

As long as one studies only a finite number of oscillators, the imaginative description of the quantum states of harmonic systems in terms of quanta is rather cute but not terribly useful or important. It is however a crucial element of relativistic and non-relativistic quantum field theories, which have an infinite number of degrees of freedom. In that case, the quanta are traditionally interpreted as particles. Photons, for example, are the quanta of the electromagnetic field and phonons are those of the vibration field. Electrons are similarly quanta of the Dirac field.

Now if you want to interpret the quanta as particles, you automatically are lead to the question that features as the title of this manuscript. One thinks of a particle as a localized object, and so it seems perfectly natural to wish to have a position operator for it, or at least some way to answer questions such as : "What is the probability of finding the particle in such and such a region of space?" In fact, as I have argued via the generalization of Knight's theorem, since the particles of quantum field theory are quanta, the situation is similar to the one we discovered already with finite systems of oscillators: the quanta cannot be perfectly localized and therefore there is no way to associate a position operator to them, and in that sense the question above does not really make any sense at all. It should be noted that whereas Knight's definition is regularly referred to in discussions of localization issues, his theorem, which is very helpful in understanding the issues at hand, seems to never be mentioned.

That quanta cannot be perfectly localized does not constitute a problem. A good notion of localized states exist: it is the one provided by considering localized excitations of the vacuum and goes back to Knight. Those states differ from the vacuum only inside a set B and there are plenty of them. They do however not form a vector subspace of the quantum Hilbert space, and no projection operator is associated with the localized excitations over a fixed set B. As I have explained, this feature of relativistic quantum field theories is also familiar from non-relativistic oscillator chains, and as such not related to relativistic invariance. If the right analogies between relativistic and non-relativistic theories are used, it is not counter-intuitive or in any way surprising.

Let me mention that a complete discussion of the localization properties of field states should also analyse notions of approximate localization, allowing for exponential or algebraic decay of the expectation values outside the set B. These notions are implicit in the physics literature, where states with exponential tails, for example, are often thought of as localized. They have been developed by several authors [1] [25] [3] [17] [48]. A complete discussion should finally also address those questions for other fields, such as complex bose fields and Fermi fields. These issues will be adressed elsewhere [45].

In conclusion, the morale of the story is this: when testing your understanding of a notion in quantum field theory, try to see what it gives for a finite system of oscillators. Examples of situations where this algorithm seems to meet with some success are the various puzzles associated with particle localization and the Reeh-Schlieder theorem and its consequences, as I have argued here. In particular, if the notion under study, when adapted to finite or infinite oscillator chains, looks funny there, it is likely to lead you astray in the context of quantum field theory as well: an example is the Newton-Wigner position operator. Of course, I am not the first one to point these analogies out. In [32], one can read: "The radiation field differs from atomic systems principally by . . . having an infinite number of degrees of freedom. This may cause some difficulties in visualizing the physical problem, but is not, in itself, a difficulty of the formalism."

References

[1] W. Amrein. Localizability for particles of mass zero. *Helv. Phys. Acta*, 42:149–190, 1969.

[2] H. Bacry. *Localizability and space in quantum physics*. Lecture Notes in Physics 308, Springer Verlag. 1988.

[3] I. Bialynicki-Birula. Exponential localization of photons. *Phys. Rev. Lett.*, 80(24):5247–5250, 1998.

[4] J. D. Björken and S. D. Drell. *Relativistic quantum fields*. Mc. Graw Hill, NY. 1965.

[5] S. De Bièvre. *Classical and quantum harmonic systems (in preparation).*

[6] S. De Bièvre. Causality and localization in relativistic quantum mechanics. In *Proceedings "Trobades Cientifiques de la Mediterrania"*, pages 234–241. 1985.

[7] S. De Bièvre. Local states of free bose fields. In *Large Coulomb Systems*, volume 695 of *Lecture Notes in Physics*, pages 17–63. 2006.

[8] V.B. Berestetskii, E.M. Lifshitz, and L.P. Pitaevskii. *Relativistic quantum theory*. Pergamon Press, Oxford. 1971.

[9] N. N. Bogoliubov and D.V. Shirkov. *Quantum fields*. The Benjamin/Cummings Publishing Company Inc. 1983.

[10] D. Buchholz and J. Yngvason. There are no causality problems for fermi's two-atom system. *Physical Review Letters 5*, 73:613–616, 1994.

[11] S.J. Chang. *Introduction to quantum field theory*. World Scientific. 1990.

[12] J.P. Dereudinger. *Theorie quantique des champs*. Presses Polytechniques et Universitaires Romandes. 2001.

[13] W. Driessler. Comments on lightlike translations and applications in relativistic quantum field theory. *Commun. Math. Phys.*, 44(2):133–141, 1975.

[14] G. Emch. *Algebraic methods in statistical mechanics and quantum field theory*. Wiley Interscience, 1972.

[15] G. Fleming and J. Butterfield. Strange positions. In *From physics to philosophy*, pages 108–165. Cambridge University Press, 1999.

[16] W. Greiner. *Relativistic quantum mechanics*. Springer. 1990.

[17] R. Haag. *Local quantum physics*. Springer. 1996.

[18] Hans Halvorson. Reeh-Schlieder defeats Newton-Wigner: on alternative localization schemes in relativistic quantum field.theory. *Philos. Sci.*, 68(1):111–133, 2001.

[19] Hans Halvorson and Rob Clifton. No place for particles in relativistic quantum theories? *Philos. Sci.*, 69(1):1–28, 2002.

[20] G. C. Hegerfeldt. Remark on causality and particle localization. *Phys. Rev. D*, 10(10):3320–3321, 1974.

[21] G. C. Hegerfeldt. Violation of causality in relativistic quantum theory? *Phys. Rev. Letters*, 54(22):2395–2398, 1985.

[22] G. C. Hegerfeldt. Causality, particle localization and positivity of the energy. In *Irreversibility and causality (Goslar, 1996)*, volume 504 of *Lecture Notes in Phys.*, pages 238–245. Springer, Berlin, 1998.

[23] K.-E. Hellwig and K. Kraus. Operations and measurements II. *Commun. Math. Phys.*, 16:142–147, 1970.

[24] S. S. Horuzhy. *Introduction to algebraic quantum field theory*, volume 19 of *Mathematics and its Applications (Soviet Series)*. Kluwer Academic Publishers Group, Dordrecht, 1990.

[25] R. Haag and J. A. Swieca. When does a quantum theory describe particles? *Commun. Math. Phys.*, 1:308–320, 1965.

[26] K. Huang. *Quantum field theory*. J. Wiley and Sons, 1998.

[27] J.M. Knight. Strict localization in quantum field theory. *Journal of Mathematical Physics*, 2(4):459–471, 1961.

[28] A. L. Licht. Strict localization. *Journal of Mathematical Physics*, 4(11):1443–1447, 1963.

[29] K. Masuda. A unique continuation theorem for solutions of the wave equation with variable coefficients. *J. Math. Analysis and Applications*, 21:369–376, 1968.

[30] K. Masuda. Anti-locality of the one-half power of elliptic differential operators. *Publ. RIMS, Kyoto University*, 8:207–210, 1972/73.

[31] F. Mandl and G.Shaw. *Quantum field theory*. Wiley. 1984.

[32] R.E. Peierls. The development of quantum field theory. In *The physicist's conception of nature*, pages 370–379. D. Reidel Publishing Cy., Dordrecht, Netherlands, 1973.

[33] M. Redhead. More ado about nothing. *Foundations of Physics 1*, 25:123–137, 1995.

[34] H. Reeh and S. Schlieder. Bemerkungen zur Unitäräquivalenz von Lorentzinvarianten Felden. *Nuovo cimento (10)*, 22:1051–1068, 1961.

[35] S. Ruijsenaars. On newton-wigner localization and superluminal propagation speeds. *Ann. Physics*, 137(1):33–43, 1981.

[36] J.J. Sakurai. *Advanced quantum mechanics*. Addison Wesley Publishing Company. 1967.

[37] S. S. Schweber. *An introduction to relativistic quantum field theory*. Row and Peterson and Company, 1961.

[38] J. Schwinger. A report on quantum electrodynamics. In *The physicist's conception of nature*, pages 413–429. D. Reidel Publishing Cy., Dordrecht, Netherlands, 1973.

[39] I. E. Segal. Quantum fields and analysis in the solution manifolds of differential equations. In *Proc. Conf. on Analysis in Function Space*, pages 129–153. M.I.T. Press, Cambridge, 1963.

[40] I. E. Segal and R. W. Goodman. Anti-locality of certain Lorentz-invariant operators. *J. Math. Mech.*, 14:629–638, 1965.

[41] G. Sterman. *An introduction to quantum field theory*. Cambridge University Press, 1993.

[42] P. Strange. *Relativistic quantum mechanics*. Cambridge University Press. 1998.

[43] R. F. Streater and A. S. Wightman. *PCT, spin and statistics and all that*. The Benjamin/Cummings Publishing Cy, Reading MA, 1964.

[44] S. Summers and R. Werner. The vacuum violates Bell's inequalities. *Physics Letters 5*, 110A:257–259, 1985.

[45] N. Tisserand. *PhD thesis (in preparation)*.

[46] R. Verch. Anti-locality and a Reeh-Schlieder theorem on manifolds. *Letters in Mathematical Physics*, 28(2):143–154, 1994.

[47] D. Wallace. Emergence of particles from bosonic quantum field theory. *preprint 2001, http://arxiv.org/abs/quant-ph/0112149*, 2001.

[48] D. Wallace. In defence of naiveté: The conceptual status of lagrangian quantum field theory. *preprint 2001, http://arxiv.org/abs/quant-ph/0112148*, 2001.

[49] E. P. Wigner. Interpretation of quantum mechanics. In *Quantum theory and measurement*, pages 260–314. Princeton University Press, Princeton, NJ, 1983.

UFR de Mathématiques et UMR P. Painlevé, Université des Sciences et Technologies de Lille, 59655 Villeneuve d'Ascq Cedex France *E-mail:* Stephan.De-Bievre@math.univ-lille1.fr

Thoughts about Anomalous Diffusion: Time-Dependent Coefficients versus Memory Functions

V. M. Kenkre and F.J. Sevilla

Relations between two natural generalizations of the standard diffusion equation, one involving memory functions and the other time-dependent coefficients, are investigated. It is shown that while the two descriptions are by no means equivalent to each other, each is equivalent to a spatially nonlocal generalization of the other. Explicit prescriptions to bridge the two formalisms are provided and illustrated in two physical transport situations. Experimental relevance of these considerations is also briefly discussed, one in the context of NMR microscopy, the other in that of transient gratings in molecular crystals.

1. Introduction and the Two Descriptions

Electrons and holes in a semiconducting device [1], interstitial atoms injected into a solid [2], molecules in a gas container, ink droplets in a glass of liquid, mice carrying the deadly Hantavirus over a landscape [3], all of these entities engage in a common activity: they diffuse. The study of diffusion has, therefore, been fundamental, important, and active in diverse disciplines. Famous thinkers who have made primary and oft-used contributions to such a study include not only the physicist Einstein [4] but the financial expert Bachelier [5], the former in his research on Brownian motion, the latter during his investigations of markets and stock movements. In the present manuscript, the authors report thoughts and calculations regarding two manners of the generalization of the fundamental process of diffusion. Such generalization becomes necessary when the mechanism of motion is more complex than in normal diffusion and may involve coherence, spatial restrictions, trapping, and similar features.

Non-equilibrium statistical mechanics in general, with transport theory as might be represented by diffusion processes as a sub-area, has benefitted from the early work [6] of Professor Gerard Emch on Master equations. One of the present authors (VMK) has mentioned some of that work in another festschrift article [7] written thirty years ago. He remembers fondly the years of overlap with Gerard as a friend and colleague in Rochester in the seventies. He thanks Gerard for much he taught him by example, including kindness, integrity and punctuality. It is with pleasure that he dedicates the present article, with the consent of his co-author (FJS), to Gerard on the occasion of his seventieth birthday.

One simple manner of describing standard diffusion is via the diffusion equation. The latter states that the time-rate of change of density P of whatever is diffusing (including the probability of a random walker) equals the product of the diffusion constant D and the Laplacian of P at the spatial location under consideration: $\frac{\partial P}{\partial t} = D\nabla^2 P$. From here onwards let us consider 1-dimensional systems for the sake of simplicity. The two manners of generalization of the diffusion equation that we wish to explore in the present paper are, respectively,

$$\frac{\partial P(x,t)}{\partial t} = D\chi(t)\frac{\partial^2 P(x,t)}{\partial x^2}, \tag{1.1}$$

$$\frac{\partial P(x,t)}{\partial t} = D\int_0^t ds\,\phi(t-s)\frac{\partial^2 P(x,s)}{\partial x^2}. \tag{1.2}$$

The first introduces time-dependence into the diffusion coefficient whereas the second injects temporal non-locality. Both reduce to standard diffusion in the respective limits $\chi(t) = 1$ and $\phi(t) = \delta(t)$.

Our interest is in studying the connections, if any, between the above mentioned two alternatives to the description of non-standard diffusion. Clearly, the quantities $\chi(t)$ and $\phi(t)$ need to be related to each other before any connection can be discussed. It is possible to show [8, 9], through a study of the underlying processes not discussed here, that a sensible relationship is

$$\chi(t) = \int_0^t ds\,\phi(s). \tag{1.3}$$

The precise source of this relation will become clearer below, but we can notice at once that it leads to *exactly* the same evolution for the mean square displacement from the two Eqs. (1.1), (1.2). Thus, multiplying each equation by x^2, integrating over x from $-\infty$ to $+\infty$, and assuming that the probability decays at $x = \pm\infty$ sufficiently fast, following standard procedures, we get the respective evolution for the mean square displacement

$$\langle x^2(t)\rangle = \langle x^2\rangle_0 + 2D\int_0^t ds\,\chi(s), \tag{1.4}$$

from the χ-formalism, and

$$\langle x^2(t)\rangle = \langle x^2\rangle_0 + 2D\int_0^t dt'\int_0^{t'} ds\,\phi(s) \tag{1.5}$$

from the ϕ-formalism. It is immediately clear that the two evolutions give identical results provided that $\chi(t)$ is the time integral of the memory $\phi(t)$, *i.e.*, that relation

(1.3) is true. We will assume that relation for the rest of the paper. Additionally, for simplicity, we will take the initial value of all moments such as $\langle x^2\rangle_0$ to vanish.

2. Higher Moments, Differences and Relations

What happens with higher moments of $P(x,t)$ in Eqs. (1.1) and (1.2)? It is straightforward to show that the χ-formalism gives, for the n-th moment (n is a positive integer) defined as $\langle x^n(t)\rangle = \int_{-\infty}^{\infty} dx x^n P(x,t)$,

$$\frac{d\langle x^n(t)\rangle}{dt} = Dn(n-1)\chi(t)\langle x^{n-2}(t)\rangle. \tag{2.1}$$

For the evolution of the same quantity, the ϕ-formalism gives

$$\frac{d\langle x^n(t)\rangle}{dt} = Dn(n-1)\int_0^t ds\, \phi(t-s)\langle x^{n-2}(s)\rangle. \tag{2.2}$$

Both formalisms connect the evolution of the n-th moment to the lower, $(n-2)$-th moment, the χ-formalism to its instantaneous value through a multiplicative factor containing $\chi(t)$, the ϕ-formalism to its values at all times s in the past through the memory function $\phi(t-s)$. Iterating Eqs. (2.1) and (2.2) over n down to $n=2$, using the explicit solutions (1.4) and (1.5) for the second moment, and focusing attention only on the even moments $\langle x^{2n}\rangle$, we have

$$\langle x^{2n}(t)\rangle = \frac{(2n)!}{n!}[D\tau]^n \tag{2.3}$$

for the χ-formalism and

$$\widetilde{\langle x^{2n}(\epsilon)\rangle} = \frac{(2n)!}{\epsilon^{n+1}}\left[D\tilde{\phi}(\epsilon)\right]^n \tag{2.4}$$

for the ϕ-formalism. In Eq. (2.3), we have defined a new time $\tau = \int_0^t ds\chi(s)$. In Eq. (2.4), tildes denote Laplace transforms and ϵ is the Laplace variable; thus, for instance, $\tilde{\phi}(\epsilon) = \int_0^\infty dt e^{-\epsilon t}\phi(t)$.

When Eq. (1.3) holds, $\phi(t)$ and $\chi(t)$ are related in Laplace domain as $\tilde{\chi}(\epsilon) = \tilde{\phi}(\epsilon)/\epsilon$. Therefore, by comparing (2.3) with (2.4) we obtain the result mentioned above that the second moments are exactly the same as predicted by the two formalisms. On the other hand, higher moments differ. Note as an illustration that the 4-th moment is given by

$$\langle x^4(t)\rangle = 12D^2\left[\int_0^t ds\ \chi(s)\right]^2 \tag{2.5}$$

from the χ-formalism, and thus is not the same as

$$\langle x^4(t)\rangle = 24D^2 \int_0^t ds \int_0^s du\ \chi(s-u)\chi(u), \tag{2.6}$$

which is the result of the ϕ-formalism, given relation (1.3). Clearly, Eqs. (2.5) and (2.6) yield different results for the fourth moment for all cases except that of standard (not anomalous) diffusion in which $\chi(t) = 1$.

The importance of demonstrating explicitly the perhaps obvious fact that the two formalisms are not equivalent to each other stems from the wide use that each has found in practical applications to situations in which diffusion is suspected to be anomalous. The χ-formalism is associated [10] with the phrase "fractional Brownian motion" if $\chi(t)$ is a power of t, while the ϕ-formalism has been called the generalized master equation (GME) approach and used widely [11] for the description of coherence. The clear difference in the higher moments along with the exact congruence of the second moment leads us to ask what the precise relation between the two formalisms is. An inspection of Eqs. (1.1)–(1.2) or of (2.1)-(2.2) shows the following. If the exact description happens to be that given by the memory function formalism, the description provided by the χ-formalism emerges as the so-called *half-Markoffian approximation*. This terminology appeared decades ago in the study of exciton motion [11]. The Markoffian approximation (see, e.g., ref. [7]) on a time non-local term such as $\int_0^t ds\phi(s)b(t-s)$ is made if the memory $\phi(s)$ varies so rapidly that the slower b may be taken out of the integral as $b(t)$. The full Markoffian approximation normally made, for instance, to convert the GME into the Pauli Master equation further assumes that $\phi(t)$ may be replaced by a δ-function times the time integral of $\phi(t)$ over all time:

$$\int_0^t ds\ \phi(s)b(t-s) \approx b(t) \int_0^\infty ds\ \phi(s).$$

The other, weaker and less-used Markoffian approximation [14] stops at taking b out of the integral, does not take the upper limit of the integral of the memory to be infinity, and has the form

$$\int_0^t ds\phi(s)b(t-s) \approx b(t) \int_0^t ds\phi(s) = b(t)\chi(t).$$

This partial Markoffian approximation, applied to the ϕ-formalism, can be said to be the content of the χ-formalism.

The above discussion should by no means imply that the χ-formalism always provides a more approximate description than does the ϕ-formalism. It is possible that the underlying dynamics of a given system may be precisely that given by Eq. (1.1). In such a case, it will be Eq. (1.2) that will be the approximation. To make this clear, let us compare the exact solutions of the two equations. Because there are no preferred points in space in the systems considered in this paper, the solution of Eq. (1.1) may be computed straightforwardly in Fourier space. It is given by the Gaussian

$$\frac{\hat{P}(k,\tau)}{\hat{P}(k,0)} = \exp\left[-Dk^2\int_0^t ds\,\chi(s)\right], \tag{2.7}$$

where k is the Fourier variable and the circumflex denotes the Fourier transform through $\hat{P}(k) = \int_{-\infty}^{\infty} dx\,P(x)e^{ikx}$. Equation (1.2), on the other hand, may be solved in the Fourier-Laplace domain as

$$\frac{\tilde{P}(k,\epsilon)}{\tilde{P}(k,0)} = \frac{1}{\epsilon + D\tilde{\phi}(\epsilon)k^2}. \tag{2.8}$$

Surely, it is impossible to find a $\phi(t)$ that would make the right hand side of Eq. (2.8) identical to the Laplace transform of the right hand side of Eq. (2.7) for *arbitrary* $\chi(t)$. The reverse is also true: it is impossible to find a $\chi(t)$ that would make the Laplace transform of the right hand side of Eq. (2.7) identical to the right hand side of Eq. (2.8) for *arbitrary* $\phi(t)$. Is there then no hope of finding a bridge between the two descriptions? One of the present authors has argued in the course of his study of stress distribution in granular materials [8] that a useful bridge might be constructed by generalizing one or the other of these two formalisms to include *spatially non-local* situations. To understand that argument, let us promote the χ-formalism by replacing the multiplicative factor $D\chi(t)$ in Eq. (1.1) by a spatial convolution:

$$\frac{\partial P(x,t)}{\partial t} = \int_{-\infty}^{\infty} dx'\,\mathfrak{D}_\chi(x-x',t)\frac{\partial^2 P(x',t)}{\partial x'^2} \tag{2.9}$$

The time rate of change of $P(x,t)$ is now connected to its second spatial derivative (Laplacian) not only at x but at all locations x' through the connecting function $\mathfrak{D}_\chi(x-x',t)$ which incorporates, in addition, the time dependence of the simpler $\chi(t)$ in a non-separable form. One recovers Eq. (1.1) from Eq. (2.9) if $\mathfrak{D}_\chi(x-x',t) = D\chi(t)\delta(x-x')$. Let us similarly generalize the ϕ-formalism by replacing the factor $D\phi(t-s)$ in Eq. (1.2) by a spatial convolution involving the new connecting function $\mathfrak{D}_\phi(x-x',t-s)$:

$$\frac{\partial P(x,t)}{\partial t} = \int_{-\infty}^{\infty} dx' \int_0^t ds\,\mathfrak{D}_\phi(x-x',t-s)\frac{\partial^2 P(x',s)}{\partial x'^2}. \tag{2.10}$$

We see that Eq. (1.2) is recovered from Eq. (2.10) if $\mathfrak{D}_\phi(x-x',t-s) = D\phi(t-s)\delta(x-x')$.

Whereas it was true that the spatially local equations of the χ- and ϕ- formalisms could not be put into equivalence, we now see that their generalizations can be.

Equating the ratio $\frac{\tilde{P}(k,\epsilon)}{\hat{P}(k,0)}$ as predicted by the two equations, we find that equivalence can be established provided

$$\int_0^\infty e^{-k^2 \int_0^t \mathfrak{D}_\chi(k,s)ds} e^{-\epsilon t} dt = \frac{1}{\epsilon + k^2 \tilde{\mathfrak{D}}_\phi(k,\epsilon)}. \tag{2.11}$$

Equation (2.11) constitutes a practical bridge to pass between the two formalisms. Assume the (spatially local) χ-formalism to be correct, i.e., that the system evolution obeys the original equation (1.1). An entirely equivalent description is then provided by the *spatially non-local* ϕ-formalism given by Eq. (2.10) in which

$$\tilde{\mathfrak{D}}_\phi(k,\epsilon) = \frac{1}{k^2} \left\{ \frac{1}{\mathcal{L}\left[e^{-Dk^2 \int_0^t \chi(s)ds}\right]} - \epsilon \right\}. \tag{2.12}$$

Thus, spatially non-locality in the ϕ-formalism is essential to describe a spatially local χ situation. Equation (2.12) is the explicit prescription through which we can find the key ϕ-quantity, viz., $\mathfrak{D}_\phi(k,\epsilon)$, for *an arbitrarily given* $\chi(t)$.

Conversely, if the (spatially local) ϕ-formalism provides the correct description, i.e., the system evolution obeys the original equation (1.2), a fully equivalent description is provided by the *spatially non-local* χ-formalism given by Eq. (2.9), the key quantiy $\mathfrak{D}(k,t)$ being computed for *any given* $\phi(t)$ by

$$\mathfrak{D}_\chi(k,t) = \frac{d}{dt}\left\{-\frac{1}{k^2} \ln\left|\mathcal{L}^{-1}\left(\frac{1}{\epsilon + k^2 D\tilde{\phi}(\epsilon)}\right)\right|\right\}. \tag{2.13}$$

In the prescriptions (2.12) and (2.13) the symbols $\mathcal{L}$ and $\mathcal{L}^{-1}$ stand, respectively, for the direct and inverse Laplace transforms, and the argument of the logarithm is the absolute value as shown.

3. Applications of the General Formalism

In this Section, we present an illustrative application of the prescription we have developed above in one physical instance in which the spatially local ϕ-formalism provides the exact description of the system. We will also mention briefly an instance of the opposite situation.

3.1. ϕ to χ: Memory Functions from a Railway-track Model

Coherence issues in exciton transport [15, 16] led to a great deal of work based on memory functions in the seventies [11, 16]. The specific form of the memory functions often arose from quantum features in the dynamics of excitons [16]. To

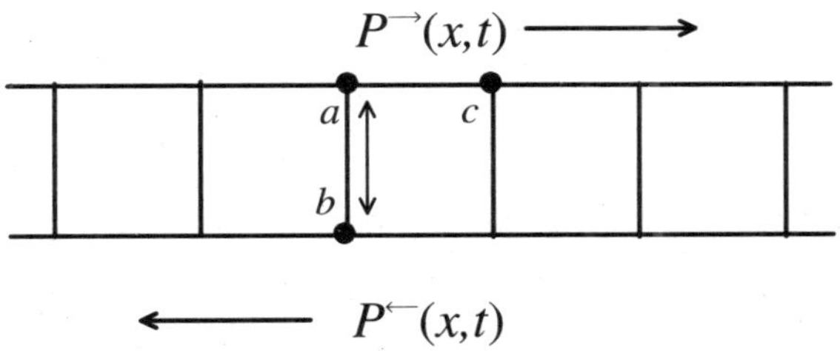

Figure 1. Pictorial description of the "railway-track" model naturally addressed via the ϕ-formalism.

dispel an incorrect notion held by some that an underlying quantum layer was essential to memory functions, one of the present authors introduced a trivially simple model that showed how memory functions could arise purely classically.[1] Because the model leads easily to Eq. (1.2), we study it here along with its equivalent spatially nonlocal χ-description.

For reasons that should be obvious, we call it the railway-track model. Let $P^{\rightarrow}$ ($P^{\leftarrow}$) be the probability that a particle moves to the right (left) with rate c ($-c$). The particle is subject to scattering at rate $\mathcal{Q}$, the only effect of scattering being to change the direction of motion from right to left and vice-versa. This system is depicted in Fig. 1 and analyzed through the coupled equations

$$\frac{\partial P^{\rightarrow}(x,t)}{\partial t} = c\frac{\partial P^{\rightarrow}(x,t)}{\partial x} + \mathcal{Q}[P^{\leftarrow}(x,t) - P^{\rightarrow}(x,t)], \tag{3.1}$$

$$\frac{\partial P^{\leftarrow}(x,t)}{\partial t} = -c\frac{\partial P^{\leftarrow}(x,t)}{\partial x} + \mathcal{Q}[P^{\rightarrow}(x,t) - P^{\leftarrow}(x,t)]. \tag{3.2}$$

Note the Markoffian nature of equations (3.1), (3.2). By defining $P(x,t) \equiv P^{\rightarrow} + P^{\leftarrow}$ and $R(x,t) \equiv P^{\rightarrow} - P^{\leftarrow}$, we get

$$R(x,t) = R(x,0)e^{-2\mathcal{Q}t} + c\int_0^t ds\, e^{-2\mathcal{Q}(t-s)}\frac{\partial}{\partial x}P(x,s). \tag{3.3}$$

For the initial condition $\partial P(x,t)/\partial t|_{t=0} = 0$, $P(x,t)$ satisfies

$$\frac{\partial P(x,t)}{\partial t} = c^2\int_0^t ds\, e^{-2\mathcal{Q}(t-s)}\frac{\partial^2 P(x,s)}{\partial x^2}. \tag{3.4}$$

[1] The utter simplicity of the model meant that there was no need to publish it–it was only discussed at conferences and private discussions. Its simplicity also means that it has been probably invoked by others before or since.

Equation (3.4) is exactly of the ϕ form of Eq. (1.2) with $D = c^2/2\mathcal{Q}$, the memory $\phi(t)$ being equal to the exponential $2\mathcal{Q}e^{-2\mathcal{Q}t}$. This trivial but clear example shows how a memory function description arises from coarsegraining involved in seeking the evolution of the *combination* $P(x,t) = P^{\rightarrow} + P^{\leftarrow}$ without interest in how much of rightward motion versus leftward motion there is. Needless to say, we have here a caricature of a system in which particles move and scatter among various velocity states, only 2 such states being considered in this caricature system.

The solution to the memory equation (3.4) with initial condition $P(x,t) = \delta(x)$ is given explicitly by [8]

$$P(x,t) = e^{-\mathcal{Q}t}\left[\frac{\delta(x+ct)+\delta(x-ct)}{2} + T(x,t)\right], \tag{3.5}$$

where $T(x,t)$ vanishes identically for $ct \le |x|$ and equals

$$T(x,t) = \left(\frac{\mathcal{Q}}{2c}\right)\left[I_0\left(\frac{\mathcal{Q}}{c}\sqrt{c^2t^2-x^2}\right) + \frac{ct}{\sqrt{c^2t^2-x^2}}I_1\left(\frac{\mathcal{Q}}{c}\sqrt{c^2t^2-x^2}\right)\right] \tag{3.6}$$

for $ct > |x|$, $I_\nu(z)$ being the modified I Bessel function of order ν and argument z. We point out in passing that the moments of $P(x,t)$ may be computed by just knowing the Laplace transform of the memory function and using (2.4). Inverting the Laplace transform, we find

$$\langle x^{2n}(t)\rangle = (ct)^{2n}\, M(n, 2n+1, -2\mathcal{Q}t), \tag{3.7}$$

where $M(a,b,z) = 1 + \frac{az}{b} + \frac{a(a+1)z^2}{b(b+1)2!} + \ldots + \frac{a(a+1)\cdots(a+n-1)z^n}{b(b+1)\cdots(b+n-1)n!} + \ldots$ is the Kummer confluent hypergeometric function of argument z [17].

Our purpose in introducing the "railway-track" model here is to examine how one may pass from the ϕ- formalism to the spatially non-local χ-formalism via our prescription (2.13). A straightforward evaluation after replacing $\tilde{\phi}(\epsilon)$ by $2\mathcal{Q}/(\epsilon + 2\mathcal{Q})$ leads to

$$\mathfrak{D}_\chi(k,t) = \left(\frac{c^2}{\mathcal{Q}}\right)\frac{\sin\left(\mathcal{Q}t\sqrt{k^2c^2/\mathcal{Q}^2-1}\right)}{\sin\left(\mathcal{Q}t\sqrt{k^2c^2/\mathcal{Q}^2-1}\right)+\sqrt{k^2c^2/\mathcal{Q}^2-1}\cos\left(\mathcal{Q}t\sqrt{k^2c^2/\mathcal{Q}^2-1}\right)}. \tag{3.8}$$

The Fourier inverse of this expression, $\mathfrak{D}_\chi(x-x',t)$, when substituted in Eq. (2.9), allows us to write the desired spatially nonlocal χ-description.

What new insights does the combination of Eqs (2.9) and (3.8) yield? One answer is that (2.9) may now be inverted explicitly as an infinite sum of local terms:

$$\frac{\partial P(x,t)}{\partial t} = D_0(t)\frac{\partial^2}{\partial x^2}P(x,t) + \frac{D_2(t)}{2!}\frac{\partial^4}{\partial x^4}P(x,t) + \frac{D_4(t)}{4!}\frac{\partial^6}{\partial x^6}P(x,t)..., \quad (3.9)$$

where the factors $D_{2n}(t)$ may be computed as $(-1)^n \left.\frac{\partial^{2n}\mathfrak{D}_\chi(k,t)}{\partial k^{2n}}\right|_{k=0}$. The first three factors are

$$D_0(t) = D(1 - e^{-2Qt}), \quad (3.10)$$

$$D_2(t) = \frac{D^2}{Q}\left[4Qte^{-2Qt} - (1 - e^{-4Qt})\right], \quad (3.11)$$

$$D_4(t) = \frac{6D^3}{Q^2}\left[2 + e^{-2Qt}(1 - 4Qt - 8Q^2t^2) - 2e^{-4Qt}(1 + 4Qt) - e^{-6Qt}\right], \quad (3.12)$$

where and henceforth we suppress the symbol c and use $D = c^2/2Q$ in addition to Q.

If we truncate (3.9) by keeping only the first term, the local χ-description emerges:

$$\frac{\partial P(x,t)}{\partial t} = D\left[\int_0^t ds\, \phi(s)\right]\frac{\partial^2}{\partial x^2}P(x,t), \quad (3.13)$$

since $D_0(t) = D\chi(t) = D\int_0^t ds\, \phi(s)$. The solution to Eq. (3.13) is given by the Gaussian

$$\frac{Q^{1/2}}{\left[2\pi D\left(2Qt - (1 - e^{-2Qt})\right)\right]^{1/2}} e^{-\frac{Qx^2}{2\pi D\left(2Qt-(1-e^{-2Qt})\right)}},$$

and leads, as mentioned above, to precisely the same $\langle x^2(t)\rangle$ as the exact solution. However, it predicts, for the *next higher* moment,

$$\langle x^4(t)\rangle = \frac{3D^2}{Q^2}\left[2Qt - (1 - e^{-2Qt})\right]^2, \quad (3.14)$$

which is only an approximation (see Fig. 1) to the exact fourth moment

$$\langle x^4(t)\rangle = \frac{6D^2}{Q^2}\left[3 - 4Qt + 2Q^2t^2 - e^{-2Qt}(3 + 2Qt)\right] \quad (3.15)$$

computed from (2.4). Any moment from the spatially-nonlocal χ-formalism may be computed from the infinite series (3.9). Note that not all the higher order terms

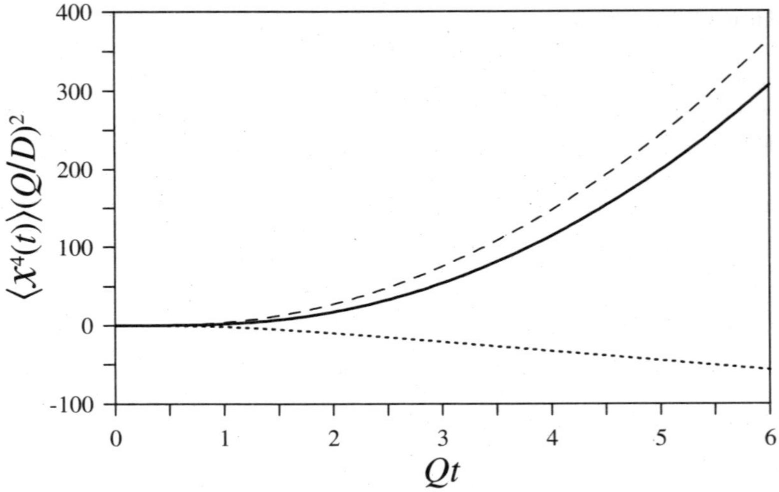

Figure 2. Fourth moment $\langle x^4(t)\rangle$ in units of $(D/\mathcal{Q})^2$ as function of the dimensionless time $\mathcal{Q}t$. We compare the exact fourth moment of (3.5) in the ϕ-formalism (solid-line) with the one given by the local χ-formalism, i.e., the "half-Markoffian" approximation (dashed-line). The exact curve lies entirely below the approximate curve. The lower curve (dotted-line) gives the correction provided by just the second term of the infinite series (3.9). The dashed and the dotted curves add up precisely to the solid curve.

are necessary to compute a given moment. Thus, the infinite series collapses into only the first two terms when computing $\langle x^4(t)\rangle$:

$$\langle x^4(t)\rangle = 12\int_0^t ds\, D_2(s) + 12\left[\int_0^t ds\, D_0(s)\right]^2 . \tag{3.16}$$

In Fig. 2 we show the fourth moment as function of time along with its half-Markoffian approximation. [2]

[2]It might be interesting to observe that the recursive relation for the even moments in the spatially-local ϕ-formalism, Eq. (2.1), the $2n$-th moment depends on the previous one $2(n-1)$-th while in the spatially-non-local χ-formalism, described by (3.9), the $2n$-th moment depends on all the previous non-vanishing moments, the explicit relation is given by

$$\frac{d}{dt}\langle x^{2n}(t)\rangle = \sum_{m=1}^{n} \frac{(2n)!}{(2n-2m)!(2m-2)!} D_{2m-2}(t)\langle x^{2n-2m}\rangle .$$

3.2. Time Dependent Diffusion Constant, a Sketch

Obviously, the local ϕ-formalism does not provide the more appropriate description in every physical system. A case in which the χ-formalism appears is in the Master equation description [18] of the effects of vibrational relaxation on intermolecular transfer of electronic excitation. Time-dependent transfer rates occur naturally there. The probability of occupation of vibrational levels as well as of site occupation, denoted by $P_m^M(t)$, where M and m refer to site and vibrational state respectively, obeys

$$\frac{d}{dt}P_m^M(t) = \sum_n [\gamma_{m,n}P_n^M(t) - \gamma_{n,m}P_m^M(t)] + \sum_N [F_m^{M,N}P_m^N(t) - F_m^{N,M}P_m^M(t)]. \tag{3.17}$$

When the use of a specific form for the relaxation rate $\gamma_{m,n}$ and the assumption of nearest-neighbor transfer rates $F_m^{M,N} = F_m\delta_{M,N}$ are made, an effective transfer equation is seen to emerge which is precisely of the form of the local χ-description. It is possible to convert that time dependence into appropriate expressions for quantities such as $\mathfrak{D}_\chi(x, x', t)$ of Eq. (2.9) and cast the problem into a spatially non-local ϕ mould. We refrain from showing any of the details here for want of space.

4. Concluding Remarks

If the phrase "anomalous diffusion" is taken to represent any process that has some basic features of standard diffusion but also has significant departures from the latter, one can state confidently that one encounters anomalous diffusion in a rich variety of physical situations. Quantum mechanical (quasi)particles, such as Frenkel excitons in photosynthetic units and photo-injected electrons in molecular crystals, obey GME's [11] whose features are in some regimes similar to those of the standard diffusion equation but in others sharply different as when coherence is substantially present. A description via memory functions [12], i.e., via the ϕ- formalism of Eq. (1.2), is natural in that case. In another extreme example of anomalous diffusion, one encounters animal movements [3] that are said to be an example of *fractional Brownian motion*, which is nothing but Eq. (1.1) representative of the χ-formalism with a power dependence of $\chi(t)$ on t [10]. There is a large number of other instances where it is not clear which of the descriptions is appropriate. Prior to 1973, it had been thought that continuous time random walks represented a description that was fundamentally different from that provided by memory functions. This viewpoint held previously (even by some of the originators of one of those descriptions) was found [7, 13, 16] to be incorrect and led to a trivially simple but important clarification. Because the demonstration of such an equivalence between methods of investigation or description can typically save much unnecessary theoretical labor, we thought it worthwhile to make that enquiry in the context of the χ- and the ϕ-formalisms. This enquiry, begun in part by one of the present authors in the context of the stress distribution of granular compacts [8], has been extended much further in the present paper.

The previous observation [8] that, as stated (i.e., in their *spatially* local form), the two descriptions cannot be generally equivalent, except in the trivial case when both describe standard diffusion, has been made transparently clear in the present paper by Eqs. (1.4), (1.5) for the mean square displacement. The latter is formally similar in the two formalisms but sharply different in content for general $\phi(t)$ or $\chi(t)$. The earlier indication [8] that each of the two formalisms is equivalent to a spatially non-local form of the other has been extended here through the complete and practical prescriptions (2.12) and (2.13). These prescriptions allow one, given an arbitrary form of $\phi(t)$ or $\chi(t)$, to obtain, at least in principle, the corresponding non-local quantities $\mathfrak{D}_\chi(x,t)$ and $\mathfrak{D}_\phi(x,t)$ in the other formalism. The prescriptions are given in k-space and the Laplace domain and are to be followed by a final Fourier-Laplace inversion if necessary.

We have also examined, in detail, a physical case in which the (spatially local) ϕ-formalism (i.e., memory functions) provides the accurate description. We have explicitly shown how our prescription developed in Eq. (3.8) is applied and how the spatially non-local character develops in the χ-formalism (see Eq. (3.9)). This is the elucidation of the railway-track model. We have also pointed out how, the reverse situation, a natural description in terms of the χ-formalism, arises in the theory of vibrational relaxation of molecular excitations in the presence of motion [18]. To avoid lengthy calculations we have refrained from showing the detail of that opposite situation.

Because the spatial shape of the propagator in the local χ-formalism is always Gaussian (for instance in fractional Brownian motion), only a generalization from the time t to $\tau = \int_0^t ds\,\chi(s)$ being necessary (see Eq. (2.7)), one may tend to believe that the χ-formalism is not as rich as the ϕ-formalism–since the latter allows more freedom in the shape of the propagator. This is, however, not an appropriate statement. Indeed, there exists a certain “symmetry in richness" as we move from the Laplace domain to the time domain. For instance, an equation such as (1.1) will appear as a convolution equation in the Laplace domain.

Since it may not be clear by inspection which of the two formalisms may be providing a correct description for a given system, we comment in passing about probing this question experimentally. We draw the attention of the reader to two observational set-ups in two widely different areas of study: transient grating observations in molecular crystals (TGO) [11, 19] and nuclear magnetic resonance microscopy (NMRM) [20, 21]. Both are sensitive not merely to the mean square displacement of the moving entities but to the entire Fourier transform of the probability density $P(x,t)$. The TGO involve crossed laser beams that are used to create a sinusoidal spatial distribution of electronic excitations. The decay of the amplitude of that distribution is measured and information is thereby extracted about the motion of the excitations. The NMRM uses the so-called pulsed-gradient spin echo technique [22]. In the limit of very short duration pulses, that technique, just as does the TGO, probes directly the Fourier transform of $P(x, t)$ and not its moments. Clearly, both experiments should be able, in principle, to discriminate between whether a local χ or local ϕ formalism is appropriate. More details will be presented elsewhere.

We have restricted attention in the present paper to spatially homogeneous (not the same as spatially non-local or local) systems only. Fourier transformation therefore diagonalizes the relevant matrices and there is no $k - k'$ interaction. If spatial locations were not all equivalent, we would have terms such as $\mathfrak{D}_\chi(x, x', t)$ and $\mathfrak{D}_\phi(x, x', t)$ in Eqs. (2.9) and (2.10) and the difference nature of the spatial kernels would be destroyed. Transformation via $\exp(ikx)$ would be useless and the problem would become quite complicated. Because situations where the diffusion coefficient itself may be spatially varying occur often in physical and biological applications, it is important to extend our analysis to incorporate them. We plan to do that in a future study.

We acknowledge fruitful discussions with Dr. Luca Giuggioli. This work was supported in part by the NSF under Grant no. INT-0336343, by NSF/NIH Ecology of Infectious Diseases under Grant no. EF-0326757 and by DARPA under Grant no. DARPA-N00014-03-1-0900.

References

[1] E.M. Conwell, High Field Transport in Semiconductors, (Academic Press, New York, 1967).

[2] V.M. Kenkre, Ceramic Transactions 21, 69 (1991); see also V.M. Kenkre, L. Skala, M. Weiser, J.D. Katz, J. Materials Science 26, 2483 (1991).

[3] L. Giuggioli, G. Abramson, V.M. Kenkre, C. Parmenter, T. Yates, J. Theoretical Biology 240, 126-135, (2006); see also V.M. Kenkre, Physica A, 356, 121-126 (2005).

[4] A. Einstein, Ann. Physik **17,** 549 (1905).

[5] L. Bachelier, Annales Scientifiques de l'École Normale Superieure **17,** 21 (1900).

[6] See, e.g., P. Martin and G. G. Emch, Helv. Phys. Acta 48, 59 (1975).

[7] V. M. Kenkre in Statistical Mechanics and Statistical Methods in Theory and Application ed. U. Landman (New York, Plenum, September 1977), pp. 441-461.

[8] V. M. Kenkre, Granular Matter 3, 23-28 (2001).

[9] V. M. Kenkre in ÒThe Granular StateÓ, MRS Symp. Proc. Vol. 627, eds. S.Sen and M.L.Hunt (MRS, Warrendale, PA 2000), pp. BB 6.5.1 - 6.5.8.; see also M. Kuś and V. M. Kenkre, unpublished results.

[10] B.B. Mandelbrot and J.W. Van Ness, SIAM Rev. **10,** 422 (1968).

[11] V.M. Kenkre, *Exciton Dynamics in Molecular Crystals and Aggregates: the Master Equation Approach*, Springer Tracts in Modern Physics, Vol. 94, ed. G. Hoehler (Springer, Berlin, 1982).

[12] See, e.g., R. W. Zwanzig, in Lectures in Theoretical Physics ed. W. E. Downs and J. Downs (Interscience, Boulder, Colo. 1961), vol. III; see also Physica 30, 1109 (1964).

[13] V.M. Kenkre, E.W. Montroll, M.F. Shlesinger, J. Stat. Phys. 9, 45 (1973); V.M. Kenkre, R.S. Knox, Phys. Rev. B9, 5279 (1974).

[14] E.W. Montroll and G.H. Weiss, J. Math. Phys. **6,** 167 (1965); E. Montroll and H. Scher, J. Stat. Phys. 9, 101 (1973).

[15] H. Haken and P. Reineker, Z. Phys. **249,** 253 (1972); M.K. Grover and R. Silbey, J. Chem. Phys. **54,** 4843 (1971).

[16] V.M. Kenkre, R.S. Knox, Phys. Rev. B9, 5279 (1974).

[17] M. Abramowitz and I.A. Stegun, *Handbook of mathematical functions* (Dover N.Y. 1972).

[18] V.M. Kenkre, Phys. Rev. A **16,** 766 (1977).

[19] M.D. Fayer, Phys. Rev. Lett. **41,** 131 (1978).

[20] P.T. Callaghan, *Principles of Nuclear Magnetic Resonance Microscopy*, Oxford Univ, Press, New York (1991).

[21] F.J. Sevilla and V.M. Kenkre, J. Phys. Condens. Matter **18,** 1-14 (2006).

[22] E.O. Stejskal, and J.E. Tanner, J Chem. Phys. **42,** 288 (1965); E.O. Stejskal, J Chem. Phys. **43,** 3595 (1965); E.O. Stejskal, and J.E. Tanner, J Chem. Phys. **49,** 1768 (1968).

Consortium of the Americas for Interdisciplinary Science and Department of Physics and Astronomy, University of New Mexico, Albuquerque, New Mexico 87131, USA

Infinite Divisibility in Euclidean Quantum Mechanics

John R. Klauder

In simple – but selected – quantum systems, the probability distribution determined by the ground state wave function is infinitely divisible. Like all simple quantum systems, the Euclidean temporal extension leads to a system that involves a stochastic variable and which can be characterized by a probability distribution on continuous paths. The restriction of the latter distribution to sharp time expectations recovers the infinitely divisible behavior of the ground state probability distribution, and the question is raised whether or not the temporally extended probability distribution retains the property of being infinitely divisible. A similar question extended to a quantum field theory relates to whether or not such systems would have nontrivial scattering behavior.

Dedicated to Gérard G. Emch
on the occasion of his 70th birthday

Introduction, Discussion & Proposition

Preliminary Details

An interesting and well studied class of quantum mechanical Hamiltonians consists of examples of the form

$$\mathcal{H} = -\tfrac{1}{2}\partial^2/\partial x^2 + V(x)\ ,$$

expressed in units where $m = \hbar = 1$ [1], where the real potential $V(x)$ is chosen so that the spectrum of $\mathcal{H}$ is nonnegative and the real, normalizable, nowhere vanishing ground state $\phi_o(x)$ has zero energy eigenvalue; such systems are referred to as "simple systems" in this article. In this case, the ground state itself determines the Hamiltonian completely since

$$\mathcal{H} = -\tfrac{1}{2}[\partial^2/\partial x^2 - \phi_o''(x)/\phi_o(x)]\ ,$$

and therefore $\phi_o(x)$ – just as much as $\mathcal{H}$ itself – can be said to determine all the properties of the system.

---

[1] For this particular dedicatee, one is tempted – following the suggestion of Victor Weisskopf – to set $e = m = c = h = 1$, despite the fact that this would give the fine structure constant an unphysical value.

Without loss of generality we can assume that $\phi_0(x)$ is normalized in the sense that

$$\int \phi_0(x)^2\,dx = 1\ ;$$

all integrals without indicated limits of integration extend from $-\infty$ to $+\infty$. In that case, $\phi_0(x)^2$ determines a probability density, and, for all real s, the expression

$$C(s) \equiv \int e^{isx}\,\phi_0(x)^2\,dx$$

defines the associated characteristic function. The information contained in the characteristic function $C(s)$ determines the probability density $\phi_0(x)^2$ and thereby the ground state $\phi_0(x)$ itself. Therefore, we can assert that the function $C(s)$ determines all the properties of the system.

The distributions in an interesting *subclass* of all probability distributions have the property of being *infinitely divisible* [1]. An infinitely divisible distribution may be characterized as follows: If X denotes a random variable the characteristic function of which is given by

$$\langle e^{isX}\rangle \equiv C(s) = \int e^{isx}\,\phi_0(x)^2\,dx\ ,$$

then, for every integer $N \geq 2$, there exists N independent identically distributed random variables, $Y_n^{(N)}$, $1 \leq n \leq N$, such that

$$X = \sum_{n=1}^{N} Y_n^{(N)}\ .$$

Alternatively stated, this property implies that the Nth root of the characteristic function $C(s)$ is again a characteristic function, i.e.,

$$C_{(N)}(s) \equiv C^{1/N}(s) = \int e^{isx}\,\rho_{(N)}(x)\,dx\ ,$$

where $\rho_{(N)}(x)$ denotes a probability density for all N. Note well that only a subclass of all probability distributions exhibits infinite divisibility.

Remark. Integer powers of characteristic functions always correspond again to characteristic functions. Therefore, for infinitely divisible distributions, the expression $C^{M/N}(s)$ again describes a characteristic functions for all positive rational numbers M/N. All characteristic functions are continuous, and it follows that in the limit as those rational numbers tend to an arbitrary positive real number, i.e., a limit such that $M/N \to r$, where $0 < r < \infty$, the result is again a characteristic function defined by $C^r(s)$. □

The general form of characteristic functions which are infinitely divisible is well known [1]. For simplicity we shall confine attention to only those distributions which are *even*, i.e., for which $V(-x) = V(x)$, leading to an even ground state $\phi_0(-x) = \phi_0(x)$, and thereby to an even characteristic function $C(-s) = C(s)$. In that case, the characteristic function $C(s)$ of an even, infinitely divisible distribution may be represented in the form [1]

$$C(s) = \exp\{-\tfrac{1}{2}as^2 - \textstyle\int[1 - \cos(sy)]\,\sigma(y)\,dy\}\,,$$

where $a \geq 0$ and $\sigma(y) \geq 0$ for all y; clearly, $\sigma(-y) = \sigma(y)$. In this expression we have allowed ourselves one simplification: namely, we have chosen an absolutely continuous measure $\sigma(y)\,dy$. Existence of this expression requires that

$$\textstyle\int[y^2/(1+y^2)]\sigma(y)\,dy < \infty\,.$$

However, it may well happen that

$$\textstyle\int\sigma(y)\,dy = \infty\,,$$

and indeed this situation is relatively common, especially in our studies.

The general form of the characteristic function for infinitely divisible distributions implies that the associated random variable X, for which $\langle e^{isX}\rangle = C(s)$, may be decomposed into a sum of *two, independent* random variables, $X = X_G + X_P$, where the random variable X_G has a Gaussian distribution (determined by a) and the random variable X_P has a Poisson distribution (determined by σ); more precisely, X_P has a compound Poisson distribution or a generalized Poisson distribution depending on whether $\int\sigma(y)\,dy$ is finite or infinite, respectively [2].

It is noteworthy that $\ln(C(s))$ is the generator of the truncated (= connected) moments of the distribution. Therefore, for infinitely divisible distributions, it follows that

$$\langle(e^{isX} - 1)^T\rangle = -\tfrac{1}{2}as^2 - \textstyle\int[1 - \cos(sy)]\,\sigma(y)\,dy\,.$$

Assuming the necessary moments exist, we learn that the truncated moments (superscript T) are given by

$$\langle X^{2p}\rangle^T = a\,\delta_{p1} + \int y^{2p}\,\sigma(y)\,dy\,,$$

and as a consequence the even order truncated moments are always nonnegative. This will be an important feature in our futher investigations.

We shall not be interested in examples that have both a Gaussian and a Poisson contribution. Indeed, we shall focus on the wide class of examples that arise from Poisson distributions alone and thus we assume that there is *no* Gaussian contribution. This means we shall set $X_G = 0$ or equivalently assume that the parameter a vanishes. Having said this, we are left to focus on that subclass of characteristic functions which can be written in the form

$$C(s) = \exp\{-\textstyle\int[1 - \cos(sy)]\,\sigma(y)\,dy\}\,,$$

for nonnegative functions σ such that

$$\textstyle\int[y^2/(1 + y^2)]\,\sigma(y)\,dy < \infty\,.$$

We shall also write

$$U(y)^2 \equiv \sigma(y)\,,$$

and we shall call the nonnegative function $U(y)\,[= U(-y)]$ the *model function*.

Just as we have asserted that the characteristic function $C(s)$ determines all the physics of our problem, we are clearly able to declare that *the model function $U(y)$ determines all the physics of our problem*. In other words, we can, without loss of generality, adopt the model function $U(y)$ as the primary input defining the problem at hand. Indeed, this view of the problem has the advantage that choosing $U(y)$ initially, and within a certain suitable class of functions, ensures that we are dealing with an even potential $V(x)$ that leads to a ground state $\phi_0(x)$ which in turn generates an infinitely divisible distribution that involves only a Poisson contribution. Of course, starting with the model function is often easier said then done!

To demonstrate that this set of conditions is not empty, it is appropriate to present a few examples.

Examples

Example 1: The first example is an idealized example that has the advantage of being analytically quite simple. It is convenient to initiate our description in terms of the ground state wave function

$$\phi_0(x) = \frac{\sqrt{a}}{\sqrt{\pi(a^2 + x^2)}}\,,$$

where $a > 0$ is a fixed parameter. In turn, this example corresponds to the potential

$$V(x) = \frac{2x^2 - a^2}{(a^2 + x^2)^2}\,.$$

The characteristic function appropriate to this example is given by

$$\begin{aligned} C(s) &= \int \frac{a\,e^{isx}}{\pi\,(a^2+x^2)}\,dx \\ &= \exp(-a\,|s|) \\ &= \exp\{-a \textstyle\int [1-\cos(sy)](1/\pi\,y^2)\,dy\}\,. \end{aligned}$$

Thus, for this example, we see that the model function reads

$$U(y) = \frac{\sqrt{a}}{\sqrt{\pi}}\,\frac{1}{|y|}\,.$$

It is true that $\int U(y)^2\,dy = \infty$, due to a divergence at $y = 0$. However, the present distribution falls very slowly as $y \to \infty$, so slowly in fact that no moments of the distribution exist.

The next example is similar to the present one except that the moments of the distribution all exist.

Example 2: For this model we start with the characteristic function of the ground state probability density in the form

$$\begin{aligned} C(s) = e^{-b\sqrt{s^2+\rho^2}+b\rho} &= \int e^{isx}\,\frac{b\rho}{\pi}\,\frac{K_1(\rho\sqrt{x^2+b^2})}{\sqrt{x^2+b^2}}\,e^{b\rho}\,dx \\ &= \exp\{-b\int [1-\cos(s\,y)]\,\frac{\rho\,K_1(\rho\,|y|)}{\pi\,|y|}\,dy\}\,, \end{aligned}$$

where $b > 0$ and $\rho > 0$ are free parameters at our disposal. Here, K_1 denotes the standard Bessel function. The ground state wave function for this example may be read off from the integrand of the characteristic function. In turn, the ground state implicitly determines the potential $V(x)$. Clearly the model function for this case is given by

$$U(y) = \frac{\sqrt{b\rho\,K_1(\rho|y|)}}{\sqrt{\pi\,|y|}}\,.$$

Since $K_1(\rho|y|) \simeq 1/(\rho|y|)$ for small argument, it follows that *near the singularity* at $y = 0$, the behavior of $U(y)$ is actually identical in the two examples when $b = a$; indeed, in the limit that $\rho \to 0$, it follows that Example 2 reduces to Example 1.

As a vast generalization of both Examples 1 and 2, we may consider characteristic functions of the general form

$$\exp\{-b\int [1-\cos(s\,y)]\,F(y)/(\pi\,y^2)\,dy\} \equiv \int e^{isx}\,G(x)\,dx\,,$$

generated by basic functions $F \in C^2$ with $F(0) = 1$, $F \geq 0$, and $\int [F(y)/(1+y^2)]\, dy < \infty$. Generally speaking, after being given an analytic expression for the basic function for such examples, we cannot analytically specify the L^1 nonnegative weight function G describing the ground state probability density; however, we know that such a probability density exists.

Euclidean Time Dependence

The Hamiltonian for our system can be used to propagate the time forward either in real time by the evolution operator $\exp(-i\mathcal{H}T)$ or in imaginary time by the evolution operator $\exp(-\mathcal{H}T)$. The latter case corresponds to Euclidean quantum mechanics, which as is well known, can be described by a stochastic process involving a stochastic variable $X(t)$, $-\infty < t < \infty$, with a distribution determined by a Feynman-Kac like probability distribution [3]. Expanding the meaning of the symbols representing expectation values, i.e., $\langle (\cdot) \rangle$, to cover the stochastic variable $X(t)$, we are led to consider correlation functions of the form

$$\int \langle X(t_1)X(t_2) \cdots X(t_p)\rangle\, f_p(t_1, t_2, \ldots, t_p)\, dt_1\, dt_2 \cdots dt_p\ ,$$

for all $p \geq 1$ and all suitable weight functions f_p. In turn, correlation functions may alternatively be described by a suitable generating functional having the meaning of a characteristic functional. Specifically, we have in mind the expression

$$E\{u\} \equiv \langle \exp[i \textstyle\int u(t)X(t)\, dt] \rangle\ ,$$

defined for all smooth test functions u. According to the tenants of Euclidean quantum mechanics, the functional $E\{u\}$ obeys all the appropriate positive-definite inequalities and continuity to satisfy the Bochner-Minlos axioms to be the functional Fourier transform of a suitable probability distribution [4].

We further observe that since the Hamiltonian is explicitly time independent, the associated stochastic process is stationary. Specifically this implies a time translation invariance of the characteristic functional, which takes the form

$$\langle \exp[i \textstyle\int u(t-\tau)X(t)\, dt] \rangle = \langle \exp[i \textstyle\int u(t)X(t)\, dt] \rangle$$

for any τ, $-\infty < \tau < \infty$.

There is a connection of the characteristic functional $E\{u\}$ involving all time and the characteristic function $C(s)$ introduced above that occurs at any one time, say at time $t = 0$. If we choose the function $u(t)$ that enters the characteristic functional as

$$u(t) = s\, \delta(t)\ ,$$

where $\delta(t)$ is a Dirac delta function, then is follows that

$$E\{u(\cdot) = s\,\delta(\cdot)\} = C(s) \;;$$

hence, the restricted evaluation of the characteristic functional E coincides with the characteristic function C defined at a single time.

We can make a similar connection with any one of the correlation functions themselves. For example, let us focus on

$$\int \langle X(t_1) \cdots X(t_4) \rangle \, f_4(t_1, \ldots, t_4)\, dt_1 \cdots dt_4$$

evaluated for

$$f_4(t_1, \ldots, t_4) = \delta(t_1) \cdots \delta(t_4) \; ,$$

leading to $\langle X(0)^4 \rangle$. We assert the equality given by

$$\langle X(0)^4 \rangle = \int x^4 \, \phi_0(x)^2 \, dx \; ,$$

i.e., the fourth moment of the ground-state probability distribution.

Now We Come to the Whole Point of the Present Paper!

By choice, we have restricted the sharp-time, ground-state probability distribution to be infinitely divisible. This is encoded into the fact that an arbitrary, positive fractional power of the characteristic function is again a characteristic function. Since the full-time characteristic functional (E) collapses to the sharp-time characteristic function (C) when the testing functions are evaluated at a sharp time (say, $t = 0$), it follows *for such a set of limited test functions* that the sharp-time restricted characteristic functional is infinitely divisible. The question arises, therefore, *whether the FULL time characteristic functional is itself infinitely divisible as well*, or whether that property is restricted to those stochastic variables with a time spread restricted to a window of finite size. It may also happen that certain distributions enjoy full time infinite divisibility following from sharp time infinite divisibility, while other distributions do not exhibit full time infinite divisibility even though they have sharp time infinite divisibility. It is clear that the question raised here is fairly complicated. The simplest possibility would seem to be to show that sharp time infinite divisibility does not imply full time infinite divisibility, and this could be determined by finding just one example for which a single property required by infinite divisibility fails to hold.

Selection of the Test Question

The structure of infinitely divisible distributions is such that the *truncated* correlation functions are positive definite functions. In particular, for such distributions, it follows that

$$\langle(\int f(t)X(t)\,dt)^4\rangle^T \equiv \langle(\int f(t)X(t)\,dt)^4\rangle - 3\langle(\int f(t)X(t)\,dt)^2\rangle^2 > 0$$

unless $f = 0$. The simplest example derives from a uniform weighting where $f(t) = 1$ in an interval $-T < t < T$, and $f(t)$ is zero elsewhere. We should like to examine the behavior of this truncated correlation function for asymptotically large T, and compare it with the behavior for extremely small T. Due to stationary of the underlying ensemble, the large T behavior diverges as $2T$; for small times, on the other hand, it behaves as $(2T)^4$ times the sharp time expectation value. To eliminate these unimportant temporal factors, it is convenient to rescale the truncated four point function by a factor $(1 + 2T)^3/(2T)^4$, which leads to the alternative form given by

$$\begin{aligned}\chi_2 &\equiv \frac{(1+2T)^3}{(2T)^4}\left[\langle(\int_{-T}^{T} X(t)\,dt)^4\rangle^T\right]\\ &= \frac{(1+2T)^3}{(2T)^4}\left[\langle(\int_{-T}^{T} X(t)\,dt)^4\rangle - 3\langle(\int_{-T}^{T} X(t)\,dt)^2\rangle^2\right].\end{aligned}$$

Although this is just one measure of the truncated four point function it is a significant one and one that is comparatively easy to evaluate. If $\chi_2 < 0$ for large T, it would establish that the full time probability functional is *not* infinitely divisible even though the sharp time probability function has been chosen (by design) to be infinitely divisible. Of course, if $\chi_2 > 0$ for all T, then no definitive conclusion can be drawn about the nature of the full time probability distribution. To proceed further, one would need to check other moments, or even change to another model function in order to test the nature of the full time probability distribution. At present, the author is not aware of any general scheme that could decide whether or not the full time probability functional is infinitely divisible whenever the sharp time probability function has been chosen to be infinitely divisible. Nevertheless, in the absence of any proof to the contrary, it seems reasonable to *conjecture* that infinite divisibility for sharp time simple quantum systems does *not* imply full time infinite divisibility for them.

Alternative Representation

It is noteworthy that the evaluation of χ_2 can be reexpressed in terms of the eigenfunctions and eigenvalues of the Hamiltonian for our simple system, and we now turn our attention to develping this alternative expression for χ_2. For convenience, let us consider those special cases where $\mathcal{H}$ has a purely discrete, nondegenerate spectrum such that

$$[-\tfrac{1}{2}\partial^2/\partial x^2 + V(x)]\phi_n(x) = E_n\phi_n(x) ,$$

where $0 = E_0 < E_1 < E_2 \cdots$ and all the (real) eigenfunctions form a complete orthonormal set of functions for which

$$\int \phi_n(x)\phi_m(x)\,dx = \delta_{nm} .$$

Here, as before, the ground state is denoted by $\phi_0(x)$. We shall also introduce these relations in the usual abstract bra-ket notation as well, namely

$$[\tfrac{1}{2}P^2 + V(Q)]|n\rangle = E_n|n\rangle ,$$

with $|n\rangle$ denoting the abstract eigenvectors, for which $\langle n|m\rangle = \delta_{nm}$. Of course, we let $|0\rangle$ denote the ground state.

By definition, the correlation functions are *symmetric* functions of their temporal arguments. However, when expressed in terms of the eigenvectors and eigenvalues, it proves convenient to define the correlation functions in a *time ordered* manner. In particular, for the time ordered two point function, this rule leads to the expression

$$\langle X(t)X(u)\rangle = \sum_{n=0}^{\infty}\langle 0|Q|n\rangle e^{-E_n(t-u)}\langle n|Q|0\rangle , \qquad t \geq u ;$$

we can of course extend this expression to all t and u values simply by replacing $(t-u)$ by $|t-u|$ on the right hand side, but we choose not to do so in order to simplify the analysis in what follows. It is clearly convenient to introduce the notational shorthand that

$$Q_{kl} \equiv \langle k|Q|l\rangle .$$

Similar expressions also exist for higher-order correlation functions. For example, for the time ordered four point function we have

$$\langle X(t)X(u)X(v)X(w)\rangle = \sum_{k,l,m=0}^{\infty} Q_{0k}\,e^{-E_k(t-u)}\,Q_{kl}\,e^{-E_l(u-v)} \times Q_{lm}\,e^{-E_m(v-w)}\,Q_{m0} , \qquad t \geq u \geq v \geq w .$$

From the assumed symmetry of the potential, i.e., $V(-x) = V(x)$, it follows that the eigenfunctions have *alternating parity*, namely, that $\phi_n(-x) = (-1)^n\phi_n(x)$. Consequently, the matrix elements of Q, such as $\langle k|Q|l\rangle = Q_{kl}$, only connect eigenvectors of *opposite* parity; in other words, if k is even, then l is odd, or vice versa. In particular, for the two point function, as presented above, only *odd* values of n lead to nonvanishing contributions. For the four point function, as presented above, only odd values of k and m contribute, while only even values of l need be considered.

To begin our construction of χ_2, we may integrate our time ordered expressions over a restricted time region and rescale the result to account for such a limited integration domain. Initially, this recipe implies that

$$\langle\!\langle (\int_{-T}^{T} X(t)\,dt)^4 \rangle\!\rangle = 24\left[\sum_{k,l,m} Q_{0k}\,Q_{kl}\,Q_{lm}\,Q_{m0}\right]\int_{-T}^{T} dt \int_{-T}^{t} du$$
$$\times \int_{-T}^{u} dv \int_{-T}^{v} dw \exp[-E_k(t-u) - E_l(u-v) - E_m(v-w)]\ .$$

Here the factor $24 = 4!$ corrects for the limited domain of integration due to time ordering. For small T, the integral above involves terms $O(T^4)$, while for large T, this integral involves terms $O(T^2)$, $O(T)$, and $o(T)$. Since we content ourselves with the integral's value for very small and very large T values, it will suffice to consider terms of the indicated types, respectively.

Small T Behavior

First, consider very small T values, or more specifically consider the situation where $E_k T \ll 1$ for all sensibly contributing energy values. In that case, we can replace all exponentials in the previous formula by unity, and therefore, to leading order

$$\frac{(1+2T)^3}{(2T)^4}\langle\!\langle (\int_{-T}^{T} X(t)\,dt)^4 \rangle\!\rangle = \sum_{k,l,m} Q_{0k}\,Q_{kl}\,Q_{lm}\,Q_{m0}\ , \qquad T \ll 1\ .$$

This result of course is equivalent to the sharp time four point moment. Second, consider the square of the two point function, which leads to the result

$$\frac{(1+2T)^3}{(2T)^4}\langle\!\langle (\int_{-T}^{T} X(t)\,dt)^2 \rangle\!\rangle^2 = \sum_{k,m=1}^{\infty} Q_{0k}\,Q_{k0}\,Q_{0m}\,Q_{m0}\ , \qquad T \ll 1\ ,$$

to leading order in T. This expression, of course, is just the square of the sharp time second moment. Finally, the truncated four point moment for small T is given by

$$\chi_2 = \sum_{k,l,m=1}^{\infty} Q_{0k}\,Q_{kl}\,Q_{lm}\,Q_{m0} - 2\sum_{k,m=1}^{\infty} Q_{0k}\,Q_{k0}\,Q_{0m}\,Q_{m0}\ ;$$

note well, that for the first term all three sums omit the ground state, and for the second term both sums omit the ground state. This change of the first term has resulted in the factor 3 becoming a factor 2 in the second term.

We can readily check this result for the harmonic oscillator, namely where $V(x) = \frac{1}{2}\omega^2 x^2$. In that case, the only nonvanishing matrix elements of interest are $Q_{01} = Q_{10} = 1/\sqrt{2\omega}$ and $Q_{12} = Q_{21} = 1/\sqrt{\omega}$. Therefore,

$$\begin{aligned}\chi_2 &= Q_{01}\, Q_{12}\, Q_{21}\, Q_{10} - 2[\,Q_{01}\, Q_{10}\,]^2 \\ &= [1/(2\omega^2) - 2/(4\omega^2)] = 0\ .\end{aligned}$$

This is the correct result: Since the ground state for the harmonic oscillator is a Gaussian, the distribution is infinitely divisible; and, as a Gaussian, all truncated moments other than the second moment vanish.

Large T Behavior

Let us now turn our attention to the evaluation of χ_2 for large T. In this case it is useful to divide the fourth moment into two distinct expressions, namely,

$$\frac{(1+2T)^3}{(2T)^4}\,\langle(\int_{-T}^{T} X(t)\,dt)^4\rangle = A + B\ .$$

In this expression,

$$A \equiv 24 \sum_{k,l,m=1}^{\infty} Q_{0k}\, Q_{kl}\, Q_{lm}\, Q_{m0}\left[\frac{1}{E_k}\,\frac{1}{E_l}\,\frac{1}{E_m}\right],$$

which is valid for $T \gg 1$, provided all E values are strictly positive, $E > 0$. This restriction leads to the omission of the term for $l = 0$ above. The omitted term, where $l = 0$, can be evaluated separately, and to leading order it follows that

$$B \equiv 24 \sum_{k,m=1}^{\infty} Q_{0k}\, Q_{k0}\, Q_{0m}\, Q_{m0}\left[T\,\frac{1}{E_m}\,\frac{1}{E_k} - \frac{1}{E_k^2}\,\frac{1}{E_m} - \frac{1}{E_k}\,\frac{1}{E_m^2}\right].$$

The two point function for large T follows from the expression given by

$$\langle(\int_{-T}^{T} X(t)\,dt)^2\rangle = \sum_{m=1}^{\infty} Q_{0k}\, Q_{k0}\left[\frac{4T}{E_k} - \frac{2}{E_k^2}\right].$$

Consequently, for very large T, we observe that

$$\begin{aligned}&\frac{(1+2T)^3}{(2T)^4}\left(\langle(\int_{-T}^{T} X(t)\,dt)^2\rangle\right)^2 \\ &\quad = \frac{1}{2T}\sum_{k,m=1}^{\infty} Q_{0k}\, Q_{k0}\, Q_{0m}\, Q_{m0}\left[\frac{4T}{E_k} - \frac{2}{E_k^2}\right]\left[\frac{4T}{E_m} - \frac{2}{E_m^2}\right]\end{aligned}$$

Finally, for large T,

$$\begin{aligned}
&\frac{(1+2T)^3}{(2T)^4} \langle\langle \int_{-T}^{T} X(t)\,dt)^4\rangle^T \\
&= 24 \sum_{k,l,m=1}^{\infty} Q_{0k}\, Q_{kl}\, Q_{lm}\, Q_{m0} \left[\frac{1}{E_k}\frac{1}{E_l}\frac{1}{E_m}\right] \\
&+24\,T \sum_{k,m=1}^{\infty} Q_{0k}\, Q_{k0}\, Q_{0m}\, Q_{m0} \left[\frac{1}{E_k}\frac{1}{E_m}\right] \\
&-24 \sum_{k,m=1}^{\infty} Q_{0k}\, Q_{k0}\, Q_{0m}\, Q_{m0} \left[\frac{1}{E_k}\frac{1}{E_m^2} + \frac{1}{E_k^2}\frac{1}{E_m}\right] \\
&-3\,\frac{1}{2T} \sum_{k,m=1}^{\infty} Q_{0k}\, Q_{k0}\, Q_{0m}\, Q_{m0} \left[\frac{4T}{E_k} - \frac{2}{E_k^2}\right]\left[\frac{4T}{E_m} - \frac{2}{E_m^2}\right],
\end{aligned}$$

which to leading order in T becomes

$$\begin{aligned}
\chi_2 = 24 \sum_{k,l,m=1}^{\infty} Q_{0k}\, Q_{kl}\, Q_{lm}\, Q_{m0} \left[\frac{1}{E_k}\frac{1}{E_l}\frac{1}{E_m}\right] \\
-24 \sum_{k,m=1}^{\infty} Q_{0k}\, Q_{k0}\, Q_{0m}\, Q_{m0} \left[\frac{1}{E_k}\frac{1}{E_m^2}\right].
\end{aligned}$$

Again, it is useful to test this expression with the harmonic oscillator. Besides the matrix elements already given previously, we need the first two excited state energy levels, $E_1 = \omega$ and $E_2 = 2\omega$. In that case, we find that

$$\chi_2 = 24\,[(1/2\omega^2)\cdot(1/2\omega^3) - (1/4\omega^2)\cdot(1/\omega^3)] = 0\,,$$

as expected.

Summary of Principal Formulas

The truncated four point function given by

$$\chi_2 = (1+2T)^3/(2T)^4\, \langle\langle \int_{-T}^{T} X(t)\,dt)^4\rangle^T$$

is expressed alternatively by

$$\chi_2 = \sum_{k,l,m=1}^{\infty} Q_{0k}\, Q_{kl}\, Q_{lm}\, Q_{m0} - 2\sum_{k,m=1}^{\infty} Q_{0k}\, Q_{k0}\, Q_{0m}\, Q_{m0}\,,$$

for *small* T, and by

$$\chi_2 = 24 \sum_{k,l,m=1}^{\infty} Q_{0k}\, Q_{kl}\, Q_{lm}\, Q_{m0} \left[\frac{1}{E_k} \frac{1}{E_l} \frac{1}{E_m} \right] - 24 \sum_{k,m=1}^{\infty} Q_{0k}\, Q_{k0}\, Q_{0m}\, Q_{m0} \left[\frac{1}{E_k} \frac{1}{E_m^2} \right],$$

for *large* T. In this expression, $Q_{kl} \equiv \langle k|Q|l\rangle$ denotes a matrix element of the position operator Q, where the states $|k\rangle$ are normalized eigenvectors of the time-independent Schrödinger equation,

$$[\tfrac{1}{2} P^2 + V(Q)]\,|k\rangle = E_k\,|k\rangle\,,$$

for the real, even potential, $V(Q)$.

Remark. The apparent difference in dimensions between the behavior for small and large T arises from our choice of a scaling factor; had we used

$$(1 + 2ET)^3/(2ET)^4$$

instead of $(1 + 2T)^3/(2T)^4$, where E is some characteristic energy level, then the two extremal expressions would have had the same dimensions. However, since our main concern is to study simple systems for which $\chi_2 > 0$ for small T with the aim of finding examples for which $\chi_2 < 0$ for large T, the formal difference in dimensionality is of little concern. □

Why Bother?

The reader may well ask why should one care about quantum systems that have infinitely divisible distributions for sharp time position variables and may – or may not – have infinitely divisible distributions for full time position variables. A brief explanation may help clarify the situation.

A Euclidean formulation for a scalar quantum field theory is characterized by a stochastic field variable we may call $\phi(x, t)$. The statistics of this variable are governed by a probability distribution, which, like our single degree of freedom examples discussed above, may be described by a characteristic functional

$$\langle e^{i \int u(x,t)\phi(x,t)\, dx\, dt} \rangle\,.$$

The sharp time expression is given simply by setting $u(x, t) = v(x)\delta(t)$ leading to

$$\langle e^{i \int v(x)\phi(x,0)\, dx} \rangle\,.$$

Interacting quantum field theories encounter divergences, and this is as true for the sharp time field expressions as it is for the full time field expressions. In a recent paper [5] it was argued that sharp time formulations involving an infinitely divisible field distribution tended to alleviate some of the principal causes of field theory divergences. From this point of view there would seem to be some merit in seeking to formulate matters so that sharp time fields had infinitely divisible distributions. Accepting such an argument opens the question of whether an infinitely divisible distribution for sharp time fields does – or does not – force the full time field distribution to be infinitely divisible.

Suppose, for the sake of argument, it was true that the full time field distribution was also infinitely divisible. By an argument of Buchholz and Yngvason [6] this would result in a theory with a unit scattering matrix. For example, consider the truncated four point function

$$\langle \phi(f_1)\phi(f_2)\phi(g_2)\phi(g_1)\rangle^T ,$$

where

$$\phi(f) \equiv \textstyle\int f(x,t)\phi(x,t)\,dx\,dt ,$$

for f a smooth test function, etc. For infinitely divisible distributions certain truncated correlation functions are nonnegative, such as

$$0 \le \langle \phi(g_1)\phi(g_2)\phi(g_2)\phi(g_1)\rangle^T .$$

Consequently, by the Schwarz inequality,

$$\begin{aligned} 0 &\le |\langle \phi(f_1)\phi(f_2)\phi(g_2)\phi(g_1)\rangle^T|^2 \\ &\le \langle \phi(f_1)\phi(f_2)\phi(f_2)\phi(f_1)\rangle^T \langle \phi(g_1)\phi(g_2)\phi(g_2)\phi(g_1)\rangle^T . \end{aligned}$$

Now, take suitable limits so that

$$\begin{aligned} &\phi(g_1) \to \phi_{in}(g_1) , \qquad && \phi(g_2) \to \phi_{in}(g_2) , \\ &\phi(f_1) \to \phi_{out}(f_1) , \qquad && \phi(f_2) \to \phi_{out}(f_2) , \end{aligned}$$

where "*in*" and "*out*" fields denote free fields appropriate to the asymptotic regime. In that case we find

$$\begin{aligned} 0 &\le |\langle \phi_{out}(f_1)\phi_{out}(f_2)\phi_{in}(g_2)\phi_{in}(g_1)\rangle^T|^2 \\ &\le \langle \phi_{out}(f_1)\phi_{out}(f_2)\phi_{out}(f_2)\phi_{out}(f_1)\rangle^T \\ &\quad \times \langle \phi_{in}(g_1)\phi_{in}(g_2)\phi_{in}(g_2)\phi_{in}(g_1)\rangle^T = 0 , \end{aligned}$$

leading to no two particle scattering. Extending this type of argument leads to a trivial theory in the sense that the scattering matrix $S = \mathbb{1}$.

If, however, an infinitely divisible sharp time field distribution did *not* imply that the full time field distribution was infinitely divisible, then triviality of the scattering matrix may be avoided.

This kind of question is extremely hard to study for a field theory. Hence, the study initiated in this paper is restricted to a single degree of freedom quantum system with the Hamiltonian form chosen to be similar in spirit to that of a traditional scalar field theory. If a positive outcome of the single degree of freedom problem emerges, it may well suggest that further study of the quantum field situation may be worthwhile.

A Proposed Problem

As a simple examination of the associated ground state reveals, neither of the two specific examples (Examples 1 & 2) suggested above are suitable for a detailed study since their spectrum is part discrete and part continuous. Unfortunately, distribution functions for general infinitely divisible characteristic functions are known for very few examples [1]. We conjecture that certain examples, such as

$$\begin{aligned} C(s) &= \exp\{-\textstyle\int[1-\cos(sy)]\,e^{-y^2}/(\pi\,|y|^\alpha)\,dy\} \\ &\equiv \int \cos(sx)\,\phi_0(x)^2\,dx\ , \end{aligned}$$

for $2 \leq \alpha < 3$, may correspond to potentials with a purely discrete spectrum which would then permit our formulas to be applied.

The implicitly defined ground state $\phi_0(x)$ above may be determined numerically. If the ground state falls to zero suitably faster than an exponential, then the spectrum of the Hamiltonian would be purely discrete. In turn, the potential associated with this ground state is

$$V(x) \equiv \phi_0''(x)/2\phi_0(x)\ ,$$

which again could be numerically determined. Even if there are small errors in the numerical determination, it is quite likely that the simple system still lies within the special class that is infinitely divisible for the ground state distribution. With the potential $V(x)$ – with symmetry enforced – now determined, standard computer programs could be used to calculate a number of eigenfunctions and eigenvalues, as well as suitable matrix elements. These quantities permit the calculation of χ_2 for large and small T values. If, for large T, $\chi_2 < 0$, the desired result will have been achieved; if, instead, $\chi_2 > 0$, then further investigations are warranted.

Of course, it may be true that sharp time infinite divisibility necessarily implies full time infinite divisibility. If this implication could be proved, then that would be an important result which would make any computation (such as outlined in the previous paragraph) completely unnecessary.

Dedication

It is a pleasure to dedicate this article to my long time friend and benefactor, Gérard Emch, on the occasion of his 70th birthday. I wish him many long years ahead of good health and productive research.

References

[1] E. Lukas, *Characteristic Functions*, 2nd Ed., (Hafner Publishing Company, New York, 1970).
[2] B. De Finetti, *Theory of Probability, Vol. 2*, (Wiley & Sons, London, 1975).
[3] B. Simon, *Functional Integration and Quantum Physics*, (Academic Press, London, 1979).
[4] I. M. Gel'fand and N. Ya. Vilenkin, *Generalized Functions, Vol. IV: Applications of Harmonic Analysis*, (Academic Press, New York, 1964).
[5] J. R. Klauder, "Isolation and Expulsion of Divergences in Quantum Field Theory", Int. J. Mod. Phys. B **10**, 1473-1483 (1996).
[6] Private communication.

Department of Physics and Department of Mathematics, University of Florida, Gainesville, FL 32611
E-mail: john.klauder@gmail.com

Adaptive Dynamics and its Application to Chaos

Masanori Ohya

Information Dynamics (ID) was proposed to find a common frame treating chaotic behaviors of several dynamical systems, which synthesizes the dynamics of state change and the complexity of system. ID enables us to attain a new concept "adaptivity" studying dynamics. In this paper, we briefly review the ID and the adaptive dynamics, and we discuss how they are used to understand chaos.

Dedicated to Professor G. G. Emch on his 70th birthday

1. Introduction

The adaptive dynamics has two aspects, one of which is the "observable-adaptivity" and another is the "state-adaptivity".

The idea of observable-adaptivity comes from the paper [21]. In that paper, we claimed that any observation will be unrelated or even contradicted to mathematical universalities such as taking limits, sup, inf, etc. Observation of chaos is a result due to taking suitable scales of, for example, time, distance or domain, and it will not be possible in the limiting cases. In other words, it is very natural to consider that observation itself plays a similar role of "noesis" of Husserl and the mode of its existence is a "being-for-itself", that is, observation itself can not exist as it is but it exists only through the results (phenomena) of objects obtained by it. Phenomena can not be phenomena without observing them, so to explain the phenomena like chaos it is necessary to find a dynamics with observation. We claimed that most of chaos are scale-dependent phenomena, so the definition of a degree measuring chaos should depends on certain scales taken and more generally it is important to find mathematics containing the rules (dynamics) of both objects and observation, which is called "Adaptive dynamics".

The idea of the state-adaptivity is implicitly started in Accardi's Chameleon dynamics [1]. This adaptivity can be used to solve a pending problem for more than 30 years whether there exists an algorithm solving a NP complete problem in polynomial time. We found such algorithms first by quantum chaos algorithm

[34, 35] and secondly by the adaptive dynamics [3] based on quantum algorithm of the SAT [33, 7].

In this paper the observable-adaptive dynamics for describing chaos is discussed based on Information Dynamics [28, 35].

There exist several approaches in the study of chaotic behavior of dynamical systems using the concepts such as (1) entropy and dynamical entropy, (2) Chaitin's complexity, (3) Lyapunov exponent (4) fractal dimension (5) bifurcation (6) ergodicity [10, 8, 1, 11, 13, 14, 22]. However these concepts are rather independently used in each case. In 1991, the present author proposed Information Dynamics [31, 29, 19] to treat such chaotic behavior of systems from a common standing point, in which a chaos degree to measure the chaos appeared in dynamical systems is defined by means of two complexities in Information Dynamics [28, 29, 35]. In particular, among several chaos degrees, the entropic chaos degree was introduced in [27] and it has been applied to several dynamical systems [27, 18, 17]. For instance, semiclassical properties and chaos degree for quantum baker's map have been considered in [16, 17].

In Section 2 of this paper, we review Information Dynamics and general chaos degree. In Section 3, we show how the entropic chaos degree is computed and a relation between the entropic chaos degree and Lyapunov exponent is clarified. In Section 4, the concept of adaptivity for studying chaos is discussed.

2. Information Dynamics and Chaos Degree

Information dynamics (ID for short) is a synthesis of the dynamics of state change and the complexity of states. It is a trial to provide a new view for the study of chaotic behavior of systems. We briefly review what ID is.

Let $(\mathcal{A}, \mathfrak{S}, \alpha(G))$ be an input (or initial) system and $(\overline{\mathcal{A}}, \overline{\mathfrak{S}}, \overline{\alpha}(\overline{G}))$ be an output (or final) system. Here $\mathcal{A}$ is a set of some objects to be observed and $\mathfrak{S}$ is a set of some means to get the observed value, $\alpha(G)$ describes a certain evolution of system. Often we have $\mathcal{A} = \overline{\mathcal{A}}$, $\mathfrak{S} = \overline{\mathfrak{S}}$, $\alpha = \overline{\alpha}$. Therefore we claim

[Giving a mathematical structure to input and output triples $\equiv$ Having a theory]

Let $(\mathcal{A}_T, \mathfrak{S}_T, \alpha_T(G_T))$ be the total system of $(\mathcal{A}, \mathfrak{S}, \alpha)$ and $(\overline{\mathcal{A}}, \overline{\mathfrak{S}}, \overline{\alpha})$, and $\mathcal{S}$ be a subset of $\mathfrak{S}$ in which we are measuring observables (e.g., $\mathcal{S}$ is the set of all KMS or stationary states in C*-system).

The dynamics of state change is described by a channel sending a state to another state Λ: $\mathfrak{S} \to \overline{\mathfrak{S}}$ (sometimes $\mathfrak{S} \to \mathfrak{S}$). Moreover ID contains two complexities, which are denoted by C and T. C is the complexity of a state φ measured from a reference system $\mathcal{S}$, in which we actually observe the objects in $\mathcal{A}$ and T is the transmitted complexity associated with a state change $\varphi \to \Lambda\varphi$, both of which should satisfy the following properties :

⟨**Axioms of complexities**⟩

(i) For any $\varphi \in \mathcal{S} \subset \mathfrak{S}$,

$$C^{\mathcal{S}}(\varphi) \geqslant 0, \ T^{\mathcal{S}}(\varphi; \Lambda) \geqslant 0$$

(ii) For any disjoint (in a proper sense) bijection $j : ex\mathcal{S} \to ex\mathcal{S}$, the set of all extremal points of $\mathcal{S}$,

$$C^{j(\mathcal{S})}(j(\varphi)) = C^{\mathcal{S}}(\varphi)$$

$$T^{j(\mathcal{S})}(j(\varphi); \Lambda) = T^{\mathcal{S}}(\varphi; \Lambda)$$

(iii) For $\Phi \equiv \varphi \otimes \psi \in \mathcal{S}_t \subset \mathfrak{S}_t$, $\psi \in \overline{\mathcal{S}} \subset \overline{\mathfrak{S}}$ (here $\otimes$ is a properly defined product)

$$C^{\mathcal{S}_t}(\Phi) = C^{\mathcal{S}}(\varphi) + C^{\overline{\mathcal{S}}}(\psi)$$

(iv) $0 \leqslant T^{\mathcal{S}}(\varphi; \Lambda) \leqslant C^{\mathcal{S}}(\varphi)$

(v) $T^{\mathcal{S}}(\varphi; id) = C^{\mathcal{S}}(\varphi)$, where "$id$" is an identity map from $\mathfrak{S}$ to $\mathfrak{S}$.

Instead of (iii), when "(iii') $\Phi \in \mathcal{S}_T \subset \mathfrak{S}_T$, put $\varphi \equiv \Phi \upharpoonright \mathcal{A}$, $\psi \equiv \Phi \upharpoonright \overline{\mathcal{A}}$ (i.e., the restriction of Φ to $\mathcal{A}$ and $\overline{\mathcal{A}}$, respectively). $C^{\mathcal{S}_t}(\Phi) \leqslant C^{\mathcal{S}}(\varphi) + C^{\overline{\mathcal{S}}}(\psi)$" is satisfied, C and T is called a pair of strong complexity. Then ID is defined as follows:

Definition 2.1. Information Dynamics is described by

$$\left(\mathcal{A}, \mathfrak{S}, \alpha(G); \overline{\mathcal{A}}, \overline{\mathfrak{S}}, \overline{\alpha}(\overline{G}); \Lambda; C^{\mathcal{S}}(\varphi), T^{\mathcal{S}}(\varphi; \Lambda)\right)$$

and some relations R among them.

Therefore, in the framework of ID, we have to

(i) mathematically determine $\left(\mathcal{A}, \mathfrak{S}, \alpha(G); \overline{\mathcal{A}}, \overline{\mathfrak{S}}, \overline{\alpha}(\overline{G})\right)$

(ii) choose Λ and R, and

(iii) define $C^{\mathcal{S}}(\varphi)$, $T^{\mathcal{S}}(\varphi; \Lambda)$.

2.1. State Change and Complexities

ID contains the dynamics of state change as its part. A state change is mathematically described by a unitary evolution, a semi-group dynamics, generally, a channeling transformation (it is simply called "channel"). Let input and output triples $(\mathcal{A}, \mathfrak{S}, \alpha(G))$ and $(\overline{\mathcal{A}}, \overline{\mathfrak{S}}, \overline{\alpha}(\overline{G}))$ be C*-dynamical systems; that is, $\mathcal{A}$ is a C^*-algebra [15] and $\mathfrak{S}$ is its state space and $\alpha(G)$ is an inner evolution of $\mathcal{A}$ with a parameter group G (or semigroup) and so is the output system. Let a channel be a mapping from $\mathfrak{S}(\mathcal{A})$ to $\overline{\mathfrak{S}}(\overline{\mathcal{A}})$.

Although there exist several complexities, one of the most fundamental pairs of C and T in quantum system is the von Neumann entropy and the mutual entropy. Other entropic complexities C and T are ε-entropy, Kolmogorov-Sinai type dynamical entropy, dynamical mutual entropy [23, 29, 19].

Here we remind that the quantum entropy and the quantum mutual entropy are examples of our complexities C and T, respectively.

Example 2.2. The entropy S and the mutual entropy I, in both classical and quantum, satisfy the conditions of the complexities C and T of ID. For a density operator ρ in a Hilbert space $\mathcal{H}$ (the case $\mathcal{A} = B(\mathcal{H})$) and a channel Λ, $C(\rho)$ is the entropy $S(\rho)$ and $T(\rho; \Lambda)$ is the mutual entropy $I(\rho; \Lambda)$:

$$C(\rho) = S(\rho) = -tr\rho \log \rho,$$

$$T(\rho; \Lambda) = I(\rho; \Lambda) = \sup \left\{ \sum_k \lambda_k S(\Lambda E_k, \Lambda \rho);\ \{E_k\} \right\},$$

where the supremum is taken over all Schatten decompositions $\{E_k\}$ of ρ; $\rho = \sum_k \lambda_k E_k$. In Shannon's communication theory in classical systems, ρ is a probability distribution $p = (p_k) = \sum_k p_k \delta_k$ and Λ is a transition probability matrix $(t_{i,j})$, so that the Schatten decomposition of ρ is unique and the compound state of ρ and its output $\overline{\rho}$ $(\equiv \overline{p} = (\overline{p}_i) = \Lambda p)$ is the joint distribution $r = (r_{i,j})$ with $r_{i,j} \equiv t_{i,j} p_j$. Then the above complexities C and T become the Shannon entropy and mutual entropy, respectively;

$$C(p) = S(p) = -\sum_k p_k \log p_k,$$

$$T(p; \Lambda) = I(p; \Lambda) = \sum_{i,j} r_{i,j} \log \frac{r_{i,j}}{p_j \overline{p}_i}.$$

We can construct several other types of entropic complexities. For instance, one pair of the complexities is

$$T(\rho; \Lambda) = \sup \left\{ \sum_k p_k S(\Lambda \rho_k, \Lambda \rho); \rho = \sum_k p_k \rho_k \right\},\quad C(\rho) = T(\rho; id)$$

where $S(\cdot, \cdot)$ is quantum relative entropy of Umegaki[37] and $\rho = \sum_k p_k \rho_k$ is a finite decomposition of ρ, over which the supremum is taken.

Example 2.3. Generalizing the entropy S and the mutual entropy I, we can construct complexities of entropy type: Let $(\mathcal{A}, \mathfrak{S}(\mathcal{A}), \alpha(G)), (\overline{\mathcal{A}}, \mathfrak{S}(\overline{\mathcal{A}}), \overline{\alpha}(\overline{G}))$ be C^* systems as before. Let $\mathcal{S}$ be a weak *-compact convex subset of $\mathfrak{S}(\mathcal{A})$ and $M_\varphi(\mathcal{S})$ be the set of all maximal measures μ on $\mathcal{S}$ with the fixed barycenter φ

$$\varphi = \int_{\mathcal{S}} \omega d\mu .$$

Moreover let $F_\varphi(\mathcal{S})$ be the set of all measures of finite support with the fixed barycenter φ. The following three pairs C and T satisfy all conditions of the complexities:

$$T^{\mathcal{S}}(\varphi;\Lambda) \equiv \sup\left\{\int_{\mathcal{S}} S(\Lambda\omega, \Lambda\varphi)d\mu; \mu \in M_\varphi(\mathcal{S})\right\}$$

$$C_T^{\mathcal{S}}(\varphi) \equiv T^{\mathcal{S}}(\varphi; id)$$

$$I^{\mathcal{S}}(\varphi;\Lambda) \equiv \sup\left\{S\left(\int_{\mathcal{S}} \omega\otimes\Lambda\omega d\mu, \varphi\otimes\Lambda\varphi\right); \mu \in M_\varphi(\mathcal{S})\right\}$$

$$C_I^{\mathcal{S}}(\varphi) \equiv I^{\mathcal{S}}(\varphi; id)$$

$$J^{\mathcal{S}}(\varphi;\Lambda) \equiv \sup\left\{\int_{\mathcal{S}} S(\Lambda\omega, \Lambda\varphi)d\mu; \mu \in F_\varphi(\mathcal{S})\right\}$$

$$C_j^{\mathcal{S}}(\varphi) \equiv J^{\mathcal{S}}(\varphi; id)$$

Here, the state $\int_{\mathcal{S}} \omega\otimes\Lambda\omega d\mu$ is the compound state exhibiting the correlation between the initial state and the final state $\Lambda\varphi$, and $S(\cdot,\cdot)$ is quantum relative entropy [6, 36, 24]. This compound state was introduced as a quantum generalization of the joint probability measure in CDS (classical dynamical system) [25, 26]. Note that in the case of $\mathfrak{S} = \mathcal{S}$, $T^{\mathcal{S}}$(resp.$C^{\mathcal{S}}$, $I^{\mathcal{S}}$, $J^{\mathcal{S}}$) is denoted by T (resp. C, I, J) for simplicity.

These complexities and the mixing $\mathcal{S}$-entropy $S^{\mathcal{S}}(\varphi)$ [32, 23], the CNT (Connes-Narnhofer-Thirring) entropy $H_\varphi(\mathcal{A})$ satisfy some relations.

Theorem 2.4. *(1)* $0 \leqslant I^{\mathcal{S}}(\varphi;\Lambda) \leqslant T^{\mathcal{S}}(\varphi;\Lambda) \leqslant J^{\mathcal{S}}(\varphi;\Lambda)$. *(2)* $C_I(\varphi) = C_T(\varphi) = C_J(\varphi) = S(\varphi) = H_\varphi(\mathcal{A})$. *(3) When* $\mathcal{A} = \overline{\mathcal{A}} = B(\mathcal{H})$, *for any density operator* ρ

$$0 \leqslant I^{\mathcal{S}}(\rho;\Lambda) = T^{\mathcal{S}}(\rho;\Lambda) \leqslant J^{\mathcal{S}}(\rho;\Lambda) .$$

2.2. Use of ID: Chaos Degree

In quantum systems, if we take $C(\rho) = S(\rho)$ =von Neumann entropy, $T(\rho;\Lambda) = I(\rho;\Lambda)$ =quantum mutual entropy and linear channel Λ, then we have

$$\begin{aligned} D(\rho;\Lambda) &= C(\Lambda\rho) - T(\rho;\Lambda) \\ &= S(\Lambda\rho) - I(\rho;\Lambda) \\ &= S(\Lambda\rho) - \sup\left\{ Tr\left(\sum_n p_n \Lambda E_n (\log \Lambda E_n - \log \Lambda\rho)\right);\ \{E_n\}\right\} \\ &= \inf\left\{\sum_n p_n S(\Lambda E_n);\ \{E_n\}\right\} = \inf\left\{\sum_n p_n C(\Lambda E_n);\ \{E_n\}\right\} \end{aligned}$$

since $S(\Lambda\rho) = -Tr\Lambda\rho\log\Lambda\rho = -Tr\left(\sum_n p_n \Lambda E_n \log \Lambda\rho\right)$ for any Schatten decomposition $\{E_n\}$ of ρ. *Therefore the above quantity $D(\rho;\Lambda)$ can be interpreted as the complexity produced through the channel Λ. We apply this quantity $D(\rho;\Lambda)$ to study chaos even when the channel describing the dynamics is not linear. $D(\rho;\Lambda)$ is called the entropic chaos degree in the sequel.*

In order to describe more general dynamics such as in continuous systems, we define the entropic chaos degree in C*-algebraic terminology. This setting will not be used in the sequel application, but for mathematical completeness we will discuss the C*-algebraic setting.

Let $(\mathcal{A}, \mathfrak{S})$ be an input C* system and $(\overline{\mathcal{A}}, \overline{\mathfrak{S}})$ be an output C* system; namely, $\mathcal{A}$ is a C* algebra with unit I and $\mathfrak{S}$ is the set of all states on $\mathcal{A}$. We assume $\overline{\mathcal{A}} = \mathcal{A}$ for simplicity. For a weak* compact convex subset $\mathcal{S}$ (called the reference space) of $\mathfrak{S}$, take a state φ from the set $\mathcal{S}$ and let

$$\varphi = \int_{\mathcal{S}} \omega d\mu_\varphi$$

be an extremal orthogonal decomposition of φ in $\mathcal{S}$, which describes the degree of mixture of φ in the reference space $\mathcal{S}$. In more detail this formula reads

$$\varphi(A) = \int_{\mathcal{S}} \omega(A) d\mu_\varphi(\omega), \quad A \in \mathcal{A}$$

The measure μ_φ is not uniquely determined unless $\mathcal{S}$ is the Schoque simplex, so that the set of all such measures is denoted by $M_\varphi(\mathcal{S})$.

Definition 2.5. The entropic chaos degree with respect to $\varphi \in \mathcal{S}$ and a channel Λ is defined by

$$D^{\mathcal{S}}(\varphi;\Lambda) \equiv \inf\left\{\int_{\mathcal{S}} S^{\mathcal{S}}(\Lambda\omega)\, d\mu;\ \mu \in M_\varphi(\mathcal{S})\right\}$$

where $S^{\mathcal{S}}(\Lambda\varphi)$ is the mixing entropy [26, 32] of a state φ in the reference space $\mathcal{S}$.

When $\mathcal{S} = \mathfrak{S}$, $D^{\mathcal{S}}(\varphi;\Lambda)$ is simply written as $D(\varphi;\Lambda)$. This $D^{\mathcal{S}}(\varphi;\Lambda)$ contains the classical chaos degree and the quantum above. The classical entropic chaos degree is the case that $\mathcal{A}$ is abelian and φ is the probability distribution of a orbit generated by a dynamics (channel) Λ; $\varphi = \sum_k p_k \delta_k$, where δ_k is the delta measure such as $\delta_k(j) \equiv \begin{cases} 1 & (k = j) \\ 0 & (k \neq j) \end{cases}$. Then the classical entropic chaos degree is

$$D_c(\varphi;\Lambda) = \sum_k p_k S(\Lambda\delta_k)$$

with the entropy S. Summarize that Information Dynamics can be applied to the study of chaos by using more general complexity $C(\varphi)$:

Definition 2.6. (1)ψ is more chaotic than φ if $C(\psi) \geqslant C(\varphi)$.

(2)When $\varphi \in \mathcal{S}$ changes to $\Lambda\varphi$, the *chaos* degree associated to this state change(dynamics) Λ is given by

$$D^{\mathcal{S}}(\varphi;\Lambda) = \inf\left\{\int_{\mathcal{S}} C^{\mathcal{S}}(\Lambda\varphi)\,d\mu ; \mu \in M_{\varphi}(\mathcal{S})\right\}.$$

Definition 2.7. A dynamics Λ produces chaos iff $D^{\mathcal{S}}(\varphi;\Lambda) > 0$.

Remark 2.8. It is important to note here that the dynamics Λ in the definition is not necessarily same as original dynamics (channel) but is one reduced from the original such that it causes an evolution for a certain observed value like orbit. However for simplicity we use the same notation here. In some cases, the above chaos degree $D^{\mathcal{S}}(\varphi;\Lambda)$ can be expressed as

$$D^{\mathcal{S}}(\varphi;\Lambda) = C^{\mathcal{S}}(\Lambda\varphi) - T^{\mathcal{S}}(\varphi;\Lambda).$$ □

3. Algorithm Computing Entropic Chaos Degree

In order to observe a chaos produced by a dynamics, one often looks at the behavior of orbits made by that dynamics, more generally, looks at the behavior of a certain observed value. Therefore in our scheme we directly compute the chaos degree once a dynamics is explicitly given as a state change of system. However even when the direct calculation does not show a chaos, a chaos will appear if one focuses to some aspect of the state change, e.g., a certain observed value. In the later case, algorithm computing the chaos degree for classical or quantum dynamics consists of the following two cases:

(1) *Dynamics is given by* $\frac{dx}{dt} = F_t(x)$ *with* $x \in I \equiv [a,b]^{\mathbf{N}} \subset \mathbf{R}^{\mathbf{N}}$: First find a difference equation $x_{n+1} = F(x_n)$ with a map F on $I \equiv [a,b]^{\mathbf{N}} \subset \mathbf{R}^{\mathbf{N}}$ into itself, secondly let $I \equiv \bigcup_k A_k$ be a finite partition with $A_i \cap A_j = \emptyset\ (i \neq j)$. Then the state $\varphi^{(n)}$ at time n of the orbit determined by the difference equation is defined by

the probability distribution $\left(p_i^{(n)}\right)$, that is, $\varphi^{(n)} = \sum_i p_i^{(n)} \delta_i$, where for an initial value $x \in I$ and the characteristic function 1_A

$$p_i^{(n)} \equiv \frac{1}{n+1} \sum_{k=m}^{m+n} 1_{A_i}\left(F^k x\right).$$

Now when the initial value x is distributed due to a measure ν on I after a proper time m, the above $p_i^{(n)}$ is given as

$$p_i^{(n)} \equiv \frac{1}{n+1} \int_I \sum_{k=m}^{m+n} 1_{A_i}\left(F^k x\right) d\nu.$$

The joint distribution $\left(p_{ij}^{(n,n+1)}\right)$ between the time n and $n+1$ is defined by

$$p_{ij}^{(n,n+1)} \equiv \frac{1}{n+1} \sum_{k=m}^{m+n} 1_{A_i}\left(F^k x\right) 1_{A_j}\left(F^{k+1} x\right)$$

or

$$p_{ij}^{(n,n+1)} \equiv \frac{1}{n+1} \int_I \sum_{k=m}^{m+n} 1_{A_i}\left(F^k x\right) 1_{A_j}\left(F^{k+1} x\right) d\nu.$$

Then the channel Λ_n at n is determined by

$$\Lambda_n \equiv \left(\frac{p_{ij}^{(n,n+1)}}{p_i^{(n)}}\right) : \text{transition probability} \Longrightarrow \varphi^{(n+1)} = \Lambda_n \varphi^{(n)},$$

and the entropic chaos degree is given by

$$D_A(x; F) = D_A\left(p^{(n)}; \Lambda_n\right) = \sum_i p_i^{(n)} S(\Lambda_n \delta_i) = \sum_{i,j} p_{ij}^{(n,n+1)} \log \frac{p_i^{(n)}}{p_{ij}^{(n,n+1)}}. \tag{3.1}$$

We can judge whether the dynamics causes a chaos or not by the value of D as

$$D > 0 \Longleftrightarrow \textit{chaotic}$$
$$D = 0 \Longleftrightarrow \textit{stable}.$$

This chaos degree was applied to several dynamical maps such logistic map, Baker's transformation and Tinkerbell map, and it could explain their chaotic characters. This chaos degree has several merits compared with usual measures such as Lyapunov exponent as explained below.

Therefore it is enough to find a partition $\{A_k\}$ such that D is positive when the dynamics produces chaos.

(2) *Dynamics is given by* $\varphi_t = F_t \varphi_0$ *on a Hilbert space*: Similarly as making a difference equation for (quantum) state, the channel Λ_n at n is first deduced from F_t, which should satisfy $\varphi^{(n+1)} = \Lambda_n \varphi^{(n)}$. By means of this constructed channel, (α) we compute the chaos degree D directly according to the definition 2 or (β) we take a proper observable X and put $x_n \equiv \varphi^{(n)}(X)$, then go back to the algorithm (1).

The entropic chaos degree for quantum systems has been applied to the analysis of quantum spin system and quantum Baker's type transformation [16, 17, 20].

Note that the chaos degree D above does depend on a partition A taken, which is somehow different from usual degree of chaos. This is a key point of our understanding of chaos, from which the idea of adaptivity comes.

Example 3.1. Logistic Map

Let us apply the entropy chaos degree (ECD) to logistic map. Chaotic behavior in classical system is often considered as exponential sensitivity to initial condition.

The logistic map is defined by

$$x_{n+1} = a x_n (1 - x_n), x_n \in [0, 1], 0 \leq a \leq 4.$$

The solution of this equation bifurcates as Fig 1.

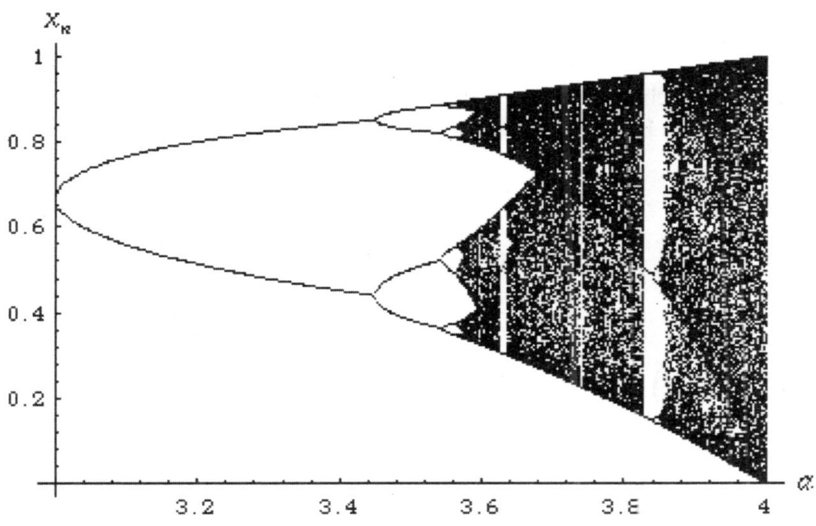

Figure 1. Logistic map

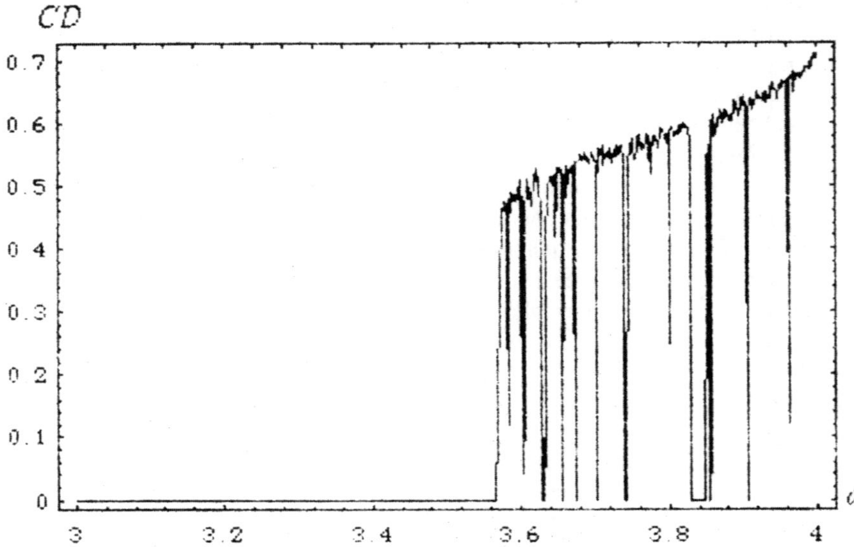

Figure 2. Entropic Chaos Degree of Logistic Map

In order to compare ECD with other measure describing chaos, we take Lyapunov exponent for this comparison: Fig. 2 and Fig. 3.

Example 3.2. Baker's transformation

We apply the chaos degree to a smooth map on R^2. Let us compute the Lyapunov exponent and the ECD for the following Baker's transformation f_a :

$$f_a\left(x^{(n)}\right) = f_a\left(x_1^{(n)}, x_2^{(n)}\right)$$

$$= \begin{cases} \left(2ax_1^{(n)}\ \tfrac{1}{2}ax_2^{(n)}\right) & \left(0 \leqslant x_1^{(n)} \leqslant 0.5\right) \\ \left(a\left(2x_1^{(n)} - 1\right), \tfrac{1}{2}a\left(x_2^{(n)} + 1\right)\right) & \left(0.5 < x_1^{(n)} \leqslant 1\right) \end{cases},$$

where $\left(x_1^{(n)}, x_2^{(n)}\right) \in [0, 1] \times [0, 1]$ and $0 \leqslant a \leqslant 1$.

The ECD of Baker's transformation is shown in Fig. 4.

The Lyapunov exponent is $\log 2a$ for Baker's transformation: Fig. 5.

Example 3.3. Tinkerbell map

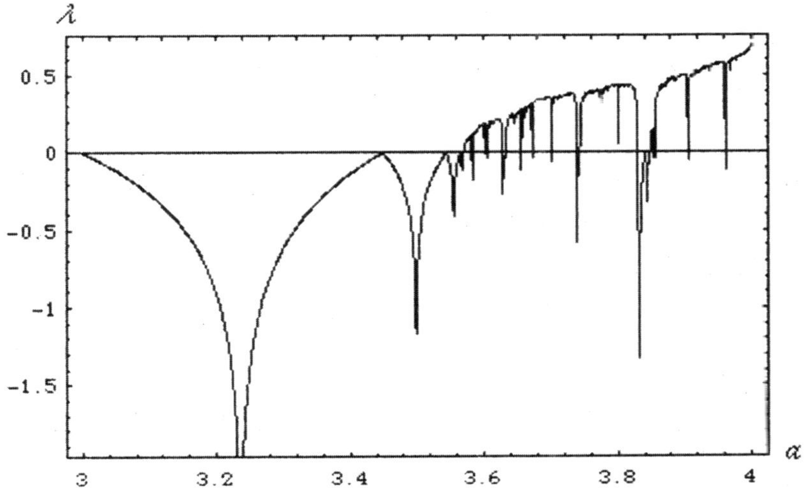

Figure 3. Lyapunov exponent

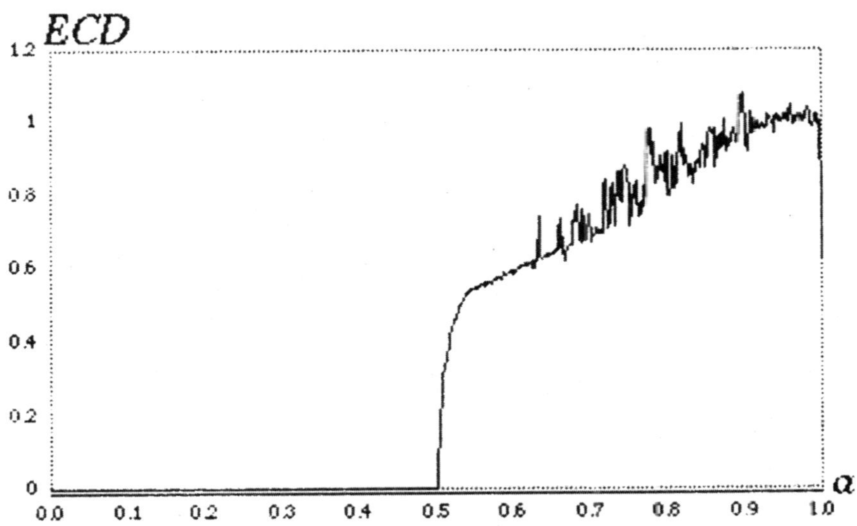

Figure 4. ECD of Baker's transformation.

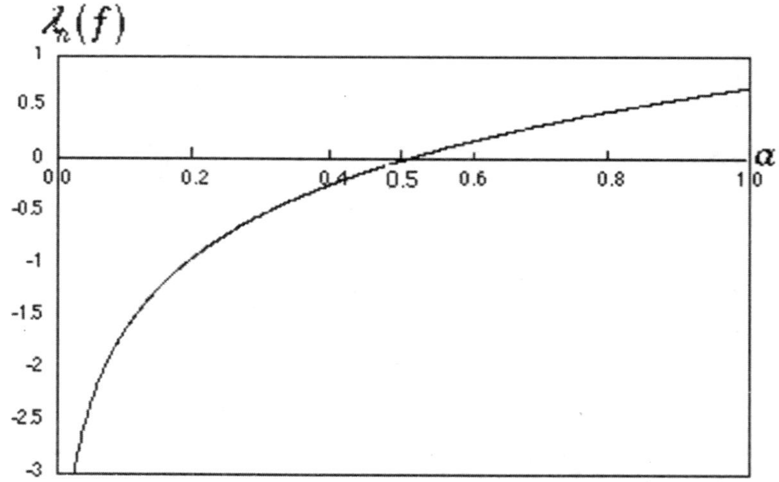

Figure 5. Lyapunov exponent of Baker's transformation

Let us compute the CD for the following two type Tinkerbell maps f_a and f_b on $I = [-1.2, 0.4] \times [-0.7, 0.3]$.

$$\begin{aligned} f_a\left(x^{(n)}\right) &= f_a\left(x_1^{(n)}, x_2^{(n)}\right) \\ &= \left(\left(x_1^{(n)}\right)^2 - \left(x_2^{(n)}\right)^2 + ax_1^{(n)} + c_2x_2^{(n)}, 2x_1^{(n)}x_2^{(n)} + c_3x_1^{(n)} + c_4x_2^{(n)}\right), \\ f_b\left(x^{(n)}\right) &= f_b\left(x_1^{(n)}, x_2^{(n)}\right) \\ &= \left(\left(x_1^{(n)}\right)^2 - \left(x_2^{(n)}\right)^2 + c_1x_1^{(n)} + c_2x_2^{(n)}, 2x_1^{(n)}x_2^{(n)} + bx_1^{(n)} + c_4x_2^{(n)}\right), \end{aligned}$$

where $\left(x_1^{(n)}, x_2^{(n)}\right) \in I, -0.4 \leqslant a \leqslant 0.9, 1.9 \leqslant b \leqslant 2.9, (c_1, c_2, c_3, c_4) = (-0.3, -0.6, 2.0, 0.5)$, and $\left(x_1^{(0)}, x_2^{(0)}\right) = (0.1, 0.1)$.

The Lyapunov exponent of Tinkerbell map is not so easy to compute, but the ECD of Tinkerbell map f_a and f_b are easily computed as: Fig. 6 and Fig. 7.

From the above example and some other maps in [18], Lyapunov exponent and chaos degree have clear correspondence, but ECD can resolve some inconvenient properties of the Lyapunov exponent as:

(1) Lyapunov exponent takes negative and sometimes $-\infty$, but ECD is always positive for all $a \geqslant 0$.

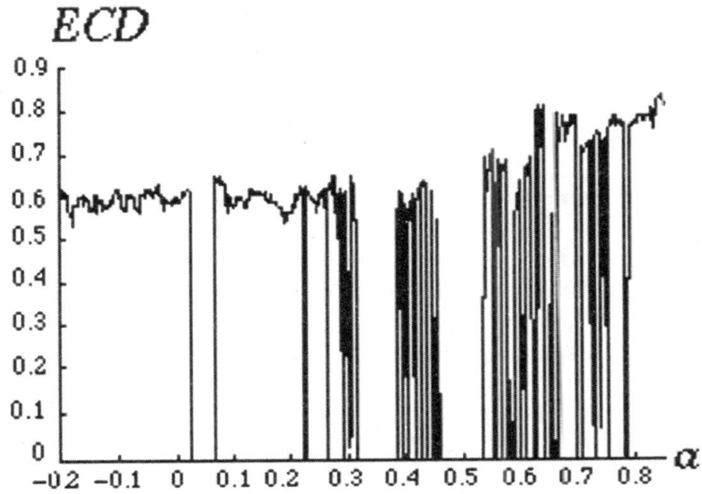

Figure 6. ECD for Tinkerbell map f_a

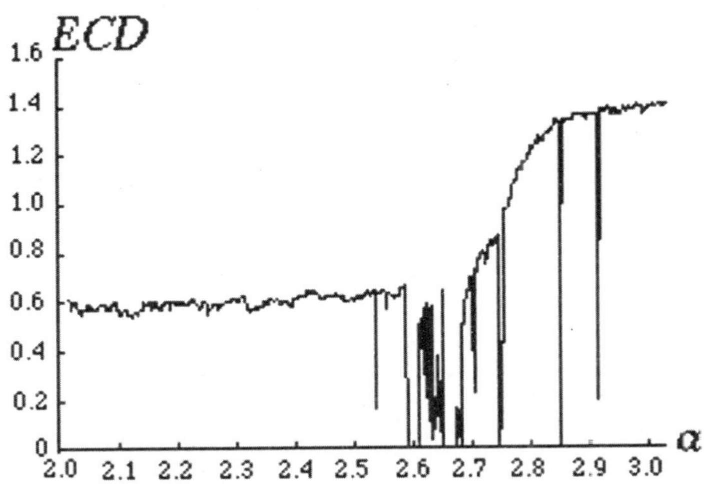

Figure 7. ECD for Tinkerbell map f_b

(2) For some map f whose Lyapunov exponent is difficult to compute (e.g., dynamics in $\mathbb{R}^n$ $(n \geqslant 2)$), the ECD of f is easily computed.

(3) Generally, the algorithm for ECD is much easier than that for Lyapunov exponent.

3.1. ECD with Memory

Here we generalize the above explained ECD to take the memory effect into account. Although the original ECD is based upon the choice of the base space $\Sigma := \{1, 2, \cdots, N\}$, we here take another choice. Σ^m, instead of Σ, is a new base space. On this base space, a probability distribution is naturally defined as

$$p_{i_0 i_1 \ldots i_m}^{(n,n+1,\ldots,n+m)} \equiv \frac{1}{n+1} \int_I \sum_{k=m}^{m+n} 1_{A_{i_0}}\left(F^k x\right) 1_{A_{i_1}}\left(F^{k+1} x\right) \cdots 1_{A_{i_m}}\left(F^{k+m} x\right) d\nu .$$

with its mathematical idealization, $p_{i_0 i_1 \ldots i_m} := \lim_{n \to \infty} p_{i_0 i_1 \ldots i_m}^{(n,n+1,\ldots,n+m)}$.The channel Λ_m over Σ^m is defined by a transition probability,

$$p_{j_0 i_1 \ldots i_{m+1}} \delta_{i_1 j_1} \cdots \delta_{i_m j_m} = p(i_1, i_2, \ldots, i_m, i_{m+1} | j_0, j_1, \ldots, j_m) p_{j_0, j_1 \ldots j_m} .$$

Thus it derives the ECD with m-steps memory effect,

$$D_m^A(x; f) = D_m^A(p; \Lambda_m) = \sum_{i_0, i_1, \ldots, i_m} p_{i_0 i_1 \ldots i_m} \log \frac{p_{i_0 i_1 \ldots i_{m-1}}}{p_{i_0 i_1 \ldots i_m}} .$$

It notes that this quantity coincides with the original ECD when $m = 1$.

This memory effect shows an interesting result, that is, the longer the memory is, the closer the ECD is to the Lyapunov exponent for its positive part [31].

Theorem 3.4. *For given f, x and A, there exists a probability space (Ω, F, ν) and a random variable g depending on f, x, A such that $\lim_{m \to \infty} D_m^A(x; f) = \int_\Omega g d\nu$ =the positive part of Lyapunov exponent.*

4. Adaptive Dynamics and Chaos Degree

In adaptive dynamics, it is essential to consider in which states and by which ways we see objects. That is, one has to select phenomena and prepare mode for observation for understanding the whole of a system. Typical adaptive dynamics are the dynamics for state-adaptive and that for observable-adaptive.

State-adaptive dynamics is that the dynamics of a system depends on a state at one instant in which the interaction is switched on, or that in a composite system the interaction depends on the instant state of at least one of sub-system. Examples of such adaptivity are seen in a compound state (or nonlinear lifting) studying quantum communication [3, 25, 26] and in an algorithm solving NP complete problem in polynomial time with stochastic limit [3].

Observable- adaptive dynamics is that the dynamics of a system depends on observables to be observed and the measurement depends on an observable chosen. Examples of this adaptivity are used to understand chaos [27, 21] and examine violation of Bell's inequality [2].

We will discuss how such adaptivities can be observed in dynamics which cause a chaos.

First of all we examine carefully when we say that a certain dynamics produces a chaos. Let us take the logistic map as an example. The original differential equation of the logistic map is

$$\frac{dx}{dt} = ax(1-x), 0 \leqslant a \leqslant 4$$

with initial value x_0 in $[0, 1]$. This equation can be easily solved analytically, whose solution (orbit) does not have any chaotic behavior. However once we make the equation above discrete such as

$$x_{n+1} = ax_n(1-x_n), 0 \leqslant a \leqslant 4. \tag{4.1}$$

This difference equation produces a chaos.

Taking the discrete time is necessary not only to make a chaos but also to observe the orbits drawn by the dynamics. Similarly as quantum mechanics, it is not possible for human being to understand any object without observing it, for which it will not be possible to trace a orbit continuously in time.

Now let us think about finite partition $A = \{A_k; k = 1, \cdots, N\}$ of a proper set $I \equiv [a, b]^{\mathrm{N}} \subset \mathbb{R}^{\mathrm{N}}$ and equi-partition $B^e = \left\{B_k^e; k = 1, \cdots, N\right\}$ of I. Here"equi" means that all elements B_k^e are identical. We denote the set of all partitions by $\mathcal{P}$ and the set of all equi-partitions by $\mathcal{P}^e$. In the section 4, we specified a special partition, in particular, an equi-partition for computer experiment calculating the ECD. Such a partition enables to observe the orbit of a given dynamics, and moreover it provides a criterion for observing chaos. There exist several reports saying that one can observe chaos in nature, which are very much related to how one observes the phenomena, for instance, scale, direction, aspect. It has been difficult to find a satisfactory theory (mathematics) to explain such chaotic phenomena. In the difference equation 4.1 we take some time interval τ between n and $n+1$, if we take $\tau \rightarrow 0$, then we have a complete different dynamics. If we take coarse graining to the orbit of x_t for time during τ; $x_n \equiv \frac{1}{\tau}\int_{(n-1)\tau}^{n\tau} x_t dt$, we again have a very different dynamics. Moreover it is important for mathematical consistency to take the limits $n \rightarrow \infty$ or N (the number of equi-partitions)$\rightarrow \infty$, i.e., making the partition finer and finer, and consider the limits of some quantities as describing chaos, so that mathematical terminologies such as"lim","sup","inf" are very often used to define such quantities. *Let us take the opposite position, that is, any observation will be unrelated or even contradicted to such limits. Observation of chaos is a result due to taking suitable scales of, for example, time, distance or domain, and it will not be possible in the limiting cases.*

It is claimed in [21] that most of chaos are scale-dependent phenomena, so the definition of a degree measuring chaos should depend on certain scales taken. Such a scale dependent dynamics is nothing but adaptive dynamics.

Taking into consideration of this view we modify the definitions of the chaos degree given in the previous sections as below.

Going back to a triple $(\mathcal{A}, \mathfrak{S}, \alpha(G))$ considered in Section 2 and we use this triple both for an input and an output systems. Let a dynamics be described by a mapping Γ_t with a parameter $t \in G$ from $\mathfrak{S}$ to $\mathfrak{S}$ and let an observation be described by a mapping $\mathcal{O}$ from $(\mathcal{A}, \mathfrak{S}, \alpha(G))$ to a triple $(\mathcal{B}, \mathfrak{T}, \beta(G))$. The triple $(\mathcal{B}, \mathfrak{T}, \beta(G))$ might be same as the original one or its subsystem and the observation map $\mathcal{O}$ may contains several different types of observations, that is, it can be decomposed as $\mathcal{O} = \mathcal{O}_m \cdots \mathcal{O}_1$.Let us list some examples of observations.

For a given dynamics $\frac{d\varphi}{dt} = F(\varphi_t)$, equivalently, $\varphi_t = \Gamma_t \varphi$, one can take several observations.

Example 4.1. Time Scaling (Discretizing): $O_\tau : t \to n$, $\frac{d\varphi}{dt}(t) \to \varphi_{n+1}$, so that $\frac{d\varphi}{dt} = F(\varphi_t) \Rightarrow \varphi_{n+1} = F(\varphi_t)$ and $\varphi_t = \Gamma_t \varphi \Rightarrow \varphi_n = \Gamma_n \varphi$. Here τ is a unit time needed for the observation.

Example 4.2. Size Scaling (Conditional Expectation, Partition): Let $(\mathcal{B}, \mathfrak{T}, \beta(G))$ be a subsystem of $(\mathcal{A}, \mathfrak{S}, \alpha(G))$, both of which have a certain algebraic structure such as C*-algebra or von Neumann algebra. As an example, the subsystem $(\mathcal{B}, \mathfrak{T}, \beta(G))$ has abelian structure describing a macroscopic world which is a subsystem of a non-abelian (non-commutative) system $(\mathcal{A}, \mathfrak{S}, \alpha(G))$ describing a micro-world. A mapping $\mathcal{O}_C$ preserving norm (when it is properly defined) from $\mathcal{A}$ to $\mathcal{B}$ is, in some cases, called a conditional expectation. A typical example of this conditional expectation is according to a projection valued measure

$$\left\{ P_k;\ P_k P_j = P_k \delta_{kj} = P_k^* \delta_{kj} \geqq 0, \sum_k P_k = I \right\}$$

associated with quantum measurement (von Neumann measurement) such that $\mathcal{O}_C(\rho) = \sum_k P_k \rho P_k$ for any quantum state (density operator) ρ. When $\mathcal{B}$ is a von Neumann algebra generated by $\{P_k\}$, it is an abelian algebra isometrically isomorphic to $L^\infty(\Omega)$ with a certain Hausdorff space Ω, so that in this case $\mathcal{O}_C$ sends a general state φ to a probability measure (or distribution) p. Similar example of $\mathcal{O}_C$ is one coming from a certain representation (selection) of a state such as one Schatten decomposition of ρ ; $\rho = \mathcal{O}_R \rho = \sum_k \lambda_k E_k$

by one-dimensional orthogonal projections $\{E_k\}$ associated to the eigenvalues of ρ with $\sum_k E_k = I$. Another important example of the size scaling is due to a finite partition of an underlining space Ω, e.g., space of orbit, defined as $\mathcal{O}_P(\Omega) = \left\{P_k; P_k \cap P_j = P_k\delta_{kj}(k, j = 1, \cdots N),\ \cup_{k=1}^N P_k = \Omega\right\}$.

4.1. Chaos Degree with Adaptivity

We go back to the discussion of the entropic chaos degree. Starting from a given dynamics $\varphi_t = \Gamma_t\varphi$, it becomes $\varphi_n = \Gamma_n\varphi$ after handling the operation $\mathcal{O}_\tau$. Then by taking proper combinations $\mathcal{O}$ of the size scaling operations like $\mathcal{O}_C$, $\mathcal{O}_R$ and $\mathcal{O}_P$, the equation $\varphi_n = \Gamma_n\varphi$ changes to $\mathcal{O}(\varphi_n) = \mathcal{O}(\Gamma_n\varphi)$, which will be written by $\mathcal{O}\varphi_n = \mathcal{O}\Gamma_n\mathcal{O}^{-1}\mathcal{O}\varphi$ or $\varphi_n^{\mathcal{O}} = \Gamma_n^{\mathcal{O}}\varphi^{\mathcal{O}}$. Then our entropic chaos degree is redefined as follows:

Definition 4.3. The entropic chaos degree of Γ with an initial state φ and observation $\mathcal{O}$ is defined by $D^{\mathcal{O}}(\varphi;\Gamma) = \int_{\mathcal{SO}} S\left(\Gamma^{\mathcal{O}}\omega^{\mathcal{O}}\right) d\mu^{\mathcal{O}}$, where $\mu^{\mathcal{O}}$ is the measure operated by $\mathcal{O}$ to a extremal decomposition measure of φ selected by of the observation $\mathcal{O}$ (its part $\mathcal{O}_R$). *The entropic chaos degree of* Γ *with an initial state* φ *is defined by* $D(\varphi;\Gamma) = \inf\left\{D^{\mathcal{O}}(\varphi;\Gamma);\mathcal{O} \in \mathcal{SO}\right\}$, *where* $\mathcal{SO}$ *is a proper set of observations naturally determined by a given dynamics.*

In this definition, $\mathcal{SO}$ is determined by a given dynamics and some conditions attached to the dynamics, for instance, if we start from a difference equation with a special representation of an initial state, then $\mathcal{SO}$ excludes $\mathcal{O}_\tau$ and $\mathcal{O}_R$.

Then one judges whether a given dynamics causes a chaos or not by the following way.

Definition 4.4. (1) A dynamics Γ is chaotic for an initial state φ in an observation $\mathcal{O}$ iff $D^{\mathcal{O}}(\varphi;\Gamma) > 0$.

(2) A dynamics Γ *is totally chaotic for an initial state* φ *iff* $D(\varphi;\Gamma) > 0$.

The idea introducing in this section to understand chaos can be applied not only to the entropic chaos degree but also to some other degrees such as dynamical entropy, whose applications and the comparison of several degrees will be discussed in [31].

In the case of logistic map, $x_{n+1} = ax_n(1 - x_n) \equiv F(x_n)$, we obtain this difference equation by taking the observation $\mathcal{O}_\tau$ and take an observation $\mathcal{O}_P$ by equi-partition of the orbit space $\Omega = \{x_n\}$ so as to define a state (probability distribution). Thus we can compute the entropic chaos degree in adaptive sense.

As an example, we consider a circle map

$$\theta_{n+1} = f_v(\theta_n) = \theta_n + \omega \quad (\text{mod } 2\pi),$$

where $\omega = 2\pi v$ $(0 < v < 1)$. If v is a rational number N/M, then the orbit $\{\theta_n\}$ is periodic with the period M. If v is irrational, then the orbit $\{\theta_n\}$ densely fills the unit circle for any initial value θ_0; namely, it is a quasiperiodic motion.

Theorem 4.5. *Let* $\mathbf{I} = [0, 2\pi]$ *be partioned into L disjoint components with equal length;* $\mathbf{I} = B_1 \cap B_2 \cap \ldots \cap B_L$.

(1) If υ *is rational number* N/M *, then the finite equi-partition* $P = \{B_k; k = 1, \cdots, M\}$ *implies* $D^{\mathcal{O}}(\theta_0; f_\nu) = 0$.

(2) If υ *is irrational, then* $D^{\mathcal{O}}(\theta_0; f_\nu) > 0$ *for any finite partition* $P = \{B_k\}$.

Note that our entropic chaos degree shows a chaos to quasiperiodic circle dynamics by the observation due to a partition of the orbit, which is different from usual understanding of chaos. However usual belief that quasiperiodic circle dynamics will not cause a chaos is not at all obvious, but is realized in a special limiting case as shown in the following theorem.

Theorem 4.6. *For the above circle map, if* υ *is irrational, then* $D(\theta_0; f_\nu) = 0$.

Such a limiting case will not take place in real observation of natural objects, so that we claim that chaos is a phenomenon depending on observations, environment or periphery, which results the adaptive definition of chaos as above.

Note here that the chaos degree and the adaptivity can be applied to understand quantum dynamics either [16, 17, 20].

References

[1] L.Accardi, *Urne e Camaleoni: Dialogo sulla realta, le leggi del caso e la teoria quantistica.* Il Saggiatore, Rome (1997)

[2] L.Accardi, K.Imafuku, M.Refoli, On the EPR-Chameleon experiment, Infinite Dimensional Analysis, Quantum Probability and Related Topics Vol. 5, No. 1 (2002) 1-2

[3] L.Accardi, M.Ohya, A Stochastic limit approach to the SAT problem", Proceedings of VLSI 2003, and Open systems and Information Dynamics (2004)

[4] L.Accardi, M.Ohya, Compound channels, transition expectations, and liftings", Appl. Math. Optim., Vol.39, 33-59 (1999)

[5] L.Accardi, M.Ohya, N.Watanabe, Note on quantum dynamical entropies Reports on mathematical physics, vol.38 n.3 457–469 (1996)

[6] H.Araki, Relative entropy of states of von Neumann Algebras, Publ.RIMS, Kyoto Univ.Vol.11, 809-833 (1976)

[7] L.Accardi, R.Sabbadini, On the Ohya–Masuda quantum SAT Algorithm, in: Proceedings International Conference UMC'01, Springer (2001)

[8] K.T.Alligood, T.D.Sauer, J.A.Yorke, Chaos-An Introduction to Dynamical Systems-, Textbooks in Mathematical Sciences, Springer (1996)

[9] R.Alicki, Quantum geometry of noncommutative Bernoulli shifts, Banach Center Publications, Mathematics Subject Classification 46L87 (1991)

[10] R.Alicki, M.Fannes, Defining quantum dynamical entropy, Lett. Math. Physics, 32, 75-82 (1994)

[11] F. Benatti, Deterministic Chaos in Infinite Quantum Systems, Springer (1993)

[12] A.Connes, H.Narnhofer, W.Thirring, Dynamical entropy of C*-algebras and von Neumann algebras, Commun.Math.Phys., 112, pp.691-719 (1987)

[13] R.L.Devaney, An Introduction to Chaotic dynamical Systems, Benjamin (1986)

[14] G.G.Emch, H.Narnhofer, W.Thirring and G.L.Sewell, Anosov actions on noncommutative algebras, J.Math.Phys., **35**, No.11, 5582-5599 (1994)

[15] G.G.Emch, Algebraic Methods in Statistical Mechanics and Quantum Field Theory, Wiley (1972)
[16] K.Inoue, M.Ohya, I.V.Volovich, Semiclassical properties and chaos degree for the quantum baker's map, J. Math. Phys., 43-2, 734-755 (2002)
[17] K.Inoue, M.Ohya, I.V.Volovich, On quantum-classical correspondence for baker's map, quant-ph/0108107 (2001)
[18] K.Inoue, M.Ohya and K.Sato, Application of chaos degree to some dynamical systems, Chaos, Soliton & Fractals, 11, 1377-1385 (2000)
[19] R.S.Ingarden, A.Kossakowski, M.Ohya, Information Dynamics and Open Systems, Kluwer Publ. Comp. (1997)
[20] K.Inoue, M.Ohya, A.Kossakowski, A Description of Quantum Chaos, Tokyo Univ. of Science preprint (2002)
[21] A.Kossakowski, M.Ohya, Y.Togawa, How can we observe and describe chaos? Open System and Information Dynamics, 10(3):221-233 (2003)
[22] A.Kossakowski, M.Ohya, N.Watanabe, Quantum dynamical entropy for completely positive maps, Infinite Dimensional Analysis, Quantum Probability and Related Topics, **2**, No.2, 267-282 (1999)
[23] N.Muraki, M.Ohya, Entropy functionals of Kolmogorov-Sinai type and their limit theorems, Letter in Mathematical Physics,36, 327-335 (1996)
[24] M.Ohya, D.Petz, Quantum Entropy and its Use, Springer-Verlag (1993)
[25] M.Ohya, On compound state and mutual information in quantum information theory, IEEE Trans. Information Theory, 29, No.5, 770-774 (1983)
[26] M.Ohya, Some aspects of quantum information theory and their applications to irreversible processes, Rep.Math.Phys., Vol.27, 19-47 (1989)
[27] M.Ohya, Complexities and their applications to characterization of chaos, International Journal of Theoretical Physics,Vol.37, No.1, 495-505 (1998)
[28] M.Ohya, State change, complexity and fractal in quantum systems, Quantum Communications and Measurement, Plenum Press, New York, 309-320 (1995)
[29] M.Ohya, Complexity and fractal dimensions for quantum states, Open Systems and Information Dynamics, 4, 141-157 (1997)
[30] M.Ohya, et al, Adaptive dynamics its use in understanding of chaos, TUS preprint
[31] M.Ohya, Information dynamics and its applications to optical communication processes, Lecture Note in Physics, **378**, 81-92 (1991)
[32] M.Ohya, Entropy Transmission in C*-dynamical systems, J.Math. Anal.Appl., 100, No.1, 222-235 (1984)
[33] M.Ohya, N.Masuda, NP problem in quantum algorithm, Open Systems and Information Dynamics, Vol.7, No.1, 33-39 (2000)
[34] M.Ohya, I.V.Volovich, New quantum algorithm for studying NP-complete problems, Rep.Math.Phys.,52, No.1,25-33 (2003) and Quantum computing and chaotic amplifier, J.Opt.B (2003)
[35] M.Ohya, I.V.Volovich, Mathematical Foundations of Quantum Information and Quantum Computation, to be published in Springer-Verlag
[36] A.Uhlmann, The 'transition probability' in the state space of a*-algebra.Rep.Math.Phys., 9:273–279 (1976)
[37] H.Umegaki, Conditional expectation in operator algebra IV, Kodai Math. Sem. Rep., 14, 59-85 (1962)

Tokyo University of Science, 278 Noda City, Chiba, Japan

The C* Axioms and the Phase Space Fomalism of Quantum Mechanics

Franklin E. Schroeck, Jr.

University of Denver, Denver, Colorado *E-mail:* fschroec@du.edu

It is proven that the phase space localization operators on the Hilbert spaces of ordinary quantum mechanics provide a set of operators that are physically motivated and form a C* algebra. Then, it is proven that the set of localization operators, when extended, are informationally complete in the original Hilbert spaces.

Dedicated to G. G. Emch.

1. Introduction

In 1972, "Algebraic Methods in Statistical Mechanics and Quantum Field Theory" was published by Gerard Emch [8], giving the axioms for a physical system in an algebraic setting using the language of Irving Segal [19], and then continuing to obtain the C*-algebraic formalism for a physical system. Of these axioms, only the fifth contained an assumption that was questionable in its physical content. Bearing in mind that for each observable A and state ϕ, one obtains a distribution of values for the observed results of measurement, we have:

> Axiom 5: For any element A in the set of observables $\mathfrak{A}$ and any non-negative integer n, there is at least one element, denoted A^n, in $\mathfrak{A}$ such that (i) the set of dispersion-free states for A^n is contained in the set of dispersion-free states for A, (ii) $< \phi; A^n > = < \phi; A >^n$ for all ϕ in the set of dispersion-free states for A. (Here $< \phi; B >$ is the expectation of B in state ϕ.)

This axiom is necessary in order to define the product of two or more observables as a member of $\mathfrak{A}$. Our objective here is to justify that axiom (as well as all the other axioms of the C* approach) for a set of observables **that are physically motivated** as well as **informationally complete** in the Hilbert spaces of ordinary quantum mechanics. In this way, we will obtain a completely physically motivated basis for the C* formalism for quantum mechanics. Then, the C* formalism may be used for

Date: June 30, 2006

2000 *Mathematics Subject Classification.* Primary 46L05, 81S10; Secondary 81Q10, 81S30.

Key words and phrases. C* algebra, phase space localization, symplectic space

more general physical situations, with an additional restriction that we will place on the algebra of observables having to do with the phase space formalism.

It is first encumbent upon us to justify why this is a problem at all. We take the position operator, Q, and the momentum operator, P, as examples. They are unbounded self-adjoint operators with a purely continuous spectrum on any of the non-relativistic Hilbert spaces in quantum mechanics. We first treat them as being generated by their spectral projections onto compact sets of their spectrum, each being a bounded self-adjoint operator. But you have a problem; each of these has a piecewise continuous spectrum, and thus has no purely discrete spectrum (eigenvalues) associated with it, much less a complete basis of eigenvectors. The same problem appears for any operator with a purely continuous spectrum in any Hilbert space. Next, we may use a theorem [22] that says that if T is a self-adjoint, bounded operator on a separable Hilbert space, then there exists a self-adjoint compact operator K such that $T + K$ has eigenvectors that span the space. Moreover, by a theorem of von Neumann [21], $\parallel K \parallel$ may be made arbitrarily small and T does not have to be bounded. But there is no physical interpretation for what K is!

A way to circumvent this came from considering the phase space formalism for quantum mechanics [16]. In this formalism, the Hilbert spaces in which quantum mechanics is done are the usual ones and on each one we may define a phase space localization operator; we shall investigate the properties of such a phase space localization operator. It will have all the properties we want.

In Section 2, we define the phase space representations for any locally compact Lie group and obtain the phase space localization operators on them. In the next section, we define the usual quantum mechanical representation spaces for the Galilei and Poincaré groups, obtain an embedding of these spaces into the appropriate phase space representation spaces, and then pull back the phase space localization operators. We also discuss the here-to-fore known informational completeness of such phase space localization operators. In Section 4, we investigate in detail the spectral properties of these pulled back phase space localization operators. We find that for any function that is L^p in the phase space, these operators have a purely discrete spectrum. In Section 5, we show that they comprise a set of operators that are physically observable and form a C* algebra. Finally, in Section 6, we show that we may construct a Hilbert space equivalent to the original one in which the localization operators form an informationally complete set.

2. Phase Space Representations of a Locally Compact Group

The majority of this section has been discussed in [6].

From *classical* experiments, one learns that classical (Newtonian) equations of motion are invariant under translations, boosts (relative velocity transformations between inertial [Galileian] reference frames), and rotations. Prior to 1887, these were invariably viewed to generate the group of Galileian transformations on spacetime. Since the Michaelson-Morley[14] experiment, and the subsequent analysis of numerous luminaries, these spacetime translations, boosts and rotations were interpreted as the generators of the group of Lorentz (or Poincaré) transformations on either energy-momentum space or on spacetime. These transformations generated the entire group from those transformations acting on an arbitrarily small neighborhood of any point. Transformations *infinitesimally* near the identity transformation form a vector space (the Lie algebra of the group) on which a non-associative operation (the Lie bracket) is defined. Thus, classical experiments reveal the kinematical groups of relevance.

The lesson learned through the efforts of mathematicians over the last 250 years is that we may use a space with a Poisson bracket, a *phase space*, to describe a classical consevative mechanics. A phase space is mathematically a symplectic manifold which possesses a closed, non-degenerate 2-form on it. Furthermore, the relevant Galilei or Poincaré group acts on this space in such a way as to preserve the Poisson bracket (acts "symplectically"). The phase space is thus a "G space", G the kinematical group. As a consequence of this set-up, "conjugate variables" are coordinates on the phase space which realize the canonical skew-symmetric form of the Poisson bracket, etc. With the experience of the Galilei and Poincaré groups, one may abstract this formulation to the setting of the action of a Lie group G on *any* phase space on which G acts symplectically.

A Lie group G generates, as above, a Lie algebra $\mathfrak{g}$; we may think of $\mathfrak{g}$ as the collection of all left-invariant vector fields on G. This process is invertible by *exponentiation* that associates an element of the group (near the identity) to any element of the Lie algebra sufficiently near the origin (zero). One may thus go from the Lie algebra to the Lie group, and *vice versa*.

In the following, it is essential that $\mathfrak{g}$ is a finite-dimensional vector space. If $\wedge$ designates the anti-symmetric tensor product on $\mathfrak{g}$ then one may form the skew-symmetric tensor algebra $*(\mathfrak{g})$ over $\mathfrak{g}$ consisting of $\mathbb{R}$, $\mathfrak{g}$, $\mathfrak{g} \wedge \mathfrak{g}$, $\mathfrak{g} \wedge \mathfrak{g} \wedge \mathfrak{g}$, etc. Let their duals be denoted by $\mathfrak{g}^*$, etc. and note that $\mathfrak{g}^*$ may be thought of as the collection of all left-invariant 1-forms on G, $(\mathfrak{g} \wedge \mathfrak{g})^*$ as the left-invariant 2-forms on G, and so on. We then define the coboundary operator δ

$$\mathbb{R} \longrightarrow^{\delta_0} \mathfrak{g}^* \longrightarrow^{\delta_1} (\mathfrak{g} \wedge \mathfrak{g})^* \longrightarrow \cdots$$

as follows. Let $\{A_i\}$ be a basis of $\mathfrak{g}$ and let $\{\omega^i\}$ be the associated dual basis of $\mathfrak{g}^*$ so that $\omega^i(A_j) = \delta^i_j$. The structure constants $C^k_{ij} \in \mathbb{R}$ of $\mathfrak{g}$, defined relative to the basis $\{A_i\}$, are determined by the Lie bracket relations: $[A_i, A_j] = \sum_k C^k_{ij} A_k$. The $\mathbb{R}$ in the sequence above can be considered to be the collection of left-invariant functions on the group G, which is assumed to be connected, so that the $\mathbb{R}$ may be thought of as the left-invariant 0-forms f on the group. We define $\delta_0 f = 0$ as an element of $\mathfrak{g}^*$. Next, thinking of the ω^i as left-invariant 1-forms we find that the Maurer-Cartan equations hold: $d\omega^k = -\frac{1}{2}\sum_{i,j} C_{ij}^k \omega^i \wedge \omega^j$. We then define

$$\delta_1 \omega^k = -\frac{1}{2}\sum_{i,j} C_{ij}^k \omega^i \wedge \omega^j$$

recognizing that this 2-form is actually in $(\mathfrak{g} \wedge \mathfrak{g})^*$. One extends this expression for δ_1 linearly, obtaining $\mathfrak{g}^* \xrightarrow{\delta_1} (\mathfrak{g} \wedge \mathfrak{g})^*$. Making use of the skew-derivation property for δ_2

$$\delta_2(\lambda \wedge \mu) \equiv (\delta_1 \lambda) \wedge \mu - \lambda \wedge (\delta_1 \mu),$$

for $\lambda, \mu \in \mathfrak{g}^*$, one defines δ_2 and proceeds inductively to define δ.

Next, let

$$Z^2(\mathfrak{g}) \equiv \left\{\omega \in (\mathfrak{g} \wedge \mathfrak{g})^* \mid \delta_2(\omega) = 0\right\}$$

denote the space of closed, left-invariant 2-forms on G, and for $\omega \in Z^2(\mathfrak{g})$, define

$$h_\omega \equiv \{\xi \in \mathfrak{g} \mid \omega(\xi, \cdot) = 0\}.$$

It turns out that h_ω is a Lie sub-algebra of $\mathfrak{g}$; so, by exponentiation, h_ω determines a subgroup H_ω of G. We must assume that H_ω is a *closed* subgroup of G to obtain a manifold:

$$\Gamma \equiv G/H_\omega.$$

It is a symplectic manifold as the 2-form ω, when factored by its kernel, is the pull-back of a non-degenerate closed 2-form on G/H_ω. That Γ is a symplectic G space follows because G acts on G/H_ω by left multiplication on left cosets: $gx = g(g_1 H_\omega) = (gg_1)H_\omega$, where $x = g_1 H_\omega$ for some g_1 in G. Since $\Gamma \equiv G/H_\omega$ is a symplectic manifold, it naturally possesses a left-invariant Liouville measure μ equal to the m-th exterior power of ω, where the dimension of Γ is equal to 2m for some integer m.

The following result (Theorem 25.1 of [10]) captures the essence of the need for the construction outlined above and is sufficient for our purposes, but only in the context of *single-particle* kinematics.

Theorem 2.1. *Any symplectic action of a connected Lie group G on a symplectic manifold M defines a G morphism, $\Psi : M \to Z^2(\mathfrak{g})$. Since the map Ψ is a G morphism, $\Psi(M)$ is a union of G orbits in $Z^2(\mathfrak{g})$. In particular, if the action of G on M is transitive, then the image of Ψ consists of a single G orbit in $Z^2(\mathfrak{g})$.*

For the Galilei group and the Poincaré group, elements of $Z^2(\mathfrak{g})$ are fixed by a choice of mass and spin. Consequently, one obtains all the *single-particle* symplectic spaces on which G acts symplectically and transitively, and one has a unified mathematical picture of kinematics in the two (Galileian and Lorentzian) cases of relevance to one-particle physics. Using the "méthode de fusion" [20], we describe multi-particle kinematics by a phase space that is a Cartesian product of the single-particle phase spaces with symplectic form equal to the "sum" of the symplectic forms on each of the single-particle factors. Thus we may start from the symplectic action of a group on classical single-particle phase space and obtain all the phase spaces (single- or multi-particle) on which G acts symplectically, in a physically meaningful way.

One may form $L^2_\mu(\Gamma)$, a Hilbert space, on which one may represent G by unitary operators $V^\alpha(g)$

$$[V^\alpha(g)\Psi](x) \equiv \alpha(h(g^{-1}, x))\Psi(g^{-1}x)$$

for $\Psi \in L^2_\mu(\Gamma)$, $x \in \Gamma$, $g \in G$, h a generalized cocycle, and α a one-dimensional representation of H; i.e., incorporate a phase factor in the left-regular representation. Then define an operator $A(f)$, for all μ-measurable f, by

$$[A(f)\Psi](x) \equiv f(x)\Psi(x).$$

When the f are characteristic functions $\chi(\Delta)$, these operators on $L^2_\mu(\Gamma)$ have a clear classically-motivated interpretation of localization observables in the phase space region Δ. The $A(f)$ are a commuting set, reflecting the classical property that the operators of position, momentum, etc. are all obtainable with precision simultaeously. For this reason and others, it will become evident that $L^2_\mu(\Gamma)$ is not a Hilbert space of fundamental importance to the description of quantum mechanical models of elementary (i.e., irreducible) single-particle systems. It will turn out that it is reducible into a direct sum (or integral) of such irreducible spaces.

3. Quantum Mechanical Representation Spaces

This section is also taken from [6].

In the case where G is one of the inhomogeneous Galilei and Lorentz groups, we know that all continuous, irreducible, unitary Hilbert space representations are obtained through the "Mackey Machine" [13] and the earlier Wigner classification [23] and that these representations are characterized by the Casimir invariants in the universal enveloping algebra of the Lie algebra. These Casimir elements are identifiable as the physical quantities of rest mass and spin (or helicity in the mass-zero case) in the case $G =$ the inhomogeneous Lorentz group. For the inhomogeneous Galilei group, the analysis of Lévy-Leblond [12] achieved a similar picture physically characterized by mass and spin.

In what follows, U will denote an irreducible unitary representation of G on an irreducible representation space, denoted $\mathcal{H}$, with inner product denoted $< \cdot, \cdot >$.

We wish to encode the entire content of the state vector $\varphi \in \mathcal{H}$ into a complex-valued function on the phase space Γ in a manner that is reversible. The goal is to be able to reconstuct the state from the complex numbers $[W^{\eta}(\varphi)](x)$ which encode it. Hence, to intertwine $\mathcal{H}$ with $L^2_{\mu}(\Gamma)$, we define a linear transformation W^{η} from $\mathcal{H}$ to $L^2_{\mu}(\Gamma)$ by

$$[W^{\eta}(\varphi)](x) \equiv \; < U(\sigma(x))\eta, \varphi >$$

for $x \in \Gamma = G/H_{\omega}$, for all $g \in G$, and for all $\varphi \in \mathcal{H}$, where η is a vector in $\mathcal{H}$ and where σ is a (Borel measurable) section

$$\sigma : G/H_{\omega} \longrightarrow G.$$

To ensure that the image of W^{η} actually lies in $L^2_{\mu}(\Gamma)$ one must exercise some care in the choice η. One first selects and fixes, once and for all, a (Borel measurable) section $\sigma : G/H_{\omega} \longrightarrow G$. Now, one says that η is *admissible* with respect to the section σ if

$$\int_{\Gamma} |< U(\sigma(x))\eta, \eta >|^2 \, d\mu(x) < \infty.$$

Assuming that η is admissible with respect to σ, one says that η is *α-admissible with respect to σ* if in addition to admissibility of η one also has

$$U(h)\eta = \alpha(h)\eta$$

for all h in H_{ω}, where α is a one-dimensional representation of H_{ω}. If η is *α-admissible with respect to σ* then we may define the mapping W^{η} from $\mathcal{H}$ to $L^2_{\mu}(\Gamma)$. [16] As we shall see, this is also enough to describe states $\varphi \in \mathcal{H}$ by their images $W^{\eta}(\varphi)$ in $L^2_{\mu}(\Gamma)$.

To illustrate that these conditions are achievable for all representations indexed by mass and spin of the Galilei and Poincaré groups, consider, for example:

(1) the case of a massive, spinless, relativistic particle ($G =$ Poincaré group) in which one finds [1] that η must be rotationally-invariant under $H_{\omega} = SU(2)$, and square-integrable over $\Gamma \equiv G/H_{\omega} \cong \mathbb{R}^6 \cong \mathbb{R}^3_{position} \times \mathbb{R}^3_{momentum}$, the classical phase space of a massive, relativistic, spinless particle,

(2) the case of a massive, relativistic particle with non-zero spin ($G =$ Poincaré group) in which one finds [3][5] that η must be rotationally invariant about the "spin axis" (but not necessarily invariant under all rotations in $SU(2)$), i.e., invariant under $H_{\omega} =$ double covering of $O(2) \cong$ stabilizer in $SU(2)$ of the spin axis, and square-integrable over $\Gamma \equiv G/H_{\omega} \cong \mathbb{R}^3_{position} \times \mathbb{R}^3_{momentum} \times S^2_{spin}$, the classical phase space of a massive, relativistic, spinning particle.

Orthogonality relations are present, insuring that the images of the W^η are orthogonal in $L^2_\mu(\Gamma)$. We have the

Theorem 3.1. [11] *Let G be a locally compact group, H a closed subgroup, $\sigma_k : G/H \to G$ any Borel sections, and U_k any unitarily inequivalent representations of G square-integrable with respect to σ_k on Hilbert space $\mathcal{H}^k$. ($k \in \{1,2\}$.) Also, let $\mathcal{H}^{k\alpha_k}$ denote the non-trivial closed subspaces of $\mathcal{H}^k$ generated by the set of α_k-admissible vectors in $\mathcal{H}^k$. Assume that the vectors $\eta_k, \xi_k \in \mathcal{H}^{k\alpha_k}$ and that $\varphi_k, \psi_k \in \mathcal{H}^k$. Then there exists a unique, positive, invertible operator C on $\mathcal{H}^{k\alpha_k}$ such that*

$$\int_{G/H} < \varphi_1, U_1(\sigma_1(x))\eta_1 >_{\mathcal{H}^1} < U_2(\sigma_2(x))\eta_2, \varphi_2 >_{\mathcal{H}^2} d\mu(x) = 0,$$

$$\begin{aligned}\int_{G/H} &< \varphi_1, U_1(\sigma_1(x))\eta_1 >_{\mathcal{H}^1} < U_1(\sigma_1(x))\xi_1, \psi_1 >_{\mathcal{H}^1} d\mu(x) \\ &= < C\eta_2, C\eta_1 >_{\mathcal{H}^1} < \varphi_1, \varphi_2 >_{\mathcal{H}^1} .\end{aligned}$$

Therefore, we have a prescription for when the representations are orthogonal in $L^2_\mu(\Gamma)$: when $< C\eta_2, C\eta_1 >_{\mathcal{H}^1} = 0$.

Note, in the case of a compact group G, the positive operator C is just a positive *constant*. In general, this does not hold on all locally non-compact groups, for example for the Poincaré group in the representations with non-zero spin. See [16, pp. 328 - 329] for a condition which guarantees that C is a constant.

We work, now, with a single choice of G, H, σ, α, and η. For the sake of simplicity we denote the closure of the image of W^η by $W^\eta(\mathcal{H}) \subset L^2_\mu(\Gamma)$. Let P^η denote the canonical projection

$$P^\eta : L^2_\mu(\Gamma) \longrightarrow W^\eta(\mathcal{H})$$

and denote by $A^\eta(f)$ the pulled back mapping [16]

$$A^\eta(f) \equiv [W^\eta]^{-1} P^\eta A(f) W^\eta : \mathcal{H} \longrightarrow \mathcal{H}.$$

This is a plausible candidate for the quantum mechanical operator that corresponds to the classical observable f. For example, for the Heisenberg group and for η = the ground state wave function of the harmonic oscillator, then $A^\eta(q) = Q$ = the position operator, and $A^\eta(p) = P$ = the momentum operator. Note that we have gone from a commuting set of $A(f)$s to a non-commuting set, the $A^\eta(f)$s.

One can prove [16] that $A^\eta(f)$ has an operator density $T^\eta(\cdot)$:

$$A^\eta(f) = \int_\Gamma f(x)T^\eta(x)d\mu(x),$$
$$T^\eta(x) \equiv |\, U(\sigma(x))\eta >< U(\sigma(x))\eta\,|,$$

and that, up to a finite renormalization of μ if necessary,

$$A^\eta(1) = 1.$$

With this set-up one can make a number of remarks:

1) We may restate the orthogonality relation in this case by replacing $|\, U(\sigma(x))\eta > < U(\sigma(x))\eta\,|$ with $T^\eta(x)$, yielding just a multiple of $< \varphi_1, \varphi_2 >_{\mathcal{H}^1}$ on the right-hand side of the second relation of Theorem 2.

2) Let ρ denote any quantum density operator; i.e., ρ is non-negative and has trace one. Then one may write $\rho = \sum \rho_i P_{\psi_i}$, the ψ_i forming an orthonormal set and P_{ψ_i} denoting the corresponding projection. Now, using the interpretation of $|< U(\sigma(x))\eta, \psi_i >|^2$ as the transition probability from ψ_i to $U(\sigma(x))\eta$, one has the quantum expectation value given by

$$Tr(\rho A^\eta(f)) = \sum_i \rho_i \int_\Gamma f(x)\,|< U(\sigma(x))\eta, \psi_i >|^2\, d\mu(x);$$

i.e., the sum over the transition probabilities. [16]

For example, when using a "screen" to detect a particle in a vector state given by ψ, one idealizes the detector (the screen) as a multi-particle quantum system consisting of identical sub-detectors. In a fixed laboratory frame of reference a sub-detector is represented by a state vector η whose phase space counterpart $W^\eta\eta$ is peaked about a reference phase space point which may be referred to as "the origin". For a fixed space-time reference frame, one may "position" a detector at all "points" of space-time (space-time events) exactly as Einstein located rods and clocks. Of course, one must now position mass spectrometers (devices that measure rest-mass in their own rest frames) and Stern-Gerlach devices at all space-time events in addition to rods and clocks. As Einstein imagined that the rods and clocks were also equipped (at all space-time coordinate events) in all inertially-related space-time reference frames, so must we imagine that our inertially-related space-time reference frames carry identical mass spectrometers and Stern-Gerlach devices in addition to rods and clocks (boosted relative to the rest "laboratory" frame). So, instead of rods and clocks situated at each space-time event and at rest in inertially-related (uniformly moving) rest frames, we must add to that *imagery* a more elaborate set of apparati. For a fixed value of momentum p there are infinitely many pairs (m, u) such that $p = mu$; of course the momentum does not alone characterize the uniform relative velocity (boost) represented by p - one requires also the rest-mass m. The totality of all such "placements" of detectors constitutes the phase-space distribution of detectors - the classical phase space frame analogous to the classical space-time (Lorentz) frame (of rods and clocks). Thus the complete detector is composed of sub-detectors each located at different "positions" (points of Γ). The sub-detector located at "position" $x \in \Gamma$, obtained from η by a kinematical placement procedure (with the same intent as Einstein's placement of identical rods and clocks at all points of spacetime), is $U(\sigma(x))\eta$. Since the probability that ψ is captured in the state given by $U(\sigma(x))\eta$ is $|< U(\sigma(x))\eta, \psi >|^2$, the formula for the expectation is justified. *One cannot improve upon this procedure when measuring, by quantum mechanical means, the distribution of the particle.*

3) Since $T^\eta(x) \geqslant 0$ and $A^\eta(1) = 1$, then $\rho_{class}(x) \equiv Tr(\rho T^\eta(x))$ is a classical (Kolmogorov) probability function [16]. Consequently,

$$\begin{aligned}
\text{quantum expectation} &= Tr(\rho A^\eta(f)) \\
&= \int_\Gamma f(x) Tr(\rho T^\eta(x)) d\mu(x) \\
&= \int_\Gamma f(x) \rho_{class}(x) d\mu(x) \\
&= \text{classical expectation.}
\end{aligned}$$

4) Since the operators $A^\eta(f)$ enjoy the feature of the same expectation as the "classical" observables f, one might ask whether these operators are sufficient to distinguish states of the quantum system.

Definition 3.2. [15]: A set of bounded self-adjoint operators $\{A_\beta \mid \beta \in I,\ I$ some index set$\}$ is **informationally complete** iff for all states ρ, ρ' such that $Tr(\rho A_\beta) = Tr(\rho' A_\beta)$ for all $\beta \in I$ then $\rho = \rho'$.

Example [15]: In spinless quantum mechanics, the set of all spectral projections for position is not informationally complete. Neither is the set of all spectral projections for momentum, nor even the union of them.

The $\{A^\eta(f) \mid f$ is measurable$\}$ (or, equivalently $\{T^\eta(x) \mid x \in \Gamma\}$) is known to be informationally complete in a number of cases and under the single additional condition on η, that $< U(g)\eta, \eta > \neq 0$ for $a.e.g \in G$:

a) spin-zero massive representations of the Poincaré group [1]

b) mass-zero, arbitrary helicity representations [4] of the Poincaré group

c) the affine group [11]

d) the Heisenberg group [11]

e) massive representations [2][16] of the inhomogeneous Galilei group.

This leaves the case of massive, non-zero spin representations [5] of the Poincaré group. We will have more to say on this in Section 6.

5) If $\mathfrak{I} = \{A_\beta \mid \beta \in I\}$ is informationally complete, then any bounded operator on $\mathcal{H}$ may be written as the closure of span($\mathfrak{I}$). [7]

6) When we specialize $A^\eta(f)$ to $f = \chi(\Delta)$, $\chi(\Delta)$ the characteristic function for the Borel set Δ, then

$$\begin{aligned} \chi(\Delta) &= \text{classical localization in } \Delta \subset \Gamma, \\ A(\chi(\Delta)) &= \text{operator on } L^2_\mu(\Gamma) \text{ localizing in } \Delta \subset \Gamma, \\ A^\eta(\chi(\Delta)) &= \text{operator on } \mathcal{H} \text{ localizing in } \Delta \subset \Gamma. \end{aligned}$$

4. Spectral Properties of the $A^\eta(f)$

These $A^\eta(f)$ have several properties [16] of relevance to us here. We first provide the

Definition 4.1. Let $\mathcal{H}$ be a Hilbert space and let A be a compact operator on $\mathcal{H}$. Let $\{\alpha_k\}$ denote the set of singular values (eigenvalues) of A. The ***n*th trace class**, $\mathcal{B}_n$, is defined to be the set of all compact operators such that $\sum_k \mid \alpha_k \mid^n < \infty$. We denote the corresponding norm by $\parallel A \parallel_{B_n} \equiv \left[\sum_k \mid \alpha_k \mid^n\right]^{1/n}$.

Then we have the

Theorem 4.2. *Let (X, Σ, μ) be a measure space, let $\mathcal{H}$ be a Hilbert space, and let $A : \Sigma \to B(\mathcal{H})$ be a positive operator valued measure. Suppose A has an operator density T such that $\parallel T_x \parallel \leqslant c$ for all $x \in X$, and $Tr(T_x) \leqslant k$ for all $x \in X$, c and k constants. Let $f \in L^p_\mu(X)$. Then $A(f) \equiv \int_X f(x) T_x d\mu(x)$ is a bounded operator that is compact with $\parallel A(f) \parallel \leqslant c^{1/p} \parallel f \parallel_p$ and $\parallel A(f) \parallel_{B_p} \leqslant r(p) \parallel f \parallel_p$ for some constant $r(p)$. In the case $p = 1$, $r(p) = k$.*

The proof is an excursion in interpolation theory. [17]

In the case at hand, we have $c = k = 1$ and $X = \Gamma = G/H$. Thus, for $A^\eta(f) \neq 0$:

a) $A^\eta(f)$ is compact for all $f \in L^p_\mu(X)$, and thus for all $f \in L^1_\mu(X)$, $1 \leqslant p < \infty$. In particular, we have that it is compact for all f equal to a characteristic function on a compact, measurable set in Γ.

b) Suppose $A^\eta(f)\varphi = \lambda\varphi$, for f in $L^1_\mu(X)$, $\lambda \sim 1$, $\parallel \varphi \parallel = 1$. Suppose also that $g \in L^1_\mu(X)$, and $\parallel f - g \parallel_1 << 1$. Then $\parallel \lambda\varphi - A^\eta(g)\varphi \parallel = \parallel A^\eta(f-g)\varphi \parallel \leqslant \parallel f - g \parallel_1 << 1$. Thus, φ is nearly an eigenfunction of $A^\eta(g)$ with the same eigenvalue as $A^\eta(f)$. The eigenvalues are close. In particular, this holds if we take g to be a characteristic function and f to be a fuzzy set function.

Now we have the

Definition 4.3. Let A and B be two effects; i.e., self-adjoint, positive operators in $\mathcal{H}$ that have spectrum in [0,1]. Then A and B are comeasurable iff we can write $A = A_1 + C$, $B = B_1 + C$, for A_1, B_1 and C effects, and $A_1 + B_1 + C$ is an effect.

In particular, for $A = A^\eta(f)$, and $B = A^\eta(g)$, then $C = A^\eta(\min\{f, g\})$. Couple that with the fact that in $\mathcal{H}$, two projections are comeasurable iff they commute and you obtain the

Theorem 4.4. [18] *The set $\{A^\eta(f) \mid 0 \leqslant f \leqslant 1,\ f\ \mu$-measurable$\}$ does not contain any two non-trivial projections.*

But the $T^\eta(x)$ are covariant as an easy proof will show. Thus the $A^\eta(f)$ are covariant:

$$\begin{aligned} U(g)A^\eta(f)U^{-1}(g) &= U(g)\int_\Gamma f(x)T^\eta(x)d\mu(x)U^{-1}(g) = A^\eta(g^{-1}.f), \\ [g.f](x) &= f(gx) \text{ for all } g \in G. \end{aligned}$$

Hence, if you have one non-trivial $A^\eta(f)$ that is a projection, you have many. Consequently,

c) The set $\{A^\eta(f) \mid 0 \leqslant f \leqslant 1,\ f\ \mu$-measurable$\}$ does not contain any non-trivial projection. [18] Hence, any non-trivial operator in the set has spectrum in (0,1). This is in spite of the informational completeness of the set $\{A^\eta(f) \mid f$ is μ-measurable$\}$. One has to be careful on this point.

Now, one may prove such things as

d) If Δ is a compact subset of Γ with a piecewise differentiable boundary, we show that $\sum(\lambda_i - \lambda_i^2)$ is small, $\{\lambda_i\}$ the eigenvalues of $A^\eta(\chi(\Delta))$. Then $A^\eta(\chi(\Delta))$ has a decreasing spectrum which starts out just below 1, remains just below 1 until it suddenly drops to values just above zero [16, pp.281-283]. Notice that $\chi(\Delta) \in L^1_\mu(X) \cap L^\infty_\mu(X)$, the additional restriction to which we referred in the introduction.

e) For all $\Delta \subset \Gamma$, $\| A^\eta(\chi(\Delta)) \| \leqslant \mu(\Delta)$.

5. The Other C* Axioms

For completeness, we discuss cursorilly the other axioms of the C* approach to physical systems. See [8] for a complete discussion. They are

Axiom 1) For each physical system Σ we can associate the triple $(\mathfrak{A}, \mathfrak{S}, <;>)$ formed by the set $\mathfrak{A}$ of all its observables, the set $\mathfrak{S}$ of all its states, and a mapping $<;>: (\mathfrak{A}, \mathfrak{S}) \to \mathbb{R}$ which associates with each pair (A, ϕ) in $(\mathfrak{A}, \mathfrak{S})$ a real number $< \phi; A >$ that we interpret as the expectation value of the observable A when the system is in the state ϕ.

Definition 5.1. For a fixed $A \in \mathfrak{A}$, we have $< \cdot; A >: \mathfrak{S} \to \mathbb{R}$. If $\mathcal{T} \subseteq \mathfrak{S}$, denote by $A \mid_{\mathcal{T}}$ the restriction $< \cdot; A >: \mathfrak{T} \to \mathbb{R}$. Declare $A \mid_{\mathcal{T}} \leqslant B \mid_{\mathcal{T}}$ whenever $< \phi; A > \leqslant < \phi; B > \forall \phi \in \mathcal{T}$. If $\mathcal{T} = \mathfrak{S}$, then we simply write $A \leqslant B$. A subset $\mathcal{T}$ is said to be **full** with respect to a subset $\mathfrak{B} \subseteq \mathfrak{A}$ iff A and B in $\mathfrak{B}$, and $A \mid_{\mathcal{T}} \leqslant B \mid_{\mathcal{T}} \Rightarrow A \leqslant B$.

Example: Let $\mathcal{H}$ be a separable Hilbert space. Let $\{\psi_i\}$ be an orthonormal basis for $\mathcal{H}$. For $\psi \in \mathcal{H}$ and $\| \psi \| = 1$, let $P_\psi \varphi = < \psi, \varphi > \psi$. Let $\mathfrak{B}_\Psi = \{\sum_i \alpha_i P_{\psi_i}, \alpha_i \in \mathbb{C}, \sum_i \mid \alpha_i \mid < \infty\}$. Then $\mathcal{T}_\Psi = \{P_{\psi_i}\}$ is full with respect to $\mathfrak{B}_\Psi$.

Axiom 2) The relation $\leqslant$ is a partial ordering relation on $\mathfrak{A}$.

Axiom 3) (i) There exist in $\mathfrak{A}$ two elements 0 and 1 such that, for all $\phi \in \mathfrak{S}$, we have $< \phi; 0 > = 0$ and $< \phi; 1 > = 1$.

(ii) For each observable $A \in \mathfrak{A}$ and any $\lambda \in \mathbb{R}$ there exists $(\lambda A) \in \mathfrak{A}$ such that $< \phi; \lambda A > = \lambda < \phi; A >$ for all $\phi \in \mathfrak{S}$.

(iii) For any pair of observables A and B in $\mathfrak{A}$ there exists an element $(A + B)$ in $\mathfrak{A}$ such that $< \phi; A + B > = < \phi; A > + < \phi; B >$ for all $\phi \in \mathfrak{S}$.

Definition 5.2. Denote the set of all dispersion-free states for the observable A by $\mathfrak{S}_A$.

Definition 5.3. A subset $\mathcal{T} \subseteq \mathfrak{S}$ is said to be **complete** if it is full with respect to the subset $\mathfrak{A}_{\mathcal{T}} \subseteq \mathfrak{A}$ defined by $\mathfrak{A}_{\mathcal{T}} = \{A \in \mathfrak{A} \mid \mathfrak{S}_A \supseteq \mathcal{T}\}$. A complete subset $\mathcal{T} \subseteq \mathfrak{S}$ is said to be **deterministic** for a subset $\mathfrak{B} \subseteq \mathfrak{A}$ whenever $\mathfrak{B} \subseteq \mathfrak{A}_{\mathcal{T}}$. A subset $\mathfrak{B} \subseteq \mathfrak{A}$ is said to be **compatible** if the set $\mathfrak{S}_{\mathfrak{B}} \equiv \cap_{B \in \mathfrak{B}} \mathfrak{S}_B$ is complete.

Example: $\mathcal{T}_\Psi$ is complete because it is full with respect to $\mathfrak{A}_{\mathcal{T}} \equiv \mathfrak{B}_\Psi$. It is moreover deterministic for any subset $\mathfrak{C} \subseteq \mathfrak{A}_{\mathcal{T}} \equiv \mathfrak{B}_\Psi$.

Example: Compatibility of $\mathfrak{B}$ in $\mathcal{H}$ is known to be given by $AB = BA \; \forall A, B \in \mathfrak{B}$.

Axiom 4) The set $\mathfrak{S}_A$ is deterministic for the one-dimensional subspace of $\mathfrak{A}$ generated by A; for any two observables A and B we have $\mathfrak{S}_{A+B} \supseteq \mathfrak{S}_A \cap \mathfrak{S}_B$, and $\mathfrak{S}_1 = \mathfrak{S}$.

Example: If in $\mathcal{H}$ we take a bounded observable A that has no eigenvalues, then $\mathfrak{S}_A$ is empty and Axiom 4 is inachievable.

Axiom 5) The axiom discussed in Section 1.

Definition 5.4. Let A and $B \in \mathfrak{A}$. $A \circ B$ is defined by $A \circ B \equiv \frac{1}{2}([A+B]^2 - A^2 - B^2)$.

Axiom 6) For any three observables A, B, and C in which A and C are compatible, $(A \circ B) \circ C - A \circ (B \circ C)$ vanishes.

Axiom 7) The norm of $A \in \mathfrak{A}$, $\| A \| \equiv \sup_{\phi \in \mathfrak{S}} |< \phi; A >|$, is finite and $\mathfrak{A}$ is topologically complete when regarded as a metric space with the distance between any two elements A and B of $\mathfrak{A}$ defined by $\| A - B \|$. $\mathfrak{S}$ is then identified with the set of all continuous positive linear functionals ϕ on $\mathfrak{A}$ satisfying $< \phi; 1 > = 1$.

Axiom 8) A sufficient condition for a set $\mathfrak{B}$ of observables to be compatible is that $\overline{\mathfrak{P}(\mathfrak{B})}$ is associative. Here $\mathfrak{P}(\mathfrak{B})$ is the set of polynomials in $\mathfrak{B}$.

Axiom 9) $\mathfrak{A}$ can be identified with the set of all self-adjoint elements of a real or complex, associative, and involutive algebra $\mathfrak{R}$ satisfying

(i) for each $R \in \mathfrak{R}$ there exists an element A in $\mathfrak{A}$ such tht $R^* R = A^2$;

(ii) $R^* R = 0$ implies $R = 0$.

We mention Axiom 10 for completeness only. It is not necessary to obtain a C* algebra.

Axiom 10) To each pair of observables A and B in $\mathfrak{A}$ corresponds an observable C in $\mathfrak{A}$ in the sense that for all $\phi \in \mathfrak{S}$, we have

$$< \phi; (A- < \phi; A > 1)^2 > \quad < \phi; (B- < \phi; B > 1)^2 > \quad \geqslant \quad < \phi; C >^2 .$$

Now, all axioms except the fourth and fifth hold in any Hilbert space construction with $< \phi; A > = Tr(\phi A)$. Consequently, consider any Hilbert space for which the phase space formalism applies. Form

$$\begin{aligned} \mathfrak{A}^+ &\equiv \text{ the algebra generated by } \mathfrak{A}_0, \\ \mathfrak{A}_0 &\equiv \{A^\eta(f) \mid f \in L^1_\mu(\Gamma) \cap L^\infty_\mu(\Gamma),\ 0 \leqslant f\}. \end{aligned}$$

Then, any $A^\eta(f) \in \mathfrak{A}^+$ is of the form $A^\eta(f) = \sum_i \lambda_i P_{\psi_i}$ for some orthonormal basis $\{\psi_i\}$. Consequently, Axioms 4 and 5 are satisfied. (One must be aware that, in $A^\eta(f) + A^\eta(h) = A^\eta(f+h)$, we may have all three of $A^\eta(f)$, $A^\eta(h)$, $A^\eta(f+h)$

possessing distinctly different eigenvectors!) Take $\mathfrak{A} \equiv \mathfrak{A}^+ - \mathfrak{A}^+$, and we have a set on which all the axioms hold. Or take $\mathfrak{R} \equiv \mathfrak{A} + i\mathfrak{A}$. Thus we have arrived at

Theorem 5.5. *Any Hilbert space in which the phase space formalism applies, including all the single particle Hilbert spaces of quantum mechanics, satisfies the axioms of the C* algebra formalism for a physical system.*

6. On the Informational Completeness of the Representation

Having a C* algebra in the phase space framework of quantum mechanics, we may now employ the G.N.S. construction [9][19] which we assume is familiar to the reader:

Choose any state ϕ in the original Hilbert space $\mathcal{H}$ in which the localization operators, $A^\eta(f)$, were defined, and form $< \phi, A >$, $A \in \mathfrak{A}$. (Note: This includes $\phi = P_\psi$ for any $\psi \in \mathcal{H}$.) Let $\mathfrak{K}_\phi = \{K \in \mathfrak{A} \mid < \phi; R^*K > = 0 \, \forall R \in \mathfrak{A}\}$, which by the Cauchy-Schwarz-Buniakowski inequality is equal to $\{K \in \mathfrak{A} \mid < \phi; K^*K > = 0\}$. Since $\mathcal{H}$ is irreducible, any vector in $\mathcal{H}$ is cyclic; so, we will take $\phi = P_\psi$, with ψ of the form $U(g^{-1})\eta$, $g \in G$. Then, abusing the notation, we have for our particular situation,

$$\begin{aligned}
\mathfrak{K}_{U(g^{-1})\eta} &= \{A^\eta(f) \in \mathfrak{A} \mid \| A^\eta(f)U(g^{-1})\eta \| = 0\} \\
&= \{A^\eta(f) \in \mathfrak{A} \mid \| A^\eta(g^{-1}.f)\eta \| = 0\}.
\end{aligned}$$

Since the set of fs is invariant under the group G, it suffices to consider just

$$\begin{aligned}
\mathfrak{K}_\eta &= \{A^\eta(f) \in \mathfrak{A} \mid \| A^\eta(f)\eta \| = 0\} \\
&= \{A^\eta(f) \in \mathfrak{A} \mid A^\eta(f)\eta = 0\}.
\end{aligned}$$

Now

$$A^\eta(f)\eta = \int f(x) < U(\sigma(x))\eta, \eta > U(\sigma(x))\eta \, d\mu(x).$$

If $< U(\sigma(x))\eta, \eta >= 0$ a.e. x for x in some compact set $\mathcal{O}$ with non-empty interior, then for all f with support in $\mathcal{O}$, $A^\eta(f)\eta = 0$. Thus $\mathfrak{K}_\eta \neq \{0\}$ in a way that is invariant under all infinitesimal transformations. If $< U(\sigma(x))\eta, \eta > \neq 0$ $a.e. x \in \Gamma$, there may be some fs such that $A^\eta(f)\eta = 0$, but $A^\eta(g.f)\eta = 0$ does not hold for all g infinitesimally in all directions. Thus, $A^\eta(f)\eta = 0$ holds only for a thin set of fs, in agreement with $A^\eta(f)$ being an integral operator.

Remark. 1) If f is always positive or always negative, then since $spec(A^\eta(f)) \subseteq (0, 1)$, resp. $\subseteq (-1, 0)$, $A^\eta(f)\eta \neq 0$.

2) The α-admissibility of η implies

$$\begin{aligned} &< U(\sigma(x))\eta, \eta > \neq 0 \, a.e. x \in \Gamma \\ \Leftrightarrow \quad &< U(g)\eta, \eta > \neq 0 \, a.e. \, g \in G; \end{aligned}$$

i.e., the same condition for obtaining informational completeness in the previously known cases for the Galilei and Poincaré groups.

3) By using the modular function we obtain: if η is admissible, then $U(g)\eta$ is admissible for all $g \in G$. Moreover, we have: if η is α-admissible, then $U(h)\eta$ is α-admissible for all $h \in H$. Thus, coupled with the results above, we have that any vector of the form $U(h)\eta$, $h \in H$, will be suitable for obtaining the results below. □

We will now obtain a representation of $\mathfrak{A}/\mathfrak{K}_\eta$, η satisfying $< U(g)\eta, \eta > \neq 0$: For R, S in $\mathfrak{A}/\mathfrak{K}_\eta$, define $(R, S) \equiv < P_\eta; R^* S >$. This turns out to be a sesquilinear form and generates a norm on $\mathfrak{A}/\mathfrak{K}_\eta$. Hence $\mathfrak{A}/\mathfrak{K}_\eta$ is a preHilbert space which has the Hilbert space $\mathcal{H}_\eta$ as its completion. Note that we are taking the completion in the topology dual to the strong or weak sense, by the cyclicity of η. This is the same topology as the topology for which informational completeness is discussed in [16].

The representation π_η of $\mathfrak{A}$ is defined by $\pi_\eta(R) : \mathfrak{A}/\mathfrak{K}_\eta \to \mathfrak{A}/\mathfrak{K}_\eta, \pi_\eta(R)S = RS$. The G.N.S. Theorem then proceeds to show that $\pi_\eta(R)$ can be extended to a bounded operator on $\mathcal{H}_\eta$. Taking $\mathfrak{A}/\mathfrak{K}_\eta$ instead of $\mathfrak{A}$ is moot when we operate on $\mathcal{H}_\eta$. The set $\{A^\eta(f)\}$ is informationally complete in this representation. But for all practical purposes, we have that $\mathcal{H}_\eta \subseteq \mathcal{H}$. Define $\mathcal{U}(g) : A^\eta(f) \mapsto A^\eta(g^{-1}.f)$, $g \in G$. $\mathcal{U}$ is an anti-representation of G on $\mathfrak{A}$. But using the covariance property of the $A^\eta(f)$, we see that this representation of the symmetry group is given by $\mathcal{U}(g)A^\eta(f) = U(g)A^\eta(f)U^{-1}(g)$; i.e. by the same U we had before. But that U is irreducible, and hence the Hilbert space obtained through the G.N.S. construction is the same as the original Hilbert space and $\pi_\eta(\mathfrak{A})\tilde{} = B(\mathcal{H})$. Consequently,

Theorem 6.1. *The set $\{A^\eta(f) \mid f \in L^1_\mu(\Gamma) \cap L^\infty_\mu(\Gamma)$, f real-valued, η α-admissible in $\mathcal{H}$ and $< U(g)\eta, \eta > \neq 0$ a.e.$g \in G\}$ is informationally complete in the G-irreducible representation space $\mathcal{H}$, for any G that is a Lie group.*

7. Conclusion

We have exhibited a set $\{A^\eta(f) \mid f$ real valued and $L^1_\mu(\Gamma) \cap L^\infty_\mu(X)\}$ of operators that have a physical meaning in any experiment in which one measures by quantum mechanical means. These $A^\eta(f)$ each have a full set of eigenvectors. They form a C* algebra, and hence form a foundation for the C*-algebraic formalism for physics in the free case. We may use the G.N.S. construction to obtain the informational completeness of $\{A^\eta(f) \mid f$ real valued and $L^1_\mu(\Gamma) \cap L^\infty_\mu(X)\}$. Generalizing to any physical system that has the phase space localization operators on it, we obtain a C* algebra and the informational completeness of these $A^{\eta}(f)$.

8. Acknowledgments

This work is done with the author's appreciation of all that he has learned from his thesis advisor and friend, Gerard G. Emch. He would also like to thank James Brooke and Paul Busch for suggesting changes to the paper that made it much more readable and accurate.

References

[1] Ali, S. T., Brooke, J. A., Busch, P., Gagnon, R., Schroeck, Jr., F. E., Can. J. Physics **66** (1988), 238–244.

[2] Ali, S. T. and Progovecki, E., Acta Appl. Math. **6** (1986), 19–45; **6** (1986) 47–62.

[3] Brooke, J. A., Schroeck, Jr., F. E., Nuclear Physics B (Proc. Suppl.) **6** (1989), 104–106.

[4] Brooke, J. A., Schroeck, Jr., F. E., J. Math. Phys.**37** (1996), 5958–5986.

[5] Brooke, J. A., Schroeck, Jr., F. E., *Phase Space Representations of Relativistic Massive Spinning Particles*, in preparation.

[6] Brooke, J. A., Schroeck, Jr., F. E., "Perspectives: Quantum Mechanics on Phase Space", International Journal of Theoretical Physics **44** (2005), 1894–1902.

[7] Busch, P., Int. J. Theor. Phys. **30** (1991), 1217–1227.

[8] Emch, G. G., *Algebraic Methods in Statistical Mechanics and Quantum Field Theory*, John Wiley and Sons, Inc., 1972.

[9] Gelfand, I., and Naimark, M. A., Mt. Sborn., N. S. **12 [54]** (1943), 197–217.

[10] Guillemin, V., and Sternberg, S., *Symplectic Techniques in Physics,* Cambridge Univ. Press, 1991.

[11] Healy, Jr., D. M. and Schroeck, Jr., F. E., J. Math. Phys. **36** (1995), 433–507.

[12] Lévy-Leblond, J.-M., J. Math. Phys. **4** (1963), 776 - 788.

[13] Mackey, G. W., Ann. Math. **55** (1952), 101–139; **58** (1953), 193–221.

[14] Michaelson, A. A., and Morley, E. W., American J. of Science **XXXI** (May, 1886), 377–386; **XXXIV** (1887), 338–339.

[15] Prugovecki, E., Int. J. Theor. Physics **16** (1977), 321–331.

[16] Schroeck, F.E., Jr., *Quantum Mechanics on Phase Space*, Kluwer Academic Publishers, 1996.

[17] Schroeck, F.E., Jr., Int. J. Theor. Phys. **28** (1989), 247–262.

[18] Schroeck, F.E., Jr., Int. J. Theor. Phys., **44** (2005), 2091–2100.

[19] Segal, I. E., Ann. Math. **48** (1947), 930–948.

[20] Souriau, J.-M., *Structure des systèmes dynamiques*, Dunod, Paris; translation: *Structure of Dynamical Systems*, Prog. Math. **149**, Birkhäuser, Boston, 1997.

[21] von Neumann, J., *Chrakterisierung des Spektrums eines Integraloperators*, Act. Sci. et Ind., Paris, 229 (1935), 11.

[22] Weyl, H., "Überbeschränkte quadratische Formen, deren Differenz Vollstetig ist", Rend. Circ. Mat. Palermo **27** (1909), 373 - 392.

[23] Wigner, E. P., Ann. Math. **40** (1939), 149–204.

Stochastic Flow on the Quantum Heisenberg Manifold

Kalyan B. Sinha

(Dedicated to Professor Gerard G. Emch on the occassion of his seventieth birthday)

1. Introduction

Following the work of Rieffel [1] on the deformation quantization of Heisenberg manifolds, a detailed study was undertaken in [2] to understand the geometry of such a manifold as a concrete example in non-commutative geometry [3]. In this article, a canonical non-commutative (quantum) stochastic flow is constructed on the quantum Heisenberg manifold which in a natural way is associated with the Dirac operator of the manifold.

First, we give a brief description of the manifold and some of its properties, more details can be found in [1] and [2]. For each $\hbar$, $\mu, \nu \in \mathbb{R}$ with $\mu^2 + \nu^2 \neq 0$ and for a given natural number c, let S^c denote the space of C^∞ functions $\varphi : \mathbb{R} \times \mathbb{T} \times \mathbb{Z} \to \mathbb{C}$ such that

a) $\varphi(x + k, y, p) = e(ckpy)\varphi(x, y, p)$ for all $k \in \mathbb{Z}$ and where we have written $e(x) = \exp(2\pi i x)$,

b) for every polynomial P on $\mathbb{Z}$ and every partial differential operator $\widetilde{X} = \dfrac{\partial^{m+n}}{\partial x^m \partial y^n}$ on $\mathbb{R} \times \mathbb{T}$, the function $P(p)(\widetilde{X}\varphi)(x, y, p)$ is bounded on $K \times \mathbb{Z}$ for any compact subset K of $\mathbb{R} \times \mathbb{T}$. For such a class of functions on the "compactified" Heisenberg manifold, we define the following non-commutative and associative product and conjugation :

$$(\varphi * \psi)(x, y, p) = \sum_q \varphi(x - \hbar(q - p)\mu, y - \hbar(q - p)\nu, q)\psi(x - \hbar q\mu, y - \hbar q\nu, p - q)$$

and

$$\varphi^*(x, y, p) = \overline{\varphi}(x, y, -p) \tag{1.1}$$

In (1.1) we note the expected properties that when $\mathfrak{h} = 0$, $(\varphi * \psi)(x, y, p) = \sum_q \varphi(x, y, q)\, \psi(x, y, p-q)$, i.e. if we look at φ and ψ as functions on $\mathbb{R} \times \mathbb{T} \times \mathbb{T}$, given as a Fourier series in the third variable, then the $*$-multiplication reduces indeed to the classical (commutative) pointwise multiplication of functions and similarly the conjugation $*$-operation is indeed just the complex conjugation. We denote this $*$-algebra given by (1.1) as $\mathcal{A}^\infty$, look at its representation in $L^2(\mathbb{R}\times\mathbb{T}\times\mathbb{T})$ given by

$$(\pi(\varphi)\xi)(x, y, p) = \sum_q \varphi(x - \mathfrak{h}(q-2p)\mu, y - \mathfrak{h}(q-2p)\nu, q)\xi(x, y, p-q), \quad (1.2)$$

and also denote by $\mathcal{A}$ and $\mathcal{N}$ the norm and weak closures of $\pi(\mathcal{A}^\infty)$ respectively. In such a case, $\mathcal{A}$ will be called the Quantum Heisenberg manifold. Here $\mathfrak{h}$ is to be interpreted as the $(2\pi)^{-1}$ times the Planck's constant in Quantum Mechanics.

It is to be noted that the (classical) Heisenberg group G acts ergodically on $\mathcal{A}$ via the following action on $\mathcal{A}^\infty$:

$$\left(L_{(r,s,t)}\varphi\right)(x, y, p) = e\left(p(t + cs(x - r))\right)\varphi(x - r, y - s, p) \quad (1.3)$$

for $(r, s, t) \in G$, and that there is a faithful normal tracial state τ on $\mathcal{N}$ given by

$$\tau(\varphi) = \int_0^1 dx \int_{T^1} dy \varphi(x, y, 0) \quad \text{on } \mathcal{A}^\infty. \quad (1.4)$$

This leads in a natural way to the following canonical Dirac operator :

$D'X = \sum_{j=1}^{3} d_{X_j} \otimes \sigma_j$, where X_j's are a basis of the Lie-algebra of G and σ's are the 3 Pauli σ_j matrices. More specifically, if we take as the basis the following three upper triangular matrices :

$$X_1 = \begin{bmatrix} 0 & 1 & 0 \\ 0 & 0 & 0 \\ 0 & 0 & 0 \end{bmatrix}, X_2 = \begin{bmatrix} 0 & 0 & 0 \\ 0 & 0 & 1 \\ 0 & 0 & 0 \end{bmatrix}, X_3 = \begin{pmatrix} 0 & 0 & c\alpha \\ 0 & 0 & 0 \\ 0 & 0 & 0 \end{pmatrix},$$

with $\alpha > 1$, then one has the following

Proposition 1.1. *Let D' be given as above. Then in $L^2([0, 1] \times [0, 1] \times \mathbb{Z})$, D' represented as $D' = \sum_{j=1}^{3} i d'_j \otimes \sigma_j$ with Dom $(D') = \{f \in L^2 | f(x, 0, p) = f(x, 1, p)$*

$f(1, y, p) = e(cpy) f(0, y, p);\ pf, \frac{\partial f}{\partial x}, \frac{\partial f}{\partial y} \in L^2\Big\} \otimes \mathbb{C}^2$ *is self-adjoint where*

$$\begin{aligned} id_1'(f) &= -i\frac{\partial f}{\partial x}, \quad id_2'(f) = -i\frac{\partial f}{\partial y} - 2\pi cpxf(x, y, p), \\ & -id_3'(f) = -2\pi cp\alpha f(x, y, p). \end{aligned} \tag{1.5}$$

It is convenient to make a unitary transformation $U : L^2([0, 1] \times [0, 1] \times \mathbb{Z}) \to L^2([0, 1] \times [0, 1] \times \mathbb{Z})$ by

$$U(f)(x, y, p) = \begin{cases} e(-cxyp) f(x, y, p) \text{ for } y < 1, \\ f(x, y, p) \text{ for } y = 1. \end{cases}$$

Under this transformation D' goes over to the self adjoint operator D given as follows :

Dom $(D) = \Big\{ f \in L^2 | f(x, 0, p) = f(x, 1, p), f(0, y, p) = f(1, y, p)$ and $pf, \frac{\partial f}{\partial x}, \frac{\partial f}{\partial y} \in L^2 \Big\} \otimes \mathbb{C}^2 \equiv \mathcal{D} \otimes \mathbb{C}^2$ with $D = \sum_j id_j \otimes \sigma_j$, where

$$id_1(f) = 2\pi cyp\ f(x, y, p) - i\frac{\partial f}{\partial x}, \quad id_2(f) = -i\frac{\partial f}{\partial y}, \quad -id_3(f) = -2\pi pc\alpha f. \tag{1.6}$$

For convenience, we shall also set

$$id_1^{(0)}(f) = -i\frac{\partial f}{\partial x}, \quad id_2^{(0)}(f) = -i\frac{\partial f}{\partial y}, \quad -id_3^{(0)}(f) = -2\pi pc\alpha f$$

and $r = -\alpha^{-1}\left(I \otimes M_y \otimes -id_3^{(0)}\right)$ on $L^2([0, 1] \times [0, 1] \times \mathbb{Z})$ with the domains of self adjointness of $d_j^{(0)}$'s being given by the appropriate periodic boundary conditions as in (1.6).

It is also clear (see for example [4]) that r is *essentially* self-adjoint on the linear span of the product domain $L^2([0, 1] \times [0, 1]) \otimes$ Dom $\left(-id_3^{(0)}\right)$; and we shall denote the self-adjoint extension also by r. The full details of the geometric properties of $\mathcal{A}$ and of K-theory on it can be found in [2] and [7]. The aim here is to construct the associated stochastic process.

2. Construction of the Stochastic Process

Clearly $d_j^{(0)}$'s, defined above, mutually commute and we set $u_t = \exp\left(i \sum_{j=1}^3 d_j^{(0)} \omega_j(t)\right)$, where $\{\omega_j\}_{j=1}^3$ are three mutually independent classical standard Brownian motions. Then u_t is a unitary operator in $\mathcal{H} \equiv L^2(\mathcal{A}, \tau) \simeq L^2([0, 1] \times [0, 1] \times \mathbb{Z})$ for almost all $\{\omega_j\}_{j=1}^3$. We also set $\alpha_t(a) = u_t a u_t^*$ for $a \in \mathcal{A}$.

Then note that though α_t is *not* a (random) automorphism of $\mathcal{A}$, $\alpha_t(a) \in \mathcal{B}(\mathcal{H})$ for almost all ω_j.

It follows from the last part of the previous section that

$$id_1 = id_1^{(0)} + r, id_2 = id_2^{(0)} \text{ and } id_3 = id_3^{(0)}. \tag{2.1}$$

Since $id_1^{(0)}$ and r (both unbounded self-adjoint operators) commute (being non-trivial in the different tensor components), it follows that $\mathcal{D}$ is the domain of essential self-adjointness of both the triplets $\{id_j^{(0)}\}$ and $\{id_j\}$.

Next we want to take into account the perturbation of the commuting triplet $\left(id_1^{(0)}, id_2^{(0)}, id_3^{(0)}\right)$ by $(r, 0, 0)$ somewhat along the lines of Evans and Hudson [5]. But the perturbation r is unbounded and hence the method of [5] cannot be applied. However, the following simple observation comes to the rescue; there is a natural "imprimitivity" between the map $e^{i\beta d_2}$ and M_y viz.

$$e^{i\beta d_2} M_y e^{-i\beta d_2} = M_{\overline{y+\beta}}, \tag{2.2}$$

where $\overline{a+b} \equiv (a+b) \bmod 1$.

Thus u_t leaves the Dom (r) invariant and

$$\alpha_t(r) = -\alpha^{-1} I \otimes M_{\overline{y+\omega_2(t)}} \otimes (id_3), \tag{2.3}$$

and $\alpha_t(r)$ is a self-adjoint operator with $\mathcal{D}$ the domain of its essential self-adjointness for almost all ω. Also, observe that $\alpha_t(r)$ and $\alpha_s(r)$ commute for all $s, t \geq 0$ and almost all ω.

Now consider the quantum stochastic differential equation [6, 7] in $\mathcal{H} \otimes \Gamma\left(L^2(\mathbb{R}_+, \mathbb{C}^3)\right)$, where $\Gamma(\mathcal{H}_0)$ is the symmetric (Boson) Fock space over the Hilbert space $\mathcal{H}_0$ (see [6] for details):

$$\begin{aligned} dU_t &= U_t \left\{ i\alpha_t(r) d\omega_1(t) - \frac{1}{2}\alpha_t(r^2) dt \right\} \\ & U_0 = I \text{ in } \mathcal{H} \otimes \Gamma(L^2(\mathbb{R}_+, \mathbb{C}^3)). \end{aligned} \tag{2.4}$$

This can be easily seen to have a unitary solution. In fact, the solution can be explicitly written down as

$$\begin{aligned} U_t &= \exp\left(i \int_0^t \alpha_s(r) d\omega_1(s) \right), \\ & \text{since } \alpha_t(r) \text{ and } \alpha_s(r) \text{ commute.} \end{aligned} \tag{2.5}$$

Finally, we set

$$\eta_t(a) = U_t \alpha_t(a) U_t^* \quad \text{for } a \in \mathcal{A}, \tag{2.6}$$

and it follows easily from the above that

$$\begin{aligned} d\eta_t(a) &= \sum_{j=1}^{3} \eta_t([id_j, a])d\omega_j(t) + \eta_t(\mathcal{L}(a))dt, \text{ where } \mathcal{L}(a) \\ &= -\frac{1}{2}\sum_{j=1}^{3}\left[d_j, \left[d_j, a\right]\right], \text{ for } a \in \mathcal{A}^\infty. \end{aligned} \tag{2.7}$$

Then we observe the following properties of the map η.

(i) Since $\{id_j\}_{j=1}^3$ are the Lie-derivations of $\mathcal{A}$ from the action of the Lie group G by (1.3) and since $\mathcal{A}^\infty$ is stable under their actions, it follows from the theory described in Chapter 8 of [7] that the differential equation (2.7) has solutions in $\mathcal{N}$ for almost all ω_j. Therefore $\eta.(.)$ maps $[0, T] \times \mathcal{A}$ into $\mathcal{N} \otimes \mathcal{B}(\Gamma(L^2(\mathbb{R}_+, \mathbb{C}^3)))$, for every $T < \infty$.

(ii) It is also clear from (2.6) that η_t is unital, multiplicative and $*$-preserving.

(iii) Thus η_t for each $t \in [0, T]$ defined by (2.6) is an unital, $*$-homomorphism from $\mathcal{A}$ into $\mathcal{N} \otimes \mathcal{B}(\Gamma(L^2(\mathbb{R}_+, \mathbb{C}^3)))$, further satisfying the quantum stochastic differential equation (2.7) on the smooth subalgebra $\mathcal{A}^\infty$.

Thus η, constructed above, describes the canonical diffusion or $*$-homomorphic stochastic flow on $\mathcal{A}$ [6,7].

References

[1] M. Rieffel : Deformation quantization of Heisenberg manifolds, *Commun. Math. Phys.* 122, 531-562 (1989).

[2] P. S. Chakraborty and Kalyan B. Sinha : Geometry on the quantum Heisenberg manifold, *Journal of Functional analysis,* 203, 425-452 (2003).

[3] A. Connes : "NonCommutative Geometry", *Academic Press Inc.* San Diego, CA (1994).

[4] M. Reed and B. Simon : Methods of Mathematical Physics, Vol.II, Academic Press, New York, 1978.

[5] M. P. Evans and R. L. Hudson : Perturbations of quantum diffusions, *J. London Math. Soc.*, 41, 373-384 (1990).

[6] K. R. Parthasarathy : "An Introduction to Quantum Stochastic Calculus", *Monographs in mathematics*, Birkhäuser Verlag, Basel (1992).

[7] Kalyan B. Sinha and Debashish Goswami: "Quantum Stochastic Processes and Non-Commutative Geometry", Cambridge Tracts in Mathematics # 169, Cambridge University Press, U.K. 2007.

Indian Statistical Institute, Delhi Centre, 7, S. J. S. Sansanwal Marg, New Delhi - 110 016, India
E-mail: kbs_jaya@yahoo.co.in